10/24/94

THINKING THROUGH TECHNOLOGY

• • • •

D0886001

• • • •

THE UNIVERSITY OF CHICAGO PRESS / CHICAGO AND LONDON

THINKING THROUGH TECHNOLOGY

• • • •

The Path between Engineering and Philosophy

CARL MITCHAM

Carl Mitcham heads the Science, Technology, and Society Program at Pennsylvania State University.

The University of Chicago Press, Chicago 60637
The University of Chicago Press, Ltd., London
© 1994 by The University of Chicago
All rights reserved. Published 1994
Printed in the United States of America
03 02 01 00 99 98 97 96 95 94   1 2 3 4 5
ISBN: 0-226-53196-1 (cloth)
        0-226-53198-8 (paper)

Library of Congress Cataloging-in-Publication Data

Mitcham, Carl.
    Thinking through technology : the path between engineering and
philosophy / Carl Mitcham.
        p.   cm.
    Includes bibliographical references and index.
    1. Technology—Philosophy. I. Title.
T14.M56 1994
601—dc20
                                                                93-44581
                                                                    CIP

⊚ The paper used in this publication meets the minimum requirements of the American National Standard for Information Sciences—Permanence of Paper for Printed Library Materials, ANSI Z39.48-1984.

Contents

• • • •

PREFATORY NOTES AND ACKNOWLEDGMENTS

• • • •

This work aspires to be a critical introduction to the philosophy of technology. It might serve as a textbook, but I also hope to make a general contribution to the interpretation of what have been termed postmodern ways of life and of the world of high-intensity artifice.

Part 1 provides a historicophilosophical overview, arguing the need to distinguish two traditions: engineering philosophy of technology, which emphasizes analyzing the internal structure or nature of technology, and humanities philosophy of technology, which is more concerned with external relations and the meaning of technology. The inclusion of illustrations is meant to emphasize the historical character of the first part.

Part 2 supplies a foundation for bridging these traditions by undertaking a humanities analysis of the broad spectrum of engineering and technology. The argument is that humanities philosophy of technology is the most philosophical tradition, but that it has failed to pay sustained or detailed attention to what really goes on in engineering and technology.

This book represents but another step in a continuing concern for the philosophical issues associated with technology. As a mid-1960s undergraduate seeking intellectual purchase on the contemporary world, it was reasonable for me to be attracted by the hypothesis that the distinguishing characteristic of our time was not so much modern science (as was often assumed) as modern technology.

Exploring this hypothesis led to the discovery of several traditions of philosophical reflection on technology and to the publication of two books on the topic: *Philosophy and Technology: Readings in the Philosophical Problems of Technology* (1972, 1983) and *Bibliography of the Philosophy of Technology* (1973, 1985). The bibliographic effort entailed by these works continued in a series of updates and special surveys and has provided the basis for some essays of historicophilosophical interpretation. Indeed, the anthologizing and bibliographing were from the start intended to prepare the way for more systematic reflection.

The present text thus attempts to realize an earlier commitment by addressing at greater length more fundamental concerns.

Earlier versions of some material in this volume can be found in the following forms:

Chapters 1 and 2: "What Is the Philosophy of Technology?" *International Philosophical Quarterly* 25, no. 1 (March 1985): 73–88.
Chapter 4: "Aspects philosophiques de la technique," *Revue Internationale de Philosophie* 41, no. 2, issue 161 (1987): 157–170.
Earlier and less complete versions of chapters 1, 2, and 4 were also used along with some quite different material in *¿Qué es la filosofía de la tecnología?* (Barcelona: Anthropos, 1989).
Chapter 5: "Philosophy and the History of Technology," in *The History of Philosophy of Technology,* ed. George Bugliarello and Dean B. Doner, pp. 163–201 (Urbana: University of Illinois Press, 1979).
Chapters 6–10: "Types of Technology," *Research in Philosophy and Technology* 1: 229–294 (Greenwich, Conn.: JAI Press, 1978).
Chapter 10: "Information Technology and the Problem of Incontinence," in *Philosophy and Technology II: Information Technology and Computers in Theory and Practice,* ed. Carl Mitcham and Alois Huning, pp. 247–255 (Boston: D. Reidel, 1986). © 1986 by D. Reidel Publishing Company. Reprinted by permission of Kluwer Academic Publishers.
Epilogue: "Three Ways of Being-with Technology," in *From Artifact to Habitat: Studies in the Critical Engagement of Technology,* ed. Gayle L. Ormiston, pp. 31–59 (Bethlehem, Pa.: Lehigh University Press, 1990).

Revisions and extensions of each of these texts have also appeared at various intervals. Permission to publish material adapted (with extensive revision) from earlier publications is gratefully acknowledged.

Since the genesis of this book has taken place over two decades, I have naturally incurred many debts, only a few of which it is possible to recognize in anything short of an autobiography. Standing out in memory, however, are Robert Mackey, Paul Durbin, and Jim Grote. Dominic Balestra did his best to protect an earlier version. Timothy Casey and

Alois Huning collaborated on research and writing that has found some place here. Durbin and Richard Buchanan, as readers for the University of Chicago Press, made helpful suggestions. Mary Paliotta read proofs and prepared the index.

For further inspiration and encouragement three others should be mentioned: Albert Borgmann, Ivan Illich, and my wife, Marylee. The book is dedicated to my mother, deceased father, and sister, to my children and grandchildren, born and unborn.

Citations policy: Classic volumes with standard pagination, and basic works in the philosophy of technology on which full information is provided by the reference list, are cited parenthetically in the text or in appropriately abbreviated form in notes. Complete references in the notes are reserved for marginal literature, which is in turn excluded from the reference list. Unless otherwise indicated, translations are mine.

INTRODUCTION

• • • •

Thinking about Technology

Technology, or the making and using of artifacts, is a largely unthinking activity. It emerges from unattended to ideas and motives, while it produces and engages with unreflected-upon objects. We make dinner, sew clothes, build houses, and manufacture industrial products. We use tools, turn on appliances, answer telephones, drive cars, listen to radios, and watch televisions. In our technological society, all this happens mostly by habit—but even in less technologically framed cultures the context of making and using is not so different, although the kinds of making and using certainly are, and artifice itself is less prevalent.

The need to think about technology is nevertheless increasingly manifest. Indeed, the inherent complexity and practical efficacy of modern technologies call forth diverse kinds of thinking—scientific and technical, of course, but also economic, psychological, political, and so forth. Within such a spectrum of approaches and issues, what does it mean to think philosophically about technology? What basic stance and distinctions characterize such thinking? Such are the principal issues to be addressed, and through them a perhaps even more fundamental question: Why try to think philosophically about technology at all? What is there about technology that is not adequately addressed by other kinds of thinking, from the scientific and technological to the psychological and political? And what are the results? What does philosophy tell us about technology?

Background and Standpoint

In the background of virtually all science and technology studies there lurks an uneasiness regarding the popular belief in the unqualified moral probity and clarity of the modern technological project. This uneasiness has been nourished not only by philosophical reflection, but also by the common experience of the citizens of technological societies over the past four decades—as all of us have been forced in divisive circumstances to address ethical issues associated with nuclear weapons and power plants, developments in information technologies from telegraphs to computers, biomedical technologies, space exploration, technological disasters, and environmental pollution. Consider, for instance, the following abbreviated and selective chronology:

1945 First atomic bomb exploded by the United States; J. Robert Oppenheimer, witnessing the test explosion in New Mexico, quotes to himself from the *Bhagavad Gita*, "I am become Death, the shatterer of worlds";[1] atomic bombs dropped on Hiroshima and Nagasaki; publication of the first issue of the *Bulletin of the Atomic Scientists*, "to explore, clarify and formulate the opinion and responsibilities of scientists in regard to the problems brought about by the release of nuclear energy" and "to educate the public."[2]

1946 First electronic computer (Electronic Numerical Integrator and Computer or ENIAC), which initiates public discussion of the possibility of artificial intelligence.

1949 Soviet Union tests its atomic bomb—to be followed by Great Britain (1952), France (1960), China (1964), and India (1974).

1950 First kidney transplant—to be followed by transplants of livers (1963), lungs (1964), hearts (1967), and so on.

1951 First hydrogen bomb exploded by the United States—to be followed by the USSR (1952), United Kingdom (1957), China (1967), and France (1968); U.S. Census Bureau buys the first commercial computer (UNIVAC).

1953 James Watson and Francis Crick discover DNA, a discovery that will become the basis of biotechnology, bioengineering, and eventually the largest biological research project in history, the international Human Genome Project (1990–present) to sequence and map the complete human genome.[3]

In this period of less than ten years nuclear energy, computers, biotechnologies and biomedical technologies all come on the world stage. Emerging from human thought, they also challenge it, as becomes apparent almost immediately:

1954 Launching of USS *Nautilus*, first nuclear submarine—to be followed by the first nuclear aircraft carrier, the USS *Enterprise* (1960).

1955 First commercial electricity from nuclear power; invention of birth control pill; Bertrand Russell and Albert Einstein issue a manifesto calling on scientists to become more involved in politics—a challenge that is taken up two years later by convening at Pugwash, Nova Scotia, the first of a series of Conferences on Science and World Affairs (subsequently known as the Pugwash Conferences).

1957 Soviet Union launches *Sputnik I*, the first artifact to orbit the earth; nuclear reactor at Windscale, England, suffers a near meltdown, creating a radioactive cloud that drifts across Western Europe; the Kyshtym dump for radioactive materials explodes in the Ural Mountains, contaminating over five hundred square miles with radioactive debris.[4]

1959 Integrated circuit invented.

1960 Laser invented.

1961 Yuri Gagarin becomes the first human being in space; thalidomide is banned in Europe after causing more than twenty-five hundred birth defects.

1962 *Mariner 2* (United States) becomes the first spacecraft to explore another planet (Venus).

This second period sees the new powers put to use within traditional human economic and political frameworks, but with increasingly conflicting results. Then comes a period of trying to adapt or alter those frameworks, punctuated by more technological disasters.

1963 Limited nuclear test ban treaty; nuclear submarine *USS Thresher* goes down at sea—to be joined by the *USS Scorpion* (1968) and at least three Soviet nuclear submarines (in 1970, 1983, and 1986).

1964 IBM makes a $10 million grant to found the Harvard University Program on Technology and Society.[5]

1965 Largest power failure in history blacks out New York City and parts of nine northeastern states—to be repeated on almost as large a scale in 1977.

1966 B-52 carrying four hydrogen bombs crashes near Palomares, Spain, contaminating a wide area with radioactivity; proposal to create a national data bank in the United States opposed by data processing professionals on ethical and political grounds.[6]

1967 The tanker *Torry Canyon* breaks apart and spills 30 million gallons of crude oil onto the beaches of southern England; oil spills subsequently become common occurrences around the world: from oil well blowout, Santa Barbara (1969) to *Exxon Valdez*, Alaska (1989); *Braer*, Shetland Islands (1993).

1968 Pope Paul VI issues *Humanae vitae* rejecting the use of artificial contraception.

1969 Neil Armstrong becomes the first human being to set foot on
the moon; U.S. Congress passes the National Environmental
Protection Act that establishes the Environmental Protection
Agency (EPA); Institute of Society, Ethics, and the Life Sciences
(known informally as the Hastings Center) founded to promote
"investigation of the ethical impact" of advances being made
"in organ transplantation, human experimentation, prenatal
diagnosis or genetic disease, the prolongation of life, and
control of human behavior";[7] Pennsylvania State University and
Cornell University initiate interdisciplinary science, technology,
and society (STS) programs—to be joined by Engineering and
Public Policy (EPP) at Carnegie-Mellon University (1970); the
Values, Technology, and Society (VTS) Program at Stanford
(1971); and related programs at a plethora of other schools from
Lehigh University to MIT; Greenpeace founded.

1970 U.S. Congress kills funding for the supersonic transport (SST).
Celebration of the first Earth Day.

1971 Founding of the Kennedy Institute of Ethics at Georgetown
University "to offer moral perspectives on the major policy
issues of our time," but with a special emphasis on bioethics
and the ethics of "in vitro fertilization, abortion, euthanasia,
genetic engineering, organ transplantation, life-sustaining
technologies, and the allocation of health-care resources."[8]

1972 Three Bay Area Rapid Transit (BART) engineers are fired for
criticizing the safety of a proposed automatic train control
system, and seven months later a BART train overruns a station
injuring five passengers, and for the first time professional
engineering societies support the whistle-blowing rights of
engineers;[9] Institute of Electrical and Electronic Engineers
(IEEE), the largest professional engineering society in the world,
establishes a Committee on the Social Implications of
Technology; pesticide DDT is banned by the EPA; U.S. Congress
passes Clean Water Act and establishes the Office of
Technology Assessment (OTA); National Science Foundation
sets up Ethics and Values in Science and Technology (EVIST)
Program to fund research; founding of the *Newsletter of the
Program on Public Conceptions of Science*, which will become the
journal *Science, Technology, and Human Values* at Harvard and
MIT and then of the Society for the Social Studies of Science (4S);
Club of Rome publishes *The Limits to Growth;* United Nations
Conference on the Human Environment (Stockholm Conference).

This ten years of new initiatives in assessment and control was one
of the most creative in science and technology policy history. But now
technological achievements and disasters enter into an almost normal

rhythm, which tests or extends public perceptions and social institutions:

1973 First spacecraft to achieve escape velocity from the solar system (*Pioneer 10*); U.S. Congress passes Endangered Species Act; first genetically engineered organism; Arab oil embargo and world energy crisis.

1974 Scientists establish a voluntary moratorium on recombinant DNA genetic engineering;[10] DC-10 crashes outside Paris, killing all 346 passengers and crew—as a result, it is later revealed, of a known fault in the engineering design of a cargo bay door;[11] Nypro chemical factory at Flixborough, United Kingdom, blows up, killing 28 workers.[12]

1976 First successful landing on Mars (*Viking 1*); three high-ranking nuclear engineers at General Electric resign to protest the dangers of nuclear power;[13] chemical plant explodes near Milan, Italy, releasing a cloud of dioxin that kills tens of thousands of animals.

1978 Soviet *Cosmos 954* with nuclear reactor aboard disintegrates over northern Canada; first test-tube baby; 2,000 residents forced to leave Love Canal, New York, because of chemical toxins; three fired BART engineers receive the first IEEE Award for Outstanding Service in the Public Interest.

1979 Partial meltdown in the nuclear reactor at Three-Mile Island; computer malfunction in the North American Air Defense Command headquarters puts United States forces on red alert.[14]

1981 Concrete skywalks at the Hyatt Regency hotel in Kansas City collapse, killing 114 people and injuring 200 more; formation of Earth First!

1982 First artificial heart implanted.

1983 Wave of computer break-ins by teenage computer hackers.[15]

1984 Union Carbide plant in Bhopal, India, explodes, killing more than 2,500 people in the worst industrial accident in history;[16] Worldwatch Institute issues its first "State of the World" report.

1985 British scientists report that thinning of stratospheric ozone has been occurring over Antarctica each spring since 1979; breakdown of Wall Street computer necessitates the borrowing of $20 billion to process stock transactions.

1986 Space shuttle *Challenger* explodes, killing seven astronauts, one a high-school science teacher; fire burns through the core of a Soviet nuclear reactor at Chernobyl, spewing lethal radioactive debris over Ukraine, Europe, and the world; giant chemical spill in the Rhine River.

1987 Montreal Protocol signed by twenty-four countries to curtail chlorofluorocarbon production that is causing stratospheric ozone depletion; World Commission on Environment and

Development, chaired by Gro Brundtland, issues its report, *Our Common Future*, trying to bridge the conflicts between environmentalists and developmentalists with a call for "sustainable development."[17]

1989 Former president Ronald Reagan, knighted in London and inducted into the French Academy in Paris, praises the democratic impact of the electronic revolution in communications and information technology.[18]

1990 Switching failure blocks half of all calls for a day on long-distance AT&T lines.

1991 Iraq sets fire to oil wells in Kuwait in an act of ecoterrorism.

1992 Earth Summit in Rio de Janeiro yields international treaty to protect biodiversity.

1993 United States House of Representatives votes 282 to 143 to stop funding the multibillion dollar superconducting supercollider.

As such a chronology shows, the late 1960s and early 1970s were a watershed in increasing consciousness of problems associated with technology and in attempts to develop mechanisms for social control. During the 1980s John Naisbitt's technological "megatrends" became Richard Lamm's "megatraumas"[19] and Charles Perrow discovered a high-tech world of "normal accidents."[20] By the 1990s it had become clear that not only would those who criticized technology have to take into account its many obvious benefits, but those who defended modern technology would have to seriously consider issues of complexity and fragility in both the environment and the technosphere and to consider the moral arguments of its critics.

There are, at the same time, reasons to be uneasy with the rush toward ethical discussions of technology as part of what has been called the "applied turn" in philosophy.[21] The philosophy of technology as currently practiced is heavily laden with such topics as environmental ethics, bioethics, nuclear ethics, computer and information ethics, development, science-technology policy studies, and global climate change.[22] Although it is true that moral problems press in upon us and demand decisions,[23] it is equally true that such decisions need to be made with as little haste and as much general understanding as possible. It is not clear to what extent philosophy can contribute directly to the effectiveness of decision making under pressure. At the very least the practical doubts of philosophers such as Socrates and Sartre—to cite two extreme cases—should raise suspicions that its unique contribution to the challenges of our time might lie elsewhere and be less direct. Certainly the need for decisiveness should not be confused with decisiveness about needs.

Efforts to integrate the general philosophical discussion of technology and specific moral issues have been singularly limited. It is remarkable, for instance, that none of the standard texts in engineering ethics contains any serious analysis of the engineering process as such.[24] The absence of theoretical analyses of technology is only slightly less pronounced in other fields of applied ethics.

The effort of this book is, in moderate contrast to prevailing inclinations, to emphasize general philosophical ideas—that is, fundamental theoretical issues dealing with technology. By standing back from the demands of practice and exploring basic philosophical questions, it aims to create more space, more open ground.[25] Through this approach it may ultimately be possible to make a more profound contribution to ethical reflection than by immediate engagement with particular moral problems. Certainly ethics is in no way rejected—and indeed, on one interpretation this book may be read as the prolegomenon to inevitably more explicit ethical reflections on technology.

Collections and Conferences

Historically, an interest in theoretical issues surrounding technology at least accompanied, if it did not wholly precede, the current ethical emphasis. Although I will say more about the ideas of the founders of the philosophy of technology and basic texts in the field, to begin it may be helpful to review some collective developments that reflect fundamental concerns and have taken place during roughly the third, pivotal period chronicled above.

The early anthologies and collections reflect an attempt to incorporate and integrate theoretical with practical issues. Although the first European collaborative work—Hans Freyer, Johannes C. Papalekas, and Georg Weippert, eds., *Technik im technischen Zeitalter* (1965)—is concerned with the "technological age," its aim is to elucidate fundamental attitudes toward this historical situation.[26] Klaus Tuchel's edited volume, *Herausforderung der Technik* (1967), likewise moves from an eighty-page essay titled "Technical Development and Social Change" to a scanning of "Documents on the Classification and Interpretation of Technology."[27]

In English, Zenon Pylyshyn's *Perspectives on the Computer Revolution* (1970), like many other collections dealing with this aspect of technology, begins with "theoretical ideas" (algorithms, automata, and cybernetics) before turning to discussions of the man-machine and machine-society relationships.[28] The Mitcham and Mackey anthology and bibliography—*Philosophy and Technology* (1972) and *Bibliography of the Philosophy of Technology* (1973)—likewise emphasize both theoretical and practical issues and their interrelations.

Broader documentation of the movement in question can be drawn from review of a series of formative conferences. The first established philosophy conference to feature a paper explicitly on the philosophy of technology was quite early, the 1911 World Congress of Philosophy, although the topic remained largely dormant among professional philosophers until after World War II. Then, in the early 1950s, in conjunction with a revived series of International Congresses of Philosophy,[29] one can identify a growing institutional effort to address technology as both a theoretical and a practical issue. Donald Brinkmann's "L'Homme et la technique," Congress XI (1953, in Brussels), for instance, focuses on alternative essential conceptions of technology and humanity. Congress XII (1958) at Venice and Padua suddenly contains a whole series of relevant papers. Congress XIII (1964) in Mexico City duplicates this situation, so that Congress XIV (1968) at Vienna introduces a special colloquium titled "Cybernetics and the Philosophy of Technology."[30] This development culminates with World Congress XV (1973) at Varna, Bulgaria, on the general theme "Science, Technology, and Man."[31]

Since then (1978, Düsseldorf, and 1982, Montreal) technology has under many guises become a regular feature of these international meetings, but with a marked shift toward ethical-political issues. World Congress XVIII (1988, Brighton, England) included sessions on, for example, ethical problems in population policy, in the treatment of animals, in contemporary medicine, and in genetic engineering, on the humanization of technology, on the dangers of nuclear war, on ecology, and on global problems in the light of systems analysis, but none on epistemological or metaphysical issues associated with technology. At Congress XIX (1993, Moscow), with a general theme of "Mankind at a Turning Point," the ethical emphasis in the philosophy of technology remains pronounced.

This same period witnesses the convening of a number of national conferences on philosophy and technology. Most notable are an Eastern European conference titled "Die marxistisch-Leninistische Philosophie und die technische Revolution" (1965)[32] and a colloquium of the International Academy of the Philosophy of Sciences in Paris in 1968, with proceedings published under the title *Civilisation technique et humanisme*.[33]

In the United States the first philosophy conference that can properly be said to take technology as its theme was a 1963 workshop "Philosophy in a Technological Culture" sponsored by the Catholic University of America (CUA). As indicated by the title, technology was approached as an issue in the philosophy of culture in a manner reflecting European intellectual concerns. Major discussions were organized around the science-technology and the technology-human nature relationships (that is, epistemology and philosophical anthropology of technology) as well as technology and ethics.

The year before the Center for the Study of Democratic Institutions and the publishers of *Encyclopaedia Britannica* convened a secular counterpart to the CUA conference under the heading "The Technological Order." Although stressing the technology-society relationship, and especially the thesis of Jacques Ellul that technology is the autonomous and defining characteristic of modern society—the English translation of Ellul's *La Technique* (1954) was being prepared under Center auspices—here the emphasis was on social theory, and there was somewhat less of an attempt than at the Catholic workshop to draw practical conclusions, make moral evaluations, or offer ethical guidance.

The first scholarly gathering to take philosophy of technology as a theme in its own right, however, and not try to sidle up to it by way of theories of culture or society was organized by Melvin Kranzberg of the Society for the History of Technology as a special symposium at the eighth annual SHOT meeting held in San Francisco in December 1965 in conjunction with a meeting of the American Association for the Advancement of Science, with proceedings published the next year in expanded form in the SHOT journal *Technology and Culture*.

In vivo this symposium consisted of papers by Joseph Agassi and Henryk Skolimowski dealing with questions of the relation between science and technology and the epistemological structure of technological thinking, respectively, followed by commentaries from J. O. Wisdom and I. C. Jarvie. The name of the symposium, "Toward a Philosophy of Technology," was taken from an unread contribution by Mario Bunge, who was prevailed on to alter his title in publication to "Technology as Applied Science." The same emphasis on theoretical issues can be found in the other two papers included in the proceedings— Lewis Mumford's "Technics and the Nature of Man" and James K. Feibleman's "Technology as Skills"—although Mumford's examination of the relation between theories of human nature and attitudes toward technology moves in the direction of ethics. As Kranzberg summarized the issues in a prefatory note, "although only in an embryonic stage, philosophy of technology already represents the variety of approaches found in older and more developed fields of philosophy. There is the questioning of technology in terms of human values; there is the attempt to define technology by distinguishing it from or by identifying it with other related fields; there is the epistemological analysis of technology; and there is the investigation of the rationale for technological developments." [34]

In 1973 a second pioneering effort exhibited this same interdisciplinary, pluralistic approach. George Bugliarello, then dean of engineering at the University of Illinois at Chicago, organized an international conference, with one day devoted to issues in the history of technology, a second to questions in the philosophy of technology, and a third to interrelationships and synthesis. Eight of the contributors to the philosophy portion of the conference continued to focus primarily on methodological, programmatic, and historicophilosophical concerns. Ethics was conspicuous by its absence, although anthropological and political theory made cameo appearances.

Despite (or perhaps because of) diverse institutional bases, none of these efforts led to an independent institutionalization of the philosophy and technology studies community. Such a step awaited the midwifery of Paul T. Durbin at the University of Delaware, who organized conferences on the philosophy of technology in 1975 and 1977.[35] These brought together a new group of scholars, with only Kranzberg having been present at both earlier ones. The weight of the discussion exhibits a slight shift, with five of the nine papers from the 1975 meeting (in the published proceedings) being strongly ethical-political in character.[36] Indeed, in his general introduction to the proceedings Durbin stresses the practical character of the existing consensus by noting that "those who see the [philosophy of technology] movement as legitimate recognize two things: (1) *There are urgent problems connected with technology and our technological culture which require philosophical clarification*, and (2) *Much that has thus far been written on these problems is inadequate— making it all the more important for serious philosophers to get involved.*" [37] Appropriately enough, then, given such a practical orientation, it was out of these conferences that there emerged three institutional structures: an occasional *Philosophy and Technology Newsletter* (1975– present),[38] an annual series titled *Research in Philosophy and Technology* (1978–present),[39] and the Society for Philosophy and Technology (SPT).

The formation of SPT nevertheless had a certain indefiniteness about it. In early 1977 Durbin, as editor of the *Philosophy and Technology Newsletter*, began to push for a formal societal organization by inviting nominations for officers. The 1977 conference considered an election but did not hold one. In mid-1979 Durbin tried again to solicit nominations, and in 1980 he conducted an election via the *Newsletter*, but the establishment of job descriptions and operational procedures remained unclear. As a result it was over a year before any substantive organizational developments took place. Still, with the formation of SPT there was created at least a nominal institutional base upon which to build wider contacts and sustained discussions.

The initial effort to take advantage of such opportunities came from Friedrich Rapp in Germany, the editor of *Contributions to a Philosophy of Technology* (1974)—an epistemologically oriented collection that reprints all the papers from the original *Technology and Culture* symposium except those by Mumford and Feibleman—and the author of *Analytische Technikphilosophie* (1978), two volumes that stress largely theoretical issues. Rapp wrote to Durbin suggesting a joint German-American conference. Held at Bad Homburg, Germany, in 1981, this initiated a series of biennial SPT meetings. The second conference was hosted by Polytechnic Institute of New York in 1983; the third by the Technological University of Twente at Enschede, the Netherlands, in 1985; the fourth by Virginia Polytechnic University in Blacksburg, Virginia, in 1987; a fifth took place in Bordeaux, France, in 1989; a sixth at the University of Puerto Rico, Mayagüez, in 1991; a seventh near Valencia, Spain, in 1993.

Although SPT has exerted a genuine effort to remain true to its origins, open to both theoretical and practical philosophy, there has been an appreciable shift toward ethical issues. The proceedings of the 1981 Bad Homburg conference, for instance, are divided into five parts, and only one is not dedicated in some form to ethical concern. The New York conference focused on theoretical and practical aspects of computers and information technology, but over two-thirds of the published papers are actually ethical-political. At Enschede the conference theme was "Technology and Responsibility," and for Blacksburg the focus was "Third World Development and Technology Transfer." The theme for the 1989 meeting was "Technology and Democracy," for 1991 "Discoveries of Technologies and Technologies of Discovery," and for 1993 "Technology and the Environment."

Appropriately enough, the practical interests of the SPT meetings in France and Spain both had more than merely discursive implications.

As an outgrowth of the Bordeaux meeting there emerged the associated Francophone Société pour la Philosophie de la Technique, with Daniel Cérézuelle as organizing secretary. Representative of the new generation of scholars who actively prepared the way for this professional group are Gilbert Hottois and Jean-Yves Goffi. Hottois's *Le Signe et la technique* (1984) is a challenging rethinking of the question of technology. Goffi's *La Philosophie de la technique* (1988) in the widely respected "Que sais-je?" series provides a balanced general introduction to the field.

In Spain, likewise, the SPT meeting was an occasion for promoting further development of a new interdisciplinary and interuniversity initiative called the Instituto de Investigaciones sobre Ciencia y Tecnología (INVESCIT).[40] As a result of its work hosting the SPT conference, INVESCIT and its program to promote the social assessment of technology projected its influence even more strongly beyond the Iberian Peninsula and into a growing network of international alliances. Moreover, José Sanmartín, the president of INVESCIT and author of two books investigating the challenge of biotechnology, *Los nuevos redentores* (1987) and *Tecnología y futuro humano* (1990), was elected the first president of SPT from outside North America.

The shift toward practical issues that has taken place within SPT and its allied associations only reflects much more profound pressures from society at large, as demonstrated by the previous chronicle. There thus continues to be a need to affirm the vitality of theory—an affirmation that can perhaps best be made not so much with specific arguments as by critically examining the historical development of the philosophy of technology and by pursuing cognitive inquiry in the presence of technological phenomena.

Themes and Variations

In defense of the theoretical stance, this book undertakes the two tasks just named, precisely to indicate the proper approach, basic conceptual distinctions, and fundamental problems within which a comprehensive philosophy of technology resides. At the very beginning it is appropriate to put forth the legitimacy and interrelation of these two tasks.

Like philosophy in general, the philosophy of technology should include at least two different but related kinds of reflection. It needs to be aware of its own history and able to articulate a set of systematically integrated issues. Without the first, it is liable to overlook insights of the past that can enrich its present; the study of history encourages respect for alternatives and guards against intellectual parochialism. Without the second, it is liable to degenerate into a hodge-podge of arguments, to be always a heap and never a whole, as Aristotle might say. Indeed, at the beginning of the history of philosophy in the West, it is the Stagarite who provides a kind of model in his pursuit of both these elements of philosophy.

The two principal parts of this book—chapters 1–5 and 6–10—thus aim to sketch out a history of the philosophy of technology and to highlight basic conceptual distinctions and associated issues. The historical component aspires, however, to be more than just a descriptive history of names, dates, and events—although it perforce includes some of that. And the conceptual analysis attempts more than simple analysis. My aim is philosophical history and substantive indication of issues, an illumination and interpretation of the chronology and concepts therein. Through reflection on the history of the philosophy of technology, I attempt to elucidate the proper philosophical approach and to point toward basic concepts; through reflection on a multitude of concepts and issues in the philosophy of technology, I make a correlated attempt to illuminate its history and point out the properly philosophical approach. These aspects are two sides of one coin, mutually informing and affirming.

Because of this mutual relation, neither the two parts nor their component chapters form a strict linear sequence. Indeed, thinking is not so much a linear, deductive process as a recursive procedure. Each part thus takes either its historical or its analytic approach, but then circles the topic as a whole in its own particular plane of reference, taking in both aspects. Part 1 stresses history, while articulating issues of significance. Part 2 stresses the articulation of conceptual distinctions, while appealing to and making use of history. In addition, each makes some attempt to hint at relations with the ethical issues that are the more prominent features of contemporary philosophy of technology.

Chapters 1 and 2 sketch the historical origins of that discipline called the philosophy of technology by distinguishing two quite different approaches: attempts by engineers and technologists themselves to create a technological philosophy, and attempts by scholars in the humanities, especially phenomenologists and others, to understand modern technology within a hermeneutic or interpretative framework. The primary aim is to call attention, first, to the thought of otherwise neglected engineer-philosophers and, second, to often ignored ideas of well-known philosophers—and to note some of the implicit arguments at issue among them. Chapter 3 then examines intermediate positions, but argues the philosophical primacy of the humanities approach.

The terminology here—"engineering philosophy of technology" versus "humanities philosophy of technology," which will on occasion be abbreviated as EPT and HPT—is chosen to emphasize two communities of discourse without prejudging the content of that discourse. Later I will comment more on this special terminology. Here it is sufficient simply to note that, despite possible uses of "humanistic" as an adjective for affairs associated with the humanities, it would be misleading to contrast engineering and "humanistic" philosophy of technology, since such a wording could connote either that engineers are not humanists in the sense of being concerned with the human (which most of them surely are) or that all members of the humanities community espouse some kind of philosophical humanism (which many of them surely do not). The terms, though clumsy—and even precisely by means of their awkwardness—are designed to keep open a special point.

The third chapter suggests but does not elaborate the full scope of questions that are part of a properly comprehensive philosophy of technology in the humanities tradition, a weakness that chapter 4 undertakes to remedy. Its playful opening compares the philosophy of technology with the philosophy of science, then it proceeds to outline a spectrum of issues ranging from the conceptual and epistemological through the ethical and political to the metaphysical. Chapter 5 returns to explicitly historicophilosophical investigations, focusing now on the period before the rise of modern technology, at the same time that it extends the themes presented by chapter 4.

Part 2 turns to more analytic tasks and seeks to furnish a conceptual framework for further exploration. The common concern of chapters 6–10 is a need fundamental in the philosophy of technology for the more careful elucidation of technology itself, in its diverse aspects, and a more intensive acquaintance on the part of students of philosophy with the self-understanding and ideas of engineers and technologists. Such a need is no doubt affirmed by the very divergences of the two communities of discourse narrated in chapters 1 and 2.

Chapter 6, by way of introduction, gives an internal summary of the state of the argument and considers some objections. By doing so it clears one stage and sets another; that is, it undertakes to move from the philosophical history of the philosophy of technology to philosophy of technology. It notes how the term "technology" is used in narrow and broad senses by engineers and by scholars in the humanities;

it defends the broader connotations but then distinguishes four modes of the manifestation of technology in the broad sense.

Chapters 7–10 explore in detail diverse categories of technology, the modes of its manifestation, suggested by the provisional analysis of chapter 6. Chapter 7 focuses on objects or artifacts, chapter 8 on technical knowledge and engineering science, chapter 9 on technological activity, and chapter 10 on technological volition. Conceptual distinctions are drawn between tools and machines; engineering knowledge is identified as entailing a distinctive epistemology; and engineering design is put forth as an activity worthy of distinctive analysis. Analysis of technology as volition returns once again to historicophilosophical considerations, while at the same time pointing toward ethical issues. Indeed, in the course of elaborating on distinctions between technology as object, as knowledge, as activity, and as volition, I raise a number of conceptual, epistemological, ethical-political, and metaphysical questions. In these chapters are numerous echoes of issues initially noted in chapter 4. Insofar as such analyses provide for the informative and helpful ordering of diverse issues related to technology, they constitute a confirmation of the very distinctions on which they are based.

The conclusion provides a brief reprise and restatement of the points developed in these ten chapters, considers the implications for technology and the humanities, and points toward further research. The epilogue offers a synthesis that, based on the analytic distinctions of part 2, returns to the historical interests of part 1 and reinterprets alternatives in the philosophy of technology.

PART ONE

· · · ·

Historical Traditions in the Philosophy of Technology

Philosophies do not spring full grown into consciousness as Pallas Athena was born from the head of Zeus. They suffer a natural and historical, not to say psychological and sociological, growth; only slowly do they develop to maturity. Even in maturity philosophies undergo change and alteration, advance and decay. Even though the period since the Industrial Revolution might well be termed the "age of technology," development of the philosophy of technology remains in its early stages; until quite recently there was little discussion that consciously saw itself as part of such a cooperative, reflective endeavor. Instead, reflection on technology tended to be subsumed within some other aspect of philosophy. The reasons are both historical and philosophical. A fitting way to introduce the philosophy of technology is thus by a brief examination of this historical and philosophical situation.

One historical complication in the birth of the philosophy of technology is that not only was it somewhat overdue, it was not even the outgrowth of a single conception. The philosophy of technology gestated as fraternal twins exhibiting sibling rivalry even in the womb.

"Philosophy of technology" can mean two quite different things. When "of technology" is taken as a subjective genitive, indicating the subject or agent, philosophy of technology is an attempt by technologists or engineers to elaborate a technological philosophy. When "of technology" is taken as an objective genitive, indicating a theme being dealt with, then philosophy of technology refers to an effort by scholars from the humanities, especially philosophers, to take technology seriously as a theme for disciplined reflection. The first child tends to be more pro-technology and analytic, the second somewhat more critical and interpretative. Before trying to decide which is more closely affiliated with philosophy itself, it is appropriate simply to observe some differences in character.

CHAPTER ONE

• • • •

Engineering Philosophy of Technology

What may be called engineering philosophy of technology has the distinction of being the firstborn of the philosophy of technology twins. It has clear historical priority in the explicit use of the phrase "philosophy of technology" and until quite recently was the only tradition to employ it. Two early anticipations of the term—"mechanical philosophy" and "philosophy of manufactures"—also point toward the overt temporal priority of engineering philosophy of technology.

Mechanical Philosophy and the Philosophy of Manufactures

"Mechanical philosophy" is a phrase of Newtonian provenance for that natural philosophy which uses the principles of mechanics to explain the world, in George Berkeley's words, as a "mighty machine."[1] Its most vigorous early exponent was the English chemist Robert Boyle—known to his contemporaries as "the restorer of mechanical philosophy," that is, of the mechanistic atomism of Democritus—whose *Mechanical Qualities* (1675) sought to explain cold, heat, magnetism, and other natural phenomena on mechanical principles. Isaac Newton, in the "Praefatio" to the first edition of his *Philosophiae naturalis principia mathematica* (1687), notes that mechanics has been wrongly limited to the manual arts, whereas he uses it to investigate the "forces of nature" and to "deduce the motions of the planets, the comets, the moon, and the sea." Indeed, he wishes he "could derive the rest of the phenomena of Nature by the same kind of reasoning from mechanical principles."[2] (That mechanical principles in the practical arts themselves called for philosophical analysis was to be argued a century later by Gaspard-François-Clair-Marie Riche de Prony in his *Mécanique philosophie*, 1799).

The eighteenth and nineteenth centuries witnessed, however, an increasing struggle over the connotations of this root metaphor—"mechanists" using it with approval and extending its application from nature to society, romantics rejecting its appropriateness in diverse contexts. In 1832, for example, an American mathematics teacher (later lawyer) named Timothy Walker (1802–1856) took it upon himself to respond to Thomas Carlyle's criticism of mechanics in *Signs of the Times* (1829). Walker did not fully appreciate Carlyle's contrast between mechanics and dynamics as poles of human action and feeling, nor could he have anticipated Carlyle's subsequent call for a reintegration of dynamics with mechanics by "captains of industry" (*Past and Present*, 1843). Instead, Walker's "Defense of Mechanical Philosophy" makes the characteristic argument that mechanical philosophy is the true means for emancipating the human mind in both thought and practice, and that through its correlate, technology, it makes democratically available the kind of freedom enjoyed only by the few in a society based on slavery.

Two years later, in 1835, the Scottish chemical engineer Andrew Ure (1778–1857) coined the phrase "philosophy of manufactures" to designate his "exposition of the general principles on which productive industry should be conducted with self-acting machines," which he contrasts to "the philosophy of the fine arts" (pp. 1 and 2). Ure's exposition includes a number of conceptual issues that have continued to concern the philosophy of technology: distinctions between craft and factory production, mechanical and chemical processes, the classification of machines, the possibility of rules for invention, and the socioeconomic implications of "automatic machinery." Because Ure's discussion is coupled with an unabashed apology for the factory system—Marx refers to him as "the Pindar of the automatic factory"[3]—his analytic side is usually overlooked. But in extending analyses made by Adam Smith and Charles Babbage,[4] Ure nevertheless advances an approach that is ancestor to operations research, systems theory, and cybernetics, as illustrated by texts such as Norbert Wiener's classic *Cybernetics* (1948) and related works.

Ernst Kapp and Technology as Organ Projection

Forty years after Ure's book, it was the German philosopher Ernst Kapp (1808–1896) who coined the phrase "Philosophie der Technik." Because Kapp is an unusual philosopher—especially unusual for a German philosopher—and the little-known originator of the term "philosophy of technology," his life and thought deserve special attention.

To begin with, his childhood was unstable, certainly less stable than that of his younger contemporary Karl Marx (1818–1883). He was the last of twelve children born to a court clerk in Ludwigstadt, Bavaria; his parents and two siblings died of typhus when he was six, and he eventually went to live with his elder brother Friedrich, who was a gymnasium teacher. This pointed him toward an academic career, and after receiving his doctorate in classical philology from the University of Bonn in 1828 with a dissertation on the Athenian state, he returned to teach under his brother in the gymnasium at Minden, Westphalia. But his interests were not limited to the classics, and in particular he was strongly influenced by the thought of both Georg W. F. Hegel (1770–1831) and Karl Ritter (1779–1859).

Along with Marx, Kapp was a left-wing Hegelian. His major scholarly study, the two-volume *Vergleichende allgemeine Erdkunde* (1845), reveals, as do Marx's economic and philosophic manuscripts from the year before, an attempt to translate Hegel's dynamic idealism into firmer materialist terms. But whereas Marx's materialism aimed to synthesize Hegel's theory of history with the new science of economics, Kapp's materialism sought to relate history to Ritter's new science of geography. Kapp's "comparative universal geography" anticipated what might today be called an environmental philosophy. On the one hand, this work stressed, like Ritter's, the formative influences of geography, especially bodies of water, on sociocultural orders. Rivers, inland seas, and oceans affect not only economies and general cultures, but political structures and military organizations. On the other hand, Kapp's adaptation of Hegelian dialectic called for the "colonization" and transformation of this environment, both externally and internally.

In a crucial section of the *Phenomenology of Spirit* (1807) Hegel analyzes the dynamics of what he presents as one of the most fundamental of social relations, the master-slave relationship. The master, to affirm his dignity and free himself from the physical environment, demands that the slave supply his needs. To do this the slaves must undertake technological work, and through work realize their own inherent dignity, independent of oppression by other human beings. Slaves can transform the world, which is thus less noble than they are. From such realization comes the drive for technological progress that can free the slave too from the physical environment and create the idea of a new society of free and equal citizens.

In the spirit of this analysis, for Kapp history is not the necessary

Ernst Kapp (1808–1896) and the house he built in the late 1840s in central Texas. Photo by Carl Mitcham.

unfolding of Absolute Idea, but the differential record of human attempts to meet the challenges of various environments—to overcome dependence on raw nature. This requires the colonization of space (through agriculture, mining, architecture, civil engineering, etc.) and of time (through systems of communication, from language to telegraph). The latter, in its perfected form, would constitute a "universal telegraphics" linking world languages, semiotics, and inventions into a global transfiguration of the earth and a truly human habitat. But this is possible only when the external colonization of the natural environment is complemented by an inner colonization of the human environment. As Hans-Martin Sass has argued, it is Kapp's theory of "inner colonization" that is the most original early idea.[5]

Because the world Kapp himself lived in was already colonized externally, Kapp devoted his own energies primarily to inner colonization in the form of politics. But when, like Marx again, he fell out with the German authorities in the late 1840s—for publishing a small volume titled *Der konstituiert Despotismus und die konstitutionelle Freiheit* (1849) he was prosecuted for sedition—and was forced to leave Germany, he chose not London (and the British Museum) but the North American frontier. Kapp immigrated to the German pioneer settlements of central Texas and simply shifted his emphasis from inner to external colonization. As he wrote to a friend at the time, "exchanging comfort for toil, the familiar pen for the unfamiliar spade," as farmer and inventor he undertook to live (quoting Goethe's *Faust*) "on free soil with free people."[6] As such, for the next two decades he led a life of close engagement with tools and machinery.

After the Civil War Kapp—who had opposed slavery, although one of his sons fought for the Confederacy—returned to Germany for a visit. But since he had become especially sick on the voyage, his physician urged him not to risk the return trip at his age, and he wound up reentering the academic world. In this capacity he revised his philosophical geography and then undertook, through reflection on his frontier experience, to formulate a philosophy of technology in which tools and weapons are understood as different kinds of "organ projections." Although this idea may have been hinted at as early as Aristotle and as late as Ralph Waldo Emerson,[7] it was certainly Kapp who, in his *Grundlinien einer Philosophie der Technik* (1877), gave it detailed and systematic elaboration. For Kapp,

the intrinsic relationship that arises between tools and organs, and one that is to be revealed and emphasized—although it is more one of unconscious discovery than of conscious invention—is that in the tool the human continually produces itself. Since the organ whose utility and power is to be increased is the controlling factor, the appropriate form of a tool can be derived only from that organ.

A wealth of spiritual creations thus springs from hand, arm, and teeth. The bent finger becomes a hook, the hollow of the hand a bowl; in the sword, spear, oar, shovel, rake, plow, and spade one observes sundry positions of arm, hand, and fingers, the adaptation of which to hunting, fishing, gardening, and field tools is readily apparent. (pp. 44–45)

Note that Kapp does not (like Emerson) think this is always a conscious process. Only after the fact, in many cases, do morphological parallels become apparent. (Indeed, chapter 9 of the *Grundlinien* is devoted to the unconscious.) And it is only on this basis that the railroad is described as an externalization of the circulatory system (chapter 7), and the telegraph as an extension of the nervous system (chapter 8). Nor is Kapp's argument limited to analogies with tools and machine networks; his book includes (chapter 10) the first philosophical reflection on the new science of mechanical engineering in the form of an analysis of Franz Reuleaux's classic *Theoretische Kinematik: Grundzüge einer Theorie des Maschinenwesens* (1875), which finds similarities between Reuleaux's description of the machine as entailing methodological limitation and the character of ethics, which also calls for principled limits on human action. Finally, even language and the state are analyzed as extensions of mental life and the res publica or *externa* of human nature (chapters 12 and 13). Well before Henri Bergson (1859–1941), Arnold Gehlen (1904–1976), and Marshall McLuhan (1911–1980), it was Kapp who articulated such ideas.[8]

As part of a sophisticated environmental philosophy, of course, Kapp's philosophy of technology to some extent transcends the framework of a strict technological philosophy. Nevertheless, the *Grundlinien* is devoid of a discussion of dialectics, and considered on its own—to some extent even in conjunction with the *Erdkunde*—it strongly projects the technological way of looking at the world into a variety of traditionally nontechnological domains. Indeed, a case could be made that the ambiguities inherent in Kapp's thought can also be found in Marxism, certainly in its late official or doctrinaire forms.

Technology and Politics according to Peter Engelmeier and Others

In the same decade when Kapp died, a minor German philosophy professor, Fred Bon (born 1871), and the Russian engineer Peter K.

Peter K. Engelmeier (1855 to ca. 1941). Photo courtesy of
Vitaly Gorokhov.

Engelmeier (1855 to ca. 1941), also began to use the term "philosophy
of technology."

Bon's *Über das Sollen und das Güte* (1898) is a treatise in neo-Kantian
ethics based on the fourfold distinction between analytic/synthetic
propositions and a priori/a posteriori knowledge. Having developed
the distinction in the critical reflection on theoretical reason or science,
Kant adapts it to an examination of practical reason or ethics, as summarized in table 1. According to Kant's *Fundamental Principles of the
Metaphysics of Morals* (1785) the imperatives or rules of skill "might also
be called technical (belonging to art)," while those of prudence could
be called "pragmatic (belonging to welfare)" (Akademie edition, p.
417). Bon simply builds on Kant's suggestion by titling his chapter on
analytic, a priori practical knowledge or rules of skill—that is, the
analysis of how means are to be chosen to achieve any given end—a
"Philosophie der Technik." The emphasis here is clearly on "philosophy of *technology*" rather than on "*philosophy* of technology."

Engelmeier, however, does not limit himself to analytic philosophy

Table 1. Fourfold Analytic/Synthetic and A Priori/A Posteriori Distinction

Relation to Experience	Relation between Subject and Predicate	
	Analytic Propositions	Synthetic Propositions
A priori knowledge	Theoretical reason: Conceptual or definitional. E.g., "Bachelors are unmarried males." Practical reason: Rules of skill. E.g., "Choose the means that can achieve the end."	Theoretical reason: Principles of knowledge. E.g., "All events have a cause." Practical reason: Commands of morality. E.g., "Act on the maxim that can be universalized."
A posteriori knowledge		Theoretical reason: Empirical knowledge. Practical reason: Counsels of prudence.

of technology. After graduating from the Moscow Imperial Technical College in 1881 as a mechanical engineer, Engelmeier quickly became an international engineer working, consulting, and studying in Russia, Germany, and France. Along with technical papers, he wrote on the economic significance of technology and the act of invention. In 1897 he even wrote a *Manual for Inventors*, published with an introduction by Leo Tolstoy, although the very next year he produced a "Critique of the Scientific and Artistic Views of Count Leo Tolstoy."[9]

When Engelmeier first uses the phrase "philosophy of technology" in a German newspaper in 1894 it is to call for the general philosophical elaboration and social application of the engineering attitude toward the world. Five years later his long, multipart article "Allgemeine Fragen der Technik" in *Dinglers Polytechnisches Journal* begins:

> Technologists or engineers [*Techniker*] generally believe they have fulfilled their social tasks when they have delivered good, cheap products. But this is only a part of their professional task. The well-educated technologists of today are not found only in factories. Highway and waterway transportation, urban economic management, etc. are already under the direction of engineers. Our professional colleagues are climbing ever higher up the social ladder; the engineer is even occasionally becoming a statesman. Yet at the same time the technologist must always remain a technologist. . . .
>
> This extension of the technical profession not only seems welcome, it is the necessary consequence of the enormous economic growth of modern society and augurs well for its future evolution.

The question then arises whether the modern technologist is sufficiently prepared to respond to the new demands. This question can hardly be answered in the affirmative, since it calls not only for governing our special fields of practical technology, but also that we try to see, with a far-reaching view, the interactions between technology and society. (p. 21)

Having set the stage, Engelmeier then proceeds to spell out the scope of a general inquiry into technology.

We must investigate what technology represents, which primary goals it pursues in its branches, what kinds of methods it uses, where its territory ends, which neighboring areas of human activity surround it, its relationship to science, art, ethics, etc. . . . [W]e should develop a total picture of technology, in which we analyze as many technical manifestations as possible . . . for technology is the spring in the great world clock of human development. (p. 21)

But as he concludes near the end of the introductory installment, the very concept of technology remains to be clarified by thinkers and technologists working together, "because what many thinkers have written about it has not been treated technically enough, and what has been written by technologists has not always been logical enough" (p. 22). Subsequently Engelmeier focuses the social function of technology, then analytic questions of the definition of technology, the machine, technological creativity, and invention.

In a 1911 paper, "Philosophie der Technik," Engelmeier restated this thesis for the World Congress of Philosophy IV. Beginning with a description of "the empire of technology" and its intensification, he considers the stages of abstraction in technology, arguing that philosophy of technology is a necessary final stage. "Technology is the inner idea of all purposeful action" (p. 591), grounded in the anthropological value of a technological will, "which springs from the utilitarian drives" (p. 592). Next year at the Moscow Imperial Technical College Engelmeier wrote the first general survey of issues in the philosophy of technology. In this four-volume, five-hundred-page work, *Filosofia tekhniki* [Philosophy of technology], he reviews the relevant ideas of previous philosophers from Aristotle through Bacon to Kapp, reports on discussions at the World Congress, and puts forth a technicist philosophy of the human being as scientist and world creator.

With the founding of the Universal Association of Engineers (abbreviated VAI in Russian) in the Soviet Union in 1917, Engelmeier began to proselytize for what in North America became the technocracy movement—the idea that business enterprises and society should be transformed and managed according to technological principles. But whereas in the United States the opposition was between business and engineering, in the Soviet Union it was between the Communist Party and the engineer.[10]

In 1927 there was a special celebration in honor of forty years' work by Engelmeier at which he delivered a lecture titled "Fifty Years of the Philosophy of Technology." That same year he helped organize a Circle on General Problems in Technology that promoted the generalization of engineering rationality. Two years later, in a journal of the Moscow Polytechnic Society affiliated with the VAI, Engelmeier published an article titled "Is Philosophy of Technology Necessary?"

> The Circle on General Problems in Technology ... refrains from any kind of propaganda. For the immediate future it has set itself the following tasks: to develop a program for the philosophy of technology [including] attempts to define the concept technology, the principles of contemporary technology, technology as a biological phenomenon, technology as an anthropological phenomenon, the role of technology in the history of culture, technology and the economy, technology and art, technology and ethics, and other social factors. (pp. 36–40)

Because Engelmeier rejected the "leading role" of the Communist Party, he fell out of favor with the Marxist authorities and could easily have been executed like Peter Palchinsky.[11] But during the 1930s he continued to pursue minor projects, apparently dying a natural death sometime in 1941.[12]

In 1914 the German chemical engineer Eberhard Zschimmer (1873–1940), who also taught at the University of Karlsruhe, became the third person to use the term "philosophy of technology," as the title of a small volume in which he defended technology against its cultural critics and proposed a neo-Hegelian interpretation of technology as "material freedom." Zschimmer's slight book went through many editions and in the 1930s was revised to reflect the ideas of National Socialism. As a result, Zschimmer's thought has been stigmatized and ignored, although it presents a cogent understanding of freedom as it can be achieved through technology, one related to that of Walker and reiterated in many contemporary engineering apologies for technological activity. That the goal of technology is human freedom achieved through and understood in terms of the material mastery of and escape from the limitations of nature has been, for instance, a common theme in the celebration of space exploration from *Sputnik I* in 1957 to the moon landing of 1969 and space shuttle operations of today.[13]

Friedrich Dessauer and Technology as Encounter with the Kantian Thing-in-Itself

The most outstanding figure in engineering-philosophy discussions during the mid-twentieth century, however, was Friedrich Dessauer (1881–1963). Dessauer—whose work ranges from *Technische Kultur?* (1908) and *Philosophie der Technik* (1927) to *Seele im Bannkreis der Technik* (1945) and *Streit um die Technik* (1956)—is also the fourth person to employ the term "philosophy of technology" as the title of a book.

Even more than Kapp, Dessauer was unusual for a German philosopher. To begin with, he was successful in business before completing his formal university education, and to the end of his life remained a devout Catholic who, as a layman, wrote numerous works on theology. But in adolescence he became fascinated with Wilhelm Röntgen's discovery of X rays, and at nineteen he dropped out of school and founded VEIFA-Werke, a company to manufacture X-ray machines.[14] As an inventor and entrepreneur he developed the techniques of deep-penetration X-ray therapy. It was related university-based research—the need for high-energy transformers to supply more powerful X-ray equipment—that in 1917 earned him a doctorate in applied physics from the University of Frankfurt.[15]

Shortly after, Dessauer received an appointment to the university and sold his company. Then in 1922, as a popular lecturer and writer, he convinced a group of industrialists to finance establishment of a research institute of biophysics, and he became its director. From 1924 he also served in the Reichstag as a Christian Democrat until 1933, when because of his opposition to Hitler, he was arrested and forced to flee the country. During the war he taught first at the University of Istanbul, then at Fribourg, Switzerland. In 1953 he returned to Germany as director of a Max Planck Institute for Biophysics. Ten years later, much of his body scarred by X-ray burns from experimental work, he died of cancer.

In his philosophy of technology Dessauer was as ecumenical as in his life; although he defended technology in the strongest possible terms, he also sought to open up a dialogue with existentialists, social theorists, and theologians. As a result, it is Dessauer's work that is most often cited when philosophers of science first acknowledged the philosophy of technology.[16] Indeed, one way to summarize Dessauer's philosophy of technology is to contrast it with standard philosophies

Friedrich Dessauer (1881–1963) as a
young man and in old age, scarred
by X-ray burns from his many experiments. Photos courtesy of Gerhard
Dessauer.

of science, which either analyze the methodologies of scientific knowledge or discuss the implications of specific scientific theories for cosmology and anthropology. For Dessauer both approaches fail to recognize the *power* of scientific-technical knowledge, which has become, through modern engineering, a new way for human beings to exist in the world. In *Philosophie der Technik*, and then again three decades later in *Streit um die Technik*—a book that restates his ideas while replying to critics and considering the arguments of others—Dessauer attempts to provide a Kantian account of the transcendental preconditions of technical power, as well as to reflect on the ethical implications of its application.

To the three Kantian critiques of scientific knowing, moral doing, and aesthetic feeling, Dessauer proposes to add a fourth—a critique of technological making. In the *Critique of Pure Reason*, Immanuel Kant (1724–1804) argues that scientific knowledge is necessarily limited to the world of appearances (the phenomenal world); it can never make unmediated contact with "things-in-themselves" (noumena). Critical metaphysics is, however, able to delineate the a priori forms of appearances and to postulate behind phenomena the existence of a noumenal reality. The *Critique of Practical Reason* (on moral doing) and the *Critique of Judgment* (concerned with aesthetic feeling) go further; they affirm the necessary existence of a "transcendent" reality beyond appearances as a precondition for the exercise of moral duty and the sense of beauty. Practical and aesthetic experience, nevertheless, fail to make positive contact with this transcendent reality; nor can the analyses of these realms of experience articulate noumenal structures.

By contrast, Dessauer argues that making, particularly in the form of technological invention, does establish positive contact with things-in-themselves. The essence of technology is encountered neither in industrial manufacture (which merely mass-produces inventions) nor in products (which are merely used by consumers), but in the act of technical creation. Before Dessauer, Engelmeier—along with the engineers Max Eyth (born 1836) and Alard DuBois-Reymond (born 1860)—had undertaken to analyze the process of technical invention.[17] For Engelmeier invention depended on the union of three key elements: will, knowledge, and skill.[18] Eyth distinguished between the creative germination of an idea, its development, and its final utilization. DuBois-Reymond likewise stressed the difference between invention as psychological event and as material artifact. All three identified the creative inspiration of the engineer with that of the fine artist, in an effort to relate engineering and the humanities. And it is significant that whenever similar efforts have been made, almost invariably they argue the unity of imagination or creativity in both the technological and the aesthetic realms. One case in point is Samuel Florman's *The Existential Pleasures of Engineering* (1976).

Dessauer acknowledges previous analyses and admits that technological creation takes place in harmony with the laws of nature and at the instigation of human purpose. Nevertheless, nature and human purpose are only necessary and not sufficient conditions for its existence. Instead there is something else, what Dessauer calls an "inner working out" (*inner Bearbeitung*) that brings the mind of the inventor into contact with a "fourth realm" of "preestablished solutions to technical problems."

It is this inner working out that makes possible the working of inventions in the real world. That this inner working out entails contact with the transcendent things-in-themselves of technical objects is confirmed for Dessauer by two facts: that the invention, as artifact, is not something previously found in the world of appearance and that, when it makes its phenomenal appearance through the inventor, it actually functions or works. An invention is not just something dreamed up, imagination without power; it arises from a cognitive encounter with the realm of preestablished solutions to technical problems. Technological invention involves "real being from ideas"—that is, the engendering of "existence out of essence," the material embodying of transcendent reality (1956, p. 234).

Although philosophers often find Dessauer's adaptation of Kant somewhat naive and unsophisticated, it can be read as an authentic extension of the Kantian project. For Kant, all reasoning is oriented toward the practical; the more practical it is, the closer experience comes to decisively transcending its phenomenal limitations. With Kant what transcendence is possible takes place in the realm of moral and aesthetic experience. Dessauer, however, locates the decisive penetration of appearances precisely in a kind of practical experience that Kant failed to recognize as worthy of serious consideration—modern technology.

Following this metaphysical analysis, Dessauer proposes a theory of the moral, almost mystical, significance of technology. Most such theories are limited to a consideration of practical benefits. For Dessauer, however, the pursuit of technology has the character of the Kantian categorical imperative or of divine command. The autonomous, world-transforming consequences of modern technology are witness to its transcendent moral value. Human beings create technology, but its power—which resembles that of "a mountain range, a river, an ice age, or planet"—goes beyond anything expected; it brings into play more than this-worldly forces. Modern technology should not be conceived simply as "the relief of man's estate" (Francis Bacon); it is instead a "participating in creation, . . . *the greatest earthly experience of mortals*" (1927, p. 66). With Dessauer, technology becomes a religious experience—and religious experience takes on technological meaning.

As mentioned, the particular form of Dessauer's argument has not survived critical scrutiny. Nevertheless both his spectrum of issues and his program for a comprehensive, interdisciplinary, and engineering-sensitive dialogue on philosophy and technology have had continuing influence. More than fifteen years after his death, for instance, Heinrich Stork in *Einführung in die Philosophie der Technik* (1977) still refers to Dessauer as the one who inaugurated a comprehensive philosophical investigation of technology. And with Heinrich Beck's much more moderate religious interpretation in *Kulturphilosophie der Technik* (1979), not only is there explicit reference to Dessauer's theory of "*homo inventor*," but the structure and evenhandedness of his argument as a whole can be seen to reflect Dessauer's expansiveness.

The Intellectual Attraction and Power of the Technical

Outside Germany, the term "philosophy of technology" has not until the 1980s been widely used, although the positive intellectual attraction and power of the technical realm have not gone philosophically unrecognized. In France one early instance can be found in *Les origines de la technologie* (1897) by the social theorist Alfred Espinas (1844–1922), who, two decades after Kapp, again emphasized the idea of technology as organ projection. Another suggestive feature of Espinas's analysis is his use of the term *technologie* and the distinction drawn between *techniques* (skills of some particular activity), *technologie* (systematic organization of some technique), and *Technologie* (generalized principles of action that would apply in many cases). Furthermore, Espinas proposed that *Technologie* (with a capital *T*) is for human making what *praxéologie* is for human action as a whole—thus introducing a specialized term that will be further exploited by the Polish philosopher Tadeusz Kotarbinski.[19] The ideas of both Espinas and Kotarbinski blend into what are now called systems theory, game theory, cybernetics, operations research, and various theories of management.

Another contribution to the engineering tradition in the philosophy of technology that exhibits even more clearly the inherent attraction of the technical realm was initiated by the French civil engineer Jacques Lafitte (1884–1966) in *Réflexions sur la science des machines* (1932), which undertakes to sketch what is termed a "mechanology," or a comprehensive analysis of technical evolution from passive machines (bowls, clothes, and houses) to active and reflexive ones (energy transformation and self-directing devices, respectively). This analysis has been deepened by Gilbert Simondon (1923–1989), a psychologist and human factors engineer, in *Du mode d'existence des objets techniques* (1958). The thrust of both works is toward a careful, analytic description of technological phenomena. With Simondon mechanology becomes a true phenomenology of machines that distinguishes between elements (parts), individuals (wholes), and ensembles (systems) as kinds of technological existence and proposes a theory of technological evolution based on detailed descriptions of developments in such artifacts as the internal combustion engine, telephone, and vacuum tube.[20]

In the Netherlands, the engineer Hendrik van Riessen began a second career in philosophy with *Filosofie en techniek* (1949), a work that continues to provide one of the most comprehensive historicophilosophical surveys of the field up through the mid-twentieth century. Van Riessen's student, the engineer-philosopher and now Dutch senator Egbert Schuurman, has made similar contributions to a philosophical analysis of the structure of modern technology, along with an appraisal of the developing philosophy of technology tradition.

In the Spanish-speaking intellectual world the thought of Juan David García Bacca (1901–1992) constitutes an even more extensive attempt to explore the philosophical and practical power of technology. Born in Spain, exiled by the Spanish Civil War to Ecuador, Mexico, and then Venezuela, García Bacca has produced a voluminous body of work that spans translations from the Greek and Latin, textbook anthologies, historical studies, and systematic treatises. His major treatises in the philosophy of science and technology emerge from research in logic, epistemology, and theology as well as extended dialogues with major figures of the Western philosophical tradition, from Plato and Aristotle through Augustine and Aquinas to Kant, Hegel, Marx, Husserl, Heidegger, Whitehead, and Russell.

In such popular works as *Elogio de la técnica* (1968) and *De magica a técnica* (1988) García Bacca utilizes the categories of both classical philosophy and modern science to present technology as the essential humanization of the world. In *Elogio*, for example, having first distinguished nature and technology in terms of the Aristotelian four causes, he presents modern technology as a humanization of the historical, intellectual, and social worlds. Although there are dangers in techno-

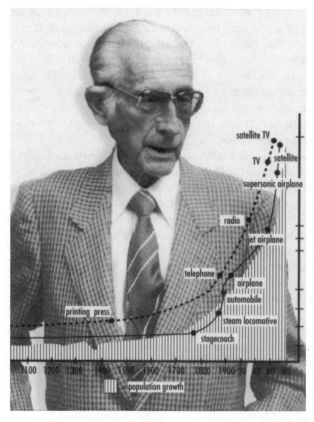

Juan David García Bacca (1901–1992) and a graph, to which he often refers, showing exponential growth in technology-related indicators. Montage by Emilie Jaffe and Matthew Jaffe, Small Time Press.

logical development, they are to be met through that same development.

In more technical treatises such as *Curso sistematico de filosofía actual* (1969) he likewise argues with a distinctive terminology drawn from multiple sources that *"science transubstantiating* [that is, technology] *is a per-objectifying and dis-alienating enterprise"* (p. 42). By "enterprise," García Bacca refers to a planned, humanly designed activity. In "per-objectifying," he uses the Latin intensifying prefix to indicate that technological objects realize their function beyond accidental utilization.

"A scale precisely reveals, makes evident, the weight of any kind of body; *objectifies* the weight—distinguishing and concretizing what was mixed up in the natural body. But the scale is a *per-object*; its aspect and functions are *per-objective*; the human, as creator, has *per-objectified* himself in it" (p. 44). And in their "dis-alienating" function, García Bacca claims that technical creations constitute "a *universe* of creatures *for* the human creator." They involve a "humanization of the natural universe, humanization of the natural man himself—insofar as he makes himself for himself a creator" (p. 47).

With García Bacca's philosophy of technology, as with Dessauer's, the role of invention and inventions is central—and ultimately supernatural. Indeed, from his perspective the human ability to create artifacts can be compared only to God's powers. The transubstantiation that medieval theologians believed took place in bread and wine through the words of the priest at Mass now takes place in a much more general way through science and technology (e.g., cave into home, stone into weapons, uranium into atomic bomb). Like God, human beings create things that can stand by themselves. A computer, for example, represents the perfect prolongation of human ways of thinking, as do industrial machines and atomic bombs. The existence of such artifacts is so independent that they can even enhance, make more powerful, degrade, or destroy their creators.

The eccentric technical enthusiasm of R. Buckminster Fuller is comparable. As engineer Buckminster Fuller writes in one of his poems, "No More Secondhand God" (1963):

I see God in
the instruments and the mechanisms that
work
reliably,
more reliably than the limited sensory departments of
the human mechanism.

(p. 4)

But the organized church
uncomprehending
the mechanical extension of man
says that such belief is pagan.

(p. 7)

In English the phrase "philosophy of technology" makes its first significant occurrence as the title for a symposium in the summer 1966 issue of *Technology and Culture*—the journal of the Society for the His-

Mario Bunge and his concept map of relations between art, technology, science, mathematics, and humanities. Montage by Emilie Jaffe and Matthew Jaffe, Small Time Press.

tory of Technology, an association (and discipline) that from its inception had strong alliances with the engineering community. The title "Toward a Philosophy of Technology" was derived from a contribution by Mario Bunge, an Argentine-Canadian philosopher with intimate knowledge of Western European discussions and strongly attracted by the logical empiricist attempt to create what he calls a scientific philosophy. For Bunge, "technophilosophy" is only an aspect of a larger effort to explain reality in scientific-technological terms and to reformulate the humanities (philosophy and ethics) along scientific and technological lines. Bunge also contrasts his approach with the "romantic wailings about the alleged evils of technology" by such persons as Heidegger and Ellul (Bunge 1979a, p. 68). His understanding of technology in the widest possible sense—with material (engineering, agriculture, medicine, etc.), social (education, industrial psychology, applied sociology, jurisprudence, administration, etc.), conceptual (information theory), and general (systems theory) branches—and his repeated outline of associated epistemological and ontological issues as well as a technoaxiology, technoethics, and technopraxiology constitutes perhaps the most comprehensive contemporary vision of engineering philosophy of technology.

Bunge's conceptions of social technology and technopraxiology are also near neighbors to what Karl Popper calls "piecemeal social engineering,"[21] which at the same time exhibits affinities with both the social pragmatism of John Dewey and the technocracy movement. Dewey, for instance, repeatedly calls for the application of science not just *to* human affairs but *in* them to make them more intelligent and to experiment with the realization of new possibilities and relationships.[22] The solution to the problems of technology is not less but more, and more comprehensive, technology. Thorstein Veblen's argument in *The Engineers and the Price System* (1921)—and even earlier in *The Instinct of Workmanship* (1914)—for the reorganization of economic (and political) life so as to free engineering principles from commercial (and political) corruption continues to exercise a certain appeal, although the once popular term "technocracy" is eschewed because of negative connotations. But one recent general theory of engineering ethics bridges these two positions with the idea of engineering as experimentation not just at the level of technical design but also at the level of social application—that is, as social experimentation.[23]

CHAPTER TWO

• • • •

Humanities Philosophy of Technology

Engineering philosophy of technology—or analyses of technology from within, and oriented toward an understanding of the technological way of being-in-the-world as paradigmatic for other kinds of thought and action—may well claim primogeniture. However, what may be called humanities philosophy of technology—or the attempt of religion, poetry, and philosophy to bring non- or transtechnological perspectives to bear on interpreting the meaning of technology—may nevertheless claim priority in the order of conception. From the origins of human history, ideas about the meaning of making activities have found expression in sacred myth, in poetry, and in philosophic discourse.[1] The attempt by Francis Bacon (1561–1626) to turn human attention toward technology and to invest human energy in its pursuit, in preference to politics and philosophy (not to mention religion and poetry), was itself undertaken by philosophical and rhetorical means. It was, we might say, the humanities that conceived technology—especially modern technology—not technology that conceived the humanities.

Although this principle—the primacy of the humanities over technologies—is the foundation for humanities philosophy of technology, it is not a principle that, especially in a highly technological culture, is self-evident or goes unchallenged. For Aristotle it was obvious that making was not an end in itself and was subordinate to various possible understandings of the good as well as to the political orders they entailed.[2] In the face of the success of Bacon's challenge to this traditional understanding and the subsequent appearance of technological societies, humanities philosophy of technology appears as a series of rear-guard attempts to defend the fundamental idea of the primacy of the nontechnical.

The modern defense of the humanities as larger and more extensive than the technological comes to the fore initially in the romantic movement. Jean-Jacques Rousseau, for instance, in his *Discourse on the Sciences and Arts* (1750), criticizes the Enlightenment idea that scientific and technological progress automatically contributes to the advancement of society by bringing about a unification of wealth and virtue. According to Rousseau, not only have "our souls been corrupted in proportion as our sciences and arts have advanced toward perfection," but "the sciences and arts owe their birth to our vices."[3] By vices Rousseau refers to selfishness and fear, with allusions no doubt to Bernard Mandeville's *Fable of the Bees* (1714), which argues that private vice (enlightened self-interest) does indeed lead to public virtue (wealth and power). "The politicians of the ancient world were always talking of morals and virtue," observes Rousseau; "ours speak only of commerce and money."[4] Romanticism affirms the significance of endeavors that transcend such limitations; it becomes fascinated by the idea of human beings outside the constrictions of civilization and the possibility of some vital faculty of mind (for the early romantics it was imagination) with access to deeper truths about reality than the rational intellect.

The subsequent romantic critique of modern technology as somehow obscuring essential elements of life is a rich and varied tradition. In the first half of the twentieth century, philosophers such as Karl Jaspers (1883–1969) and Gabriel Marcel (1889–1973) point up problematic aspects of modern technological society. For present purposes, however, it is convenient to concentrate on four not usually associated representatives of the romantic tradition who make a strong case for humanities philosophy of technology: Lewis Mumford (1895–1988), José Ortega y Gasset (1883–1955), Martin Heidegger (1889–1976), and Jacques Ellul (1912–1994).

Lewis Mumford: The Myth of the Machine

Like Dessauer, Mumford was excited in adolescence by electronics, did not take a standard university education—despite four years of college he was never awarded his bachelor's degree—and has made his way in philosophy as an outsider. But in contrast to Dessauer, his focus was from the humanities and he has been a persistent critic of technology in the American tradition of worldly romanticism that extends from Ralph Waldo Emerson to John Dewey. The tradition is worldly in its concern with the ecology of the environment, the harmonies of urban life, the preservation of nature, and a sensitivity to organic realities. It is romantic in insisting that material nature is not

Lewis Mumford. Drawing by Dirk Leach.

the final explanation of organic activity, at least in its human form. The basis of human action is mind and the human aspiration for creative self-realization.[5]

In 1930 Mumford published a short article arguing that the machine should be considered in terms of "its psychological as well as its practical origins" and appraised in aesthetic and ethical terms as much as technical ones.[6] This proposal led to an invitation to teach an extension course, "The Machine Age," at Columbia University, which was followed by an exhaustive research tour of European technical museums and libraries. The fruit was *Technics and Civilization* (1934), whose bibliography reflects a prescient knowledge of Ure, Zschimmer, Veblen, and Dessauer. Here Mumford first employs his understanding of human nature to provide an extensive analysis of the broad sweep of mechanical civilization and to write, in the process, a classic in the history of technology.

But despite his resultant reputation as a historian, Mumford's interest is not simply historical. The first two chapters of his book describe, in turn, the psychological and cultural origins, then the material and efficient causes, of technology; afterward he outlines a linear history of machine technics, dividing it into three "over-lapping and interpenetrating phases" of intuitive technics using water and wind (to about 1750), empirical technics of coal and iron (1750 to 1900), and scientific technics of electricity and metal alloys (1900 to the present). The last third of his book, however, undertakes an evaluation of the contemporary social and cultural reactions. As he summarizes, "We have observed the limitations the Western European imposed upon himself in order to create the machine and project it as a body outside his personal will: we have noted the limitations the machine has imposed upon men through the historic accidents that accompanied its development. We have seen the machine arise out of the denial of the organic and the living, and we have in turn marked the reaction of the organic and the living upon the machine" (p. 433).

If the machine is an organ projection, for Mumford it is so only by limitation. He might well sympathize with the interpretation of Ernst Cassirer (1928), who argues that along with "analogical" technologies that function as extensions of human organs, there are also "mimetic" technologies that imitate nature, and "purely symbolic" technologies.[7] Emerging from technical abstraction, purely symbolic technologies— what today are called information technologies—are liberated from both analogy and mimesis.

Much of Mumford's voluminous writing since has been an expansion and commentary on his pioneering early work, culminating three and a half decades later in *The Myth of the Machine* (2 vols., 1967, 1970). In this restatement of his case, Mumford argues that although the human being is rightly engaged in worldly activities, he or she is properly understood not as *Homo faber* but as *Homo sapiens*. It is not making but thinking, not the tool but the mind, that is the basis of humanity. As Mumford says in more than one essay, the human essence is not making but finding or interpreting. "What we know of the world comes to us mainly by interpretation, not by direct experience, and the very vehicle of interpretation itself is a product of that which must be explained: it implies man's organs and physiological aptitudes, his feelings and curiosities and sensibilities, his organized social relations and his means for transmitting and perfecting that unique agent of interpretation, language."[8] The importance of this hermeneutic activity can scarcely be overemphasized: "If all the mechanical inventions of the last five thousand years were suddenly wiped away, there would be a catastrophic loss of life; but man would still be human. But if one took away the function of interpretation, ... the whole round earth would fade away more swiftly than Prospero's vision [and] man would sink into a more helpless and brutish state than any animal; close to paralysis."[9]

Against what Mumford considers a technical-materialist image of humanity, he maintains that technology in the narrow sense of tool making and using has not been the main agent in human development, not even in technology itself. All human technical achievements are, Mumford maintains, "less for the purpose of increasing food supply or controlling nature than for utilizing his own immense organic resources ... to fulfill more adequately his superorganic demands and aspirations." The elaboration of symbolic culture through language, for instance, "was incomparably more important to further human development than the chipping of a mountain of hand-axes" (1967, p. 8). For Mumford, the human being "is pre-eminently a mind-making, self-mastering, and self-designing animal" (1967, p. 9).

On the basis of this philosophical anthropology, Mumford constructs a distinction between two basic kinds of technology: polytechnics and monotechnics. Poly- or biotechnics is the primordial form of making; at the beginning (logically but also to some extent historically), technics was "broadly life-oriented, not work-centered or power-centered" (1967, p. 9). This is the kind of technology that is in harmony with the polymorphous needs and aspirations of life, and it functions in a democratic manner to realize a diversity of human potentials. In contrast, monotechnics or authoritarian technics is "based upon scientific intelligence and quantified production, directed mainly toward economic expansion, material repletion, and military superiority" (1970, p. 155)—in short, toward power.

Although modern technology is a primary example of monotechnics, this authoritarian form did not begin with the Industrial Revolution. Its origins go back five thousand years to the discovery of what Mumford calls the "megamachine"—that is, rigid, hierarchical social organization. Standard examples of the megamachine are large armies or organized work crews such as those that built the Pyramids and the Great Wall of China. The megamachine often brings with it striking material benefits, but at the expense of a dehumanizing limitation of human endeavors and aspirations. A large army can conquer territory and extend power, but only by enforcing among its soldiers a discipline that either does away with family life, play, poetry, music, and art or rigorously subordinates them to military ends. The consequence is the "myth of the machine," or the notion that megatechnics is both irresistible and ultimately beneficent. This is a myth and not reality because the megamachine can be resisted, and it is not ultimately beneficial. Mumford's work as a whole is an attempt to demythologize and delimit megatechnics and thereby to initiate a radical reorientation of mental attitudes that would transform monotechnical civilization. As he says in an earlier essay, "to save technics itself we shall have to place limits on its heretofore unqualified expansion." [10]

An important feature of Mumford's work, however, is that his negative criticisms of monotechnics are complemented by positive studies of art and urban life, culminating with *The City in History*, which won a National Book Award for 1961. *Technics and Civilization* is itself designated the first in a four-volume "renewal of life" series, and the second volume states the case for a technology modeled on patterns of human biology and a "biotechnic economy." [11] In *Art and Technics* (1952), midway between *Technics and Civilization* and *The Myth of the Machine*, Mumford contrasts art as symbolic communication of the inner life of the mind with technics as power-manipulation of external objects. Mumford is clearly not arguing for a simpleminded rejection of any and all technology. Instead, he seeks to make a reasoned distinction between two kinds of technology, one that is in accord with human nature, and another that is not. His aim is not to discard the Promethean myth of human beings as tool-using animals, but to "supplement" it with that of Orpheus as "man's first teacher and benefactor." The animal became human "not because he made fire [a] servant, but because he found it possible, by means of his symbols, to express fellowship and love, to enrich [a] present life with vivid memories of the past and formative impulses toward the future, to expand and intensify those moments of life that had value and significance." [12] Technology is thus to be promoted when it contributes to and enhances what Mumford calls this "personal" aspect of existence, not when it restricts and narrows human life with a focus on power.

José Ortega y Gasset. Drawing by Dirk Leach.

José Ortega y Gasset: Meditation on Technics

Ortega is the first professional philosopher to address the question of technology, which he does in a series of university lectures delivered in 1933 in Spain and then first published in 1935 in Buenos Aires in the newspaper *La Nación*. The first authorized book publication occurred four years later. Ortega thus raises the issue of technology about the same time as Mumford and in the context of a philosophical anthropology that, although it exhibits some similarity to Mumford's, is certainly of greater metaphysical depth.

The importance that Ortega himself places on his "Meditations on Technics," which has not (even among Ortega scholars) been accorded the attention it deserves, is indicated by its very title. Two decades earlier, in his first book, *Meditations on Quixote* (1914)—with its allusion to and criticism of Descartes—Ortega puts forth a new understanding of what it is to be human with the formula "Yo soy yo y mi circunstancia" (I am I plus my circumstances). Before publishing this inaugural work he had spent two years in Germany and had come into contact with the phenomenology of Edmund Husserl. In criticizing the Husserlian analysis of consciousness Ortega developed a version of existential intentionality or "real human life" as the coexistence of the ego and its circumstances—a view that would later become more widely associated with the thought of the early Heidegger of *Being and Time* (1927).

Ortega's book, like Descartes's meditations at the beginning of the modern period, proposes to bring about a revolution in philosophy. Yet Ortega's meditation is not on "first philosophy." It is instead on a figure of central importance for anyone existing in Spanish circumstances. His meditation is not rationalist but existential, although Ortega himself eschews the term "existentialism" in favor of "ratiovitalism." *Meditations on Quixote*—a rational reflection on real life—further announces that it is the first of a series of meditations. But only a very few other works with this title follow. Among the most substantial is "Meditations on Technics."

According to Ortega, technics is necessarily involved with what is to be human. Ortega's philosophy of technology rests on the idea of human life as entailing a relationship with circumstances—not, however, in a passive manner, but as an active response to and creator of those circumstances. "I am I plus my circumstances"—meaning "I" is not to be identified with just itself (idealism) or with just its circumstances (materialist empiricism), but with both and their interaction. The opening sections of his "Meditations on Technics" are designed to develop this metaphysical thesis. Human nature, unlike that of a rock, tree, or animal, is not something given by existence; rather, it is something people must create for themselves. A person's "life does not correspond with the profile of organic necessities" (p. 323)[13] but projects beyond itself.

This self-interpretative, self-creative undertaking proceeds through two distinct stages. First, there is the creative imagination of a project or attitude toward the world that the person desires to realize. Second, there is the material realization of that project, since once we have imagined what we want to become, what we want to make ourselves—whether this is a gentleman, bodhisattva, or hidalgo (to use Ortega's own historical illustrations)—there are certain technical requirements for its realization. And of course, because these requirements will differ according to the project to be realized—the gentleman requires a water closet, unlike both bodhisattva and hidalgo—there are as many different kinds of technics as there are human projects.

For Ortega the human being actually might, to some extent, be defined as *Homo faber,* provided *faber* is not restricted to material fabrication but includes spiritual creativity. "This invented life, invented as the invention of a novel or a work of the theater, is what a person calls human . . . and this a person makes himself beginning with the invention of it" (pp. 334–335). Inner invention precedes and provides the basis for external invention. Technics, again, may even be thought of as a kind of human projection, but not on strictly natural or organic foundations (as with Kapp or Gehlen). There is a break or a rupture between the human and the world.

Near the end of his life, at a conference in Darmstadt, Germany, in 1951, Ortega returned to this theme with a story titled "The Myth of Humanity outside Technics." In this presentation he begins by affirming from external observation that the human being is indeed "a technical being" (p. 618).[14] But why should this be? The reason, he argues, is that the human being is not part of nature but has an *idea,* an *interpretation* of nature. Although there is only extremely limited scientific knowledge about the origins of such a being, since science explains only how things arise within or as part of nature, it is possible to construct a myth of how the human might have been in nature and outside technics and then was transformed into a being outside nature and within technics.

Ortega imagines a prehuman species that simply accepts whatever is given to it by nature. Its members do not think about anything other than what simply happens; they are happy, content. Then through some genetic mutation this animal develops an inner life of multiple fantasies, so that a member of the new species "has *to choose, to select*" between fantastic possibilities (p. 622). This new animal is essentially what the Latins called *eligens,* from which are derived the words *intellegens* and *intellegentia,* that is "intelligent." Such intelligence gives rise to an *insatisfaction,* a discontent with the world, to the desire to create a new world, and thus to technics.

"Meditation on Technics" begins with a metaphysical argument (sections 1–5) which is then illustrated with references to technics of different historical periods that exhibit more or less equal status as technics (sections 6 and 7). There is nevertheless some truth in the myth of human beings existing outside technics. This truth is the common notion that modern technics is the epitome of technology. Indeed, following his historical illustrations in "Meditation on Technics" Ortega develops a history of technology that argues this thesis (sections 8–12).

To present this thesis Ortega outlines the evolution of technology, dividing it into three main periods similar to those found in Mumford: the technics of chance, the technics of the craftsman, and the technics of the technician or engineer. The difference between these three is in the way one discovers the means to realize the project one has chosen to become—that is, in the "technicity" (el tecnicismo) of technical thinking. In the first period, there are no methods or techniques at all, and a technics must be discovered simply by chance. In the second, certain technics have become conscious and are passed from one generation to the next by a special class, the artisans. Still, there is no systematic or conscious study called technology; technics is simply a skill, not a science. It is only in the third period, with the development of that analytic way of thinking associated with the rise of modern science, that the technics of the technician or engineer— scientific technics, "technology" in the literal sense—comes to be. Discovering the technical means for realizing any end itself becomes a self-conscious scientific method or technique. "The technicity of modern technics is radically different from that which inspired all previous technics" and indeed is "a new way for the mind to operate that manifests itself both in technics and even more in pure [or scientific] theory" (p. 371). In our time, as Ortega puts it, humanity has "la técnica" (that is, technology) before "a technics." People can know how to realize any project they might choose even before they choose some particular project.

The perfection of scientific technics leads, for Ortega, to a uniquely modern problem: the drying up or withering away of the imaginative or wishing faculty, an aboriginal faculty that accounts for the invention of human ideals in the first place. In the past people were mainly conscious of things they were unable to do, of their limitations and restrictions. After willing some project, a person had to expend years of energy in solving the technical problems involved in its realization. Now, however, with the possession of a general method for discovering the technical means to realize any projected ideal, people often lose the

ability to will any ends at all. In the absurd logic of a "brilliant construction engineer," Alexei Kirilov argues (in Dostoyevsky's *The Possessed*) that human beings are completely free and anything is possible,
but nothing is required—except suicide, as a definitive demonstration
of freedom.[15] And such suicide need not be the result of an explicit
act. Human beings have become so entranced with their new technology that they have forgotten that "to be a technician and only a technician means to be able to be everything and consequently not to be
anything determinate" (p. 366). In the hands of technicians alone,
people devoid of the imaginative faculty, technics is "an empty form—
like the most formalistic logic; it is unable to determine the content of
life" (p. 366). Scientific technicians are dependent on a source they cannot master by reducing it to scientific or technical terms. Because of
this, Ortega provocatively suggests, the West may be forced to turn to
the technicians of Asia.

Martin Heidegger: The Question concerning Technology

Heidegger's philosophy of technology is not easily summarized, although it has features in common with Mumford's and, at a deeper
level, with Ortega's. Like Mumford, Heidegger adopts a strategy of
distinguishing between two kinds of technology and, without rejecting
technology in any general sense, trying to enclose modern technology
within a more expansive framework. Like Ortega, Heidegger approaches the issue of technology from the perspective of what he terms
fundamental ontology and ultimately raises issues about the historical
destiny of the West.

In approaching Heidegger's discussion of technology, however, there
are two points to keep in mind. First, Heidegger is to some extent a
philosopher in the Socratic tradition of raising questions rather than
providing answers. He thinks that more than anything else questions,
difficulties, or problems are what philosophy is all about. He has no
desire to resolve questions like the positivists or to dissolve problems
after the manner of Ludwig Wittgenstein and some other analytic philosophers. In truth, Heidegger is inordinately suspicious of all answers
or solutions. Second, the overriding question for Heidegger concerns
Being. Now, exactly what this question is has been much debated. Heidegger himself has worded it differently at different points in his life.
Originally it was the question of the meaning of Being; then it became
the question of the truth of Being; later it was the question of the place

Martin Heidegger. Drawing by Dirk Leach.

of Being. Later still he resorts to an archaic spelling of the German *Sein* (*Seyn*) or simply crosses it out (B̶e̶i̶n̶g̶). Yet he insists that it is the same question he is trying to formulate.

It is significant that of the three works Heidegger titled "The Question of . . ."—as distinct from actually being questions themselves (which accounts for at least another five works)—one is called "The Question of Being" (1955), another "The Question of the Thing" (1967), and a third "The Question concerning Technology" (1954). This suggests a need to examine the "question concerning technology" especially in relation to the "question of Being" and perhaps even the "question of the thing." It may also be that these other two questions concerning the thing and technology can help illuminate the fundamental question of Being.

Technology is a question or an issue in at least three senses. The first concerns the whatness or essence of what we call technology. This is the initial focus of Heidegger's work on the technology question in an essay that arose out of four lectures first delivered in 1949. Heidegger rejects traditional answers to the question "What is technology?"—answers describing technology as a neutral means or a human activity. In contrast to the instrumental view of technology as neutral human means, Heidegger argues that technology is a kind of truth or revealing, and that modern technology in particular is a revealing that sets up and challenges nature to yield a kind of energy that can be independently stored and transmitted.

To clarify this characterization of modern technology as a revealing that has the special character of a "setting-upon" and "challenging-forth," Heidegger contrasts the traditional windmill or waterwheel with an electric power plant. Each harnesses the energy of nature and puts it to work to serve human ends. Yet the windmill and waterwheel remain related to nature in a way that makes them, Heidegger suggests, similar to works of art. First, of course, they are dependent on the earth in ways that modern technology is not, simply because they only transfer motion. If the wind is not blowing or the water not running, nothing can be done. Second, even as structures they generally tend to fit into a landscape, intensifying and deepening its character, often revealing and throwing into relief geographic features that would otherwise be easily overlooked. The windmill stands forth on the plains like a lighthouse, calling attention to a small oasis and highlighting by its upright posture the stark flatness of this region of the earth.

A coal-fired electric power plant, by contrast, unlocks basic physical energies and then stores them up in abstract, nonsensuous form. It does not just transmit motion; it transforms, or releases and then transforms it. From prehistoric times until the Industrial Revolution the materials and forces human beings worked with remained fairly constant: timber, stone, wind, falling water, animals. But modern technology proceeds to exploit the earth in a new way—extracting stored-up energy in the form of coal, then transforming it into electricity that can be re-stored and kept ready for distribution or use at human will. "To open up, to transform, to store, to distribute, to switch are modes of revealing" that are characteristic of modern technology (p. 18).[16] Moreover, an electric power plant seldom fits into or complements the natural landscape. Large dams flood canyons and cover over rapids; not only do nuclear reactors contaminate the environment with heat and radiation, their location is determined by urban utilities, and they have a form that is hostage to internal structural calculations, so that they exhibit the same character wherever they are set down upon the landscape.

It is this latter fact that connects the question of technology to the question of the thing. Heidegger argues that technological processes, unlike traditional techniques, never create things in the genuine sense. The atomic bomb, Heidegger says elsewhere, only makes explicit what has already happened, the destruction of all thinghood.[17] In place of unique things like the potter's earthenware jug, modern technology generates a world of what Heidegger calls *Bestand*—"resources," "standing reserve," "stock" objects that are available to be used and consumed. The world of modern artifacts always stands ready and available to be manipulated, consumed, or discarded. This is not just because of mass production, but because of the kinds of articles that are mass-produced. *Bestand* consists of objects with no inherent value apart from human use. Like plastic, their whole form is dependent on human decisions about what they will be used for and how they will be decorated or packaged.

Note, too, how this ties in with what Heidegger has argued elsewhere about the relation between modern science and technology. Modern science is characterized by an objectification of the natural world, the re-presentation of the world in mathematical terms that necessarily leave out of account its earthiness, thus setting up the possibility for producing objects without true individuality or thinghood. Instead of describing technology as applied science, Heidegger suggests science is more accurately called theoretical technology.

At this juncture, however, Heidegger poses the question of technology in a second sense: Who or what brings about the technological revealing of the world as pure object? Is it, as the positivist or anthropological view of technology suggests, merely the result of human agency? Is modern technology the simple consequence of a personal or collective human decision? Not according to Heidegger. For Heidegger what lies behind or beneath modern technology as a revealing that sets up and challenges the world is what he calls *Ge-stell*.

Ge-stell names, to use Kantian language, the transcendental precondition of modern technology. With this term Heidegger admits to taking a common word, which in its normal unhyphenated form means something like "stand," "frame," or "rack," and giving it a deeper philosophical meaning. "*Ge-stell* refers to the gathering together of the setting-up that sets up human beings, that is, challenges them, to reveal reality, by the mode of ordering, as *Bestand*" or resource (p. 20). The root *stell* alludes to *stellendes* (setting-up). "*Ge-stell* refers to the mode of revealing that rules in the essence of modern technology and is not itself anything technological" (p. 20). The *Ge-stell* or framework is not another part of technology; it is the attitude that is at the foundation of, yet wholly present within, modern technological activity. It is, simply put, the technological attitude toward the world.

From one point of view, *Ge-stell* is an impersonal cognitive framework. But according to Heidegger, in what is undoubtedly his most provocative argument, *Ge-stell* is fundamentally what might be termed an intersubjective not to say impersonal volition. Not only does *Ge-stell* "set up" and "challenge" the world—a description that already hints at volitional elements—it also sets upon and challenges human beings to set upon and challenge the world. Ultimately, it is not just human needs and aspirations that give rise to modern technology. "The essence of modern technology starts human beings upon the way of that revealing through which reality everywhere, more or less distinctly, becomes resource" (p. 24). Heidegger wants to say, perhaps, that the very fact that reality leaves itself open to technological manipulation to some extent calls forth such manipulation. Reality must bear some responsibility for its own exploitation, in the same way that a householder who leaves a door unlocked to some extent invites burglary.

Such an idea now raises the question of technology in still a third sense: This "challenging *Ge-stell* not only conceals a former mode of revealing [i.e., art or craft and its bringing-forth of things], but it conceals revealing itself and with it That [capitalized] wherein unconcealment, that is, truth, happens" (p. 27). Nature or Being hides itself. This is the deepest sense in which modern technology presents itself as a problem or a question. It is also at this level that the relation between the question of technology and the question of Being comes to the fore. When Heidegger speaks of the "That [capitalized] wherein unconcealment . . . comes to pass" he refers to Being. Modern technology not only covers over or obscures the thinghood in things, it also covers over or obscures the Being of beings, and ultimately itself. Technology cannot be understood in terms of technology.

Heidegger's idea can be rephrased in words adapted from Socrates. According to Socrates, dogmatic opinion such as that exhibited by Euthyphro and Ion, not to mention Polus and Thrysamachus, obscures the truth, yet not because it is formally false. Socrates ultimately agrees with Thrysamachus that justice is "the interest of the stronger," provided "stronger" is correctly understood. Thrysamachus's doctrine obscures the truth not because it is inherently false but because it is so easily misunderstood. At the same time, Socrates does not claim to have the truth in any substantive sense. His wisdom consists merely in knowing he does not know. Indeed, it is precisely Socrates' studied or cultivated ignorance—also known as irony—that makes him open to the truth. Modern technology, in Heidegger's view, can be characterized as a kind of reified dogmatism. It is so certain about how to construct this or fabricate that. It has an efficient method or procedure that excludes all other methods or procedures. And in this it does not recognize its own limits; it does not know itself.

(Parenthetically, one can appreciate Heidegger's point without necessarily buying his substantive theory about the character of Being as an *Ereignis* or event that is ever undergoing historical changes in its worldly manifestations. One can simply say that an overwhelming involvement in the material level tends to detract from metaphysical or spiritual reality. Technology is a kind of existential rejection of the metaphysical or spiritual—in the sense of not paying attention—in the same ways that any dogma, precisely in its worldly powerfulness, rejects or ignores the more subtle affairs of mind and heart.)

But what is the way out of this difficulty? How can one respond to this reified dogmatism with the deepest part of one's self? The proper response is decidedly not, says Heidegger, simply to try to get rid of technology, to reject its rejection. Technology "will not be rejected and certainly not smashed" (p. 38). "Technology, whose essence is Being itself, will never allow itself to be overcome by human beings. That would mean, after all, that humanity was the master of Being" (p. 38). The overcoming of technology is more like "what happens when, in the human realm, one overcomes pain" (p. 38). The overcoming of technology must be lived through, extended and deepened, the way grief or pain can be lived through to the point that it becomes an observed grief or pain and thus in some mysterious way is set aside or transcended.

When we suffer or are in pain, we are simply too close to what we are experiencing; we need distance, some self-knowledge, appreciation of who we really are and of our limitations. But this is acquired not through rejection or repression of the pain; it comes only with time and through naming the source of our pain by asking questions and talking about it, rendering our suffering or recalling its background of happiness in poetry and art, sitting quietly and experiencing its presence—or rather what is immediately and unobtrusively there, just on the other side of the curtain of our disturbed feelings—gradually standing back and becoming detached from the tossed surface of our conscious calculations.

It is remarkable that, as if to provide a positive counterpoint to his negative critique of technology, in other works Heidegger mentions just these kinds of experiences: questioning; art and poetry; *Denken*, or meditative, nondiscursive thinking; *Gelassenheit*, or detached acceptance. But at the end of the essay "The Question concerning Technology" he places the emphasis, appropriately enough, on questioning alone. "For questioning," Heidegger writes, "is the piety of thinking" (p. 36). There is, in the end, a sense in which technology must be questioned, and indeed invites its own questioning, in the same way Euthyphro's self-certainty almost begs for Socrates to punch holes in it. And it is this questioning of technology, or the attempt to enclose technological certitude within philosophical questioning, that is at the core of Heidegger's philosophy of technology.

Excursus on Ortega and Heidegger

Having discussed first Ortega and then Heidegger, consider briefly some relations between the two. On the one hand, there are many similarities in the thought of these two philosophers on the issue of technology. At the most superficial level, they are the first two professional philosophers to explicitly address the issue of technology. They also do so within the framework of an existential phenomenology that emphasizes the primacy of practical over theoretical concerns, is sensitive to the issues of freedom and destiny, and recognizes historical or life-world distinctions between different kinds of technology. (Although Ortega distinguishes three periods in the history of technology to Heidegger's two, Ortega's technologies of chance and of the craftsman can easily be interpreted as subdivisions within Heidegger's ancient as opposed to modern technology.) Both assert the deep affinity between humanity and technology while denying that the human is exhausted by the technological or that the essence of technology can be grasped through the technological. Both reject the definition of technology as applied science and view modern science as inherently technological. Finally, both see dangers in too much technology.

On the other hand, whereas Heidegger explicitly rejects the idea of technology as a neutral means—what he also calls the anthropological approach to technology—Ortega seems to affirm such a view. For Heidegger, technology as a form of truth is therefore a means for the revelation of Being, but one that hides its own essence. For Ortega, by contrast, technology is a means for the realization of some human project, although a project that gets hidden within an ever enlarged and penetrating technological effectiveness. For Heidegger technology is relativized by being associated with a regional (or limited) ontology; for Ortega any particular technics denotes a regional (or specialized) anthropology. As one astute commentator summarizes the difference, whereas Heidegger presents the human "as a means of access to the mysterious ground of all that is, as an opening or clearing for Being," Ortega is "content to transform human life itself into the radical reality or foundation." [18]

It is nevertheless crucial to note that for Ortega and Heidegger the projection of the human into the world is not a "natural" or "organic" activity as it is with, say, Kapp or Gehlen. Human technics—as opposed to animal technics such as spiderwebs, bird's nests, and beaver dams—derive from a radical rupture in the organic or natural world. As Ortega says in "Ensimismamiento y alteración," the long essay that introduces "Meditación de la técnica" and is posthumously incorporated into *El hombre y la gente* (1957),

> human beings are technical, are capable of modifying their environment to fit their sense of convenience because they take advantage of every respite that things allow in order to retire within themselves, to enter into themselves and form ideas about the world, about things and their relations to them, to forge a plan of attack upon circumstances, in short, to construct an inner world. From this inner world they emerge and return to the outside. But they return ... with *selves* they did not have before ... in order to impose their wills and designs, to realize in the outside world their ideas, to mold the planet according to the preferences of their interiority.[19]

At the same time, this interior world reveals no transcendent solutions to technical problems (Dessauer) nor even Being as *Ereignis* (Heidegger), but only itself, the human reality of estranged worldliness.

> Far from losing themselves in this return to the world, on the contrary, human beings carry themselves into the other, project themselves energetically, masterfully, upon things, that is, convert the other—the world—little by little into the human. Humanity humanizes the world, injects it, impregnates it with its

own ideal substance, and it is possible to imagine that, one day in the distant future, this terrible external world will become so saturated with the human that our descendants will be able to traverse it as today we move about within our most intimate selves—it is possible to imagine that the world, without ceasing to be, will become converted into something like a materialized soul, and, as in Shakespeare's *Tempest*, the winds will blow at the bidding of Ariel, the elf of Ideas.[20]

With such a suggestion, however, Ortega comes close to transforming a humanities philosophy into an engineering philosophy of technology.

As a further but related aside, one can consider the problem of Heidegger's commitment to National Socialism in contrast to Ortega's antifascism. As Michael Zimmerman (1990) has shown in abundant detail, Heidegger developed a philosophy of technology that unites a reactionary modernism with a view of the historicity of Being. As a result of this union, some critics have argued an essential relation between Heidegger's metaphysics and Nazism. The example of Ortega could, however, serve to qualify such a judgment. Ortega, too, argues a historicist metaphysics and historicist philosophical anthropology, while developing a nuanced critique of many of the weaknesses of culture under the influence of industrial technology—but Ortega was at the same time a resolutely progressive modernist.

Jacques Ellul: Technology as the Wager of the Century

During the same period when Heidegger was formulating the question concerning technology, Jacques Ellul was developing a systematic analysis of "la Technique" as the most important societal phenomenon of the modern world. According to Ellul, capital is no longer the dominant force it was in the nineteenth century; instead it is "technology," which he defines as "the *totality of methods rationally arrived at and [aiming at] absolute efficiency* (for a given stage of development) in *every* field of human activity."[21]

Indeed, it is Ellul's aim to offer for the twentieth century the same kind of orientation toward essentials that Marx's *Das Kapital* (1867) once provided. As Ellul says in a later autobiographical reflection on that period during which he began studies that would culminate in *La Technique* (1954): "I was certain . . . that if Marx were alive in 1940 he would no longer study economics or the capitalist structures but technology. I thus began to study technology using a method as similar as

Jacques Ellul. Drawing by Dirk Leach.

possible to the one Marx used a century earlier to study capitalism" (1981a, p. 155). Furthermore, all the work conceived during that period

> was intended to be, with few exceptions, part of the detailed analysis of this technological society. For example, *La Technique* [1954] studies this society as a whole; *Propagande* [1962] examines the technical means that serve to alter opinion and transform the individual; *L'Illusion politique* [1965] is the study of what politics becomes in a technological society; *Métamorphose du bourgeois* [1967] looks at the social classes in a technological society. My two books on revolution pose the question of what kind of revolution is possible in a technological society. . . . And finally, *L'Empire du Non-Sens* [1980] is the study of what art becomes in the technological milieu.[22]

La Technique, translated into English as *The Technological Society* (1964), provides the fundamental analysis by distinguishing between what he calls "technical operations" and "the technical phenomenon." Technical operations are many, traditional, and limited by the diverse contexts in which they occur; the technical phenomenon—or "la Technique"—is one, and constitutes that uniquely modern form of making and using artifacts that tends to dominate and incorporate into itself all other forms of human activity. With the technical phenomenon or the comprehensive pursuit of efficiency, "technique has taken over the totality of human activities, not only those of productive activity" (1954, p. 2).

In his "characterology" of modern technology Ellul identifies it as artificial, self-augmenting, universal, and autonomous. It replaces the natural milieu with one increasingly fabricated by human beings. As the common phrase has it, "The solution to the problems of technology is not less but more technology." It is progressively the same everywhere and seems to increase according to its own laws. These characteristics are manifested in economics, politics, and even in areas now conceived in technological terms as "human resources." Medicine, education, sports, and entertainment also become subject to input-output, cost-benefit analysis in search of "the one best way" to achieve results (p. 75; the phrase is in English in the original).

The contrast between technical operations and the technical phenomenon resembles that between biotechnics and monotechnics in Mumford. Technical operations include the technics of chance and craft technics of Ortega, while the technical phenomenon includes his technics of the technician. The challenge of the technical phenomenon is precisely that it resists incorporation into or subordination to nontechnical attitudes and ways of thinking. It explains other actions as forms of itself and thereby transforms them into itself. It constitutes, as it were, the social manifestation of Heidegger's *Ge-stell.*

Mumford provides a formal contrast between these two ways of being technological and argues the superiority of polytechnics with an ideal of humanistic pluralism not unlike that espoused by Marx, for whom it is desirable "to do one thing today and another tomorrow, to hunt in the morning, fish in the afternoon, rear cattle in the evening, criticize after dinner, just as I have a mind, without ever becoming hunter, fisherman, shepherd or critical critic."[23] Ortega probes the philosophical-anthropological foundations of the possibility of any technology. Heidegger stresses the epistemological-ontological character of modern technology. Ellul, however, elucidates the "characterology" of the technical phenomenon in terms of seven general characteristics of modern technology: rationality, artificiality, self-directedness, self-supporting growth, indivisibility, universality, and autonomy. These characteristics are further explored in chapters dealing with how they are manifested in and transform the economy, the state, and what Ellul calls human technologies (in education, work, advertising, recreation, sports, and medicine).

Ellul's view, especially as elaborated by Langdon Winner in *Autonomous Technology* (1977), has sometimes been termed a "technological determinism." Recent historical and sociological criticism of technological determinism, arguing that technologies are as much "social constructions" as they are technical constructions, can easily be read as challenging Ellul's characterology with a kind of technological relativism. For instance, Trevor Pinch and Wiebe Bijker have shown that the evolution of the bicycle from an unstable device with a large peddle-driven front wheel to the much more stable machine with two equal wheels and a chain-linked rear-wheel drive was anything but linear.[24] There were many fits and starts in all sorts of diverse technical directions, and the bicycle was subject to considerable interpretive flexibility among various socioeconomic groups. But for Ellul—who argues for technological determinism in only a highly qualified sense, that is, under certain social conditions—that numerous sociocultural and economic factors are taken up into the technical process, and that at different times and places the search for the "one best" solution to technical problems can yield superficially different results, in no way undermines the comprehensiveness of the technical phenomenon.[25]

For Ellul technological determinism, insofar as it exists, is the result of a societal bet. Indeed, in contrast to Heidegger, the deep questioning of this new way of being-in-the-world is to recognize it as the bet or wager of the century. What is happening with technology is not some unqualified conquest of nature but the replacement of the natural milieu with the technical milieu. The modern gamble concerns whether this new milieu, in contrast with the natural milieu, will be better or ultimately even possible. To some extent the wager is the opposite of Pascal's; it bets on the human ability to know and control or to act with good intentions. Such a bet is no sure thing. Indeed, in one of his more recent books he speaks of the "technological bluff" and describes even the philosophy of the absurd as reflecting and infected by the technical milieu.

To throw this wager or secular faith into the boldest possible relief, Ellul places it in dialectical contrast with biblical faith. As a dialectical contrast to *La Technique,* for instance, Ellul writes *Sans feu ni lieu* (1975, although written much earlier). Whereas technology is the attempt of human beings to create their home in this world, the Bible denies that they are ever truly at home here (see Matt. 8:20 and Luke 9:58). In his richly detailed biblical studies Ellul is able to propose a more explicit alternative to the technology of the technician than does either Ortega or Heidegger. Like Mumford, he invests considerable imagination in the alternative to the technological way of being in the world. But unlike Mumford, this alternative is not just an alternative technology, the aesthetically pleasing urban landscape. The biblical view of the city is quite different from technical *and* aesthetic ones.

Although he is sometimes accused of leaving things as much up in the air as either Ortega (with his implicit hope for a new creative burst of culture) or Heidegger (with his desire to accept and work through the destiny of modern technology), Ellul does provide concrete guidance even for those who do not inhabit the city of faith. At the same time he expresses more affinities with Heidegger than with Ortega. But in place of *Gelassenheit* or detachment, Ellul argues for an ethics of nonpower that would sharply delimit technical practice.

> An ethics of nonpower—the root of the affair—is obviously that human beings agree not to do everything they are able to do. Nevertheless, there is no more ... divine law to oppose technology from the outside. It is thus necessary to examine technology from the inside and to recognize the impossibility of living with it, indeed of just living, if one does not practice an ethics of nonpower. This is the fundamental option.... [W]e must search systematically and willingly for nonpower, which of course does not mean accepting impotence ... , fate, passivity, etc. (1983, p. 16)

Thus not only will such an ethics of nonpower seek to set limits, it will pursue freedom (from technology) and thereby introduce new tensions and conflicts into the technical world. It will even turn the practice of transgression (taking drugs, breaking sexual taboos, etc.) against that very technical phenomenon that makes possible the typically modern transgressions. It will turn off television sets, drive cars at slower speeds, and turn away from overconsumption and environmental pollution, all of which can engender new ways of speaking and listening, building and inhabiting, thinking—which in turn can be nourished by and promote not only the freedom to question but also a certain countertechnical wager.

CHAPTER THREE

• • • •

From Engineering to Humanities Philosophy
of Technology

The basic contrast of chapters 1 and 2 is between the philosophy of
technology as developed by persons representing the engineering and
the humanities traditions. Different in both historical origins and basic
orientations, engineering and humanities approaches to the philosophy of technology are necessarily somewhat at odds. Although one in
their concern with technology, they differ over how this concern is to
be articulated. The challenge of difference is judgment.

The Two Philosophies in Tension: A Dialogue

Engineering philosophy of technology beings with the justification of
technology or an analysis of the nature of technology itself—its concepts, its methods, its cognitive structures and objective manifestations. It then proceeds to find that nature manifested throughout human affairs and, indeed, even seeks to explain both the nonhuman
and the human worlds in technological terms. Culture is a form of
technology (Kapp); the state and economy should be organized according to technological principles (Engelmeier and Veblen); religious
experience is united with technological creativity (Dessauer and
García Bacca). For engineers, the "of technology" in the phrase "philosophy of technology" is a subjective genitive, indicating the subject or
agent. Engineering philosophy of technology might even be termed a
technological philosophy, one that uses technological criteria and paradigms to question and to judge other aspects of human affairs, and
thus deepen or extend technological consciousness.

Humanities, or what might also be called hermeneutic philosophy
of technology, seeks by contrast insight into the meaning of technology—its relation to the transtechnical: art and literature, ethics and
politics, religion. It typically beings with nontechnical aspects of the

human world and considers how technology may (or may not) fit in or correspond. In its attempt to appreciate the nontechnical aspects of human experience and to bring nontechnical criteria to bear on the questioning of technology, it reinforces an awareness of the nontechnological. For humanities scholars, the "of technology" in the phrase "philosophy of technology" is an objective genitive, indicating the object being dealt with; humanities philosophy of technology refers to efforts by philosophers and others to take technology seriously as a theme for reflection—from some nontechnological point of view. Technology can be interpreted as a special myth (Mumford), in relation to human self-definition (Ortega), as posing ontological questions (Heidegger), or as a risk-fraught attempt at total control (Ellul). But in each case technology is related to a nontechnological dimension of reality.

The word "hermeneutics" comes into play in this context because of the central place interpretation occupies in all such humanities reflection. Hermeneutics, in its original development (Schleiermacher and Dilthey), was an attempt to reach out for sympathetic *understanding* via humanities disciplines rather than for logical *explanation* via scientific and technological ones.[1] The hermeneutic or interpretative enterprise is pervaded by personal, interpersonal, and historically conditioned elements, and thus tenuously articulated within a human world of fluctuating intersubjective consensus. One way to define the modern scientific-technological project is to say that it rests upon a firm but narrow agreement about how to construct that limited form of understanding known as explanation. Humanities philosophy of technology, in its interpretations and speculations, subsists within a diverse but fragile lifeworld in opposition to the hard-edged presence of economic analysis and utilitarian logic characteristic of engineering emphases in the philosophy of technology.

In some sense, of course, it is unfair to appropriate the term "humanities" for nonengineering philosophy of technology. Certainly engineers commonly think of themselves as humanists—although this is not precisely the same as those who practice the humanities. Indeed, they pursue their profession expressly because they view it as humanizing—which, again, is not exactly the same as "humanitizing" (to coin a term). Their efforts are necessarily grounded in some conception of the human. For engineer philosophers of technology, however, this self-understanding is commonly taken as given, accepted in a largely unproblematic manner. Qua engineers, they do not question it, and they commonly regard questions raised by others as distracting or beside the point. After all, has history not led to an increasingly world-wide acceptance of scientific engineering and technology? Do not all peoples, irrespective of their cultures, finally opt for or adopt modern technology as soon as they can? As a result, engineer philosophers of technology typically undertake to translate other human pursuits into their language, to view the larger human world in technological terms. Indeed, technical terminology has itself become a kind of Esperanto of the emerging world technoculture.

Humanities philosophy of technology, however, does approach the human as a question, even as the most fundamental question—perhaps one that cannot ever be definitively answered. As such, whenever representatives of the humanities—who are commonly called "humanists" because of their commitment to the primacy of this question—come in contact with new or different languages, their impulse is not just to translate into some already known idiom, but to try to learn, interpret, and understand them. Translation, even of the most sophisticated sort, tends to leave a residue of untranslated and untranslatable meaning. Technical Esperanto is seen as a language without roots in a particular time or place, a kind of free-floating postcultural means not of poetry but of minimal communication. Aspects of the human are obscured and diminished.

In the present instance, however, it is perhaps ironic to speak of a hermeneutic philosophy of technology, just because this approach to the philosophy of technology so often seems to reject learning the new language of technology. Doesn't philosophy of technology in the subjective genitive really pay more attention to technology than philosophy of technology in the objective genitive? Humanities philosophy of technology too often seems to be a philosophy of antitechnology and to close itself off in romantic subjectivity from technological aspects of the human—aspects that are fundamental constituents of the contemporary techno-lifeworld, if not of the human world at all times and places.

Technological philosophy can criticize humanities philosophy of technology for being too speculative or based on too narrow if not unempirical foundations. Humanities thinkers do not understand what they talk about, say engineer-philosophers. C. P. Snow, for example, once contrasted the scientific and literary intellectuals, arguing that the former know more about Shakespeare than the latter know about thermodynamics.[2] Engineering curricula require engineering students to take humanities courses, but how many engineering courses are humanities students required to take?

Humanities thinkers reply (no doubt with some uneasiness), that a commonsense or ordinary experiential acquaintance with the technological should be a firm enough basis for understanding its meaning, and that becoming mired in the specialized details of technology and its many processes tends to obscure relationships to nontechnological aspects of the human. Their rejection of technological discourse may even be to some extent an act of self-defense, in the face of a language that bears down with the force of history and appears ready to reduce all others to interesting but nonessential dialects. The dominating power and attractive glamour of modern technology, they may suggest, threaten to deprive culture of depth by distracting the critical faculties from moral issues and subverting those articulated and often geographically grounded diversities that historical destiny has placed in the care of the humanities.

Engineering philosophers of goodwill might nevertheless respond that humanities concern for critical analysis and moral sensitivity easily cloaks irrationality and bad judgment. Engineer-philosophers can also appeal to their philosophical counterparts in the humanities to admit—based on a recognition of the historical character of the interpretative enterprise—that the commonsense understanding is historically conditioned. Today the ordinary person is better acquainted with the details and principles of science and technology than were the experts of premodern times. Humanities scholars themselves use computers and understand the basic principles of flight. It is obvious that there has been a deepening of commonsense acquaintance with the technological, so surely there exists some basis for a limited rapprochement between the competitive claims for expertise and ordinary knowledge. Interdisciplinary research that links engineering and humanities disciplines in the common investigation of issues and problems points in the same direction.

Two Attempts at Reconciliation

At least two attempts to bridge these two traditions in the philosophy of technology deserve notice. One emerged within the engineering community, another within the philosophical community.

The "Mensch und Technik" Committee of the VDI

After World War II in Germany, engineering-related philosophy of technology experienced a period of sustained, systematic growth— one that in part can be attributed to the strong German philosophical tradition and to guilt about the role engineers had played in the war.

As engineer-architect Albert Speer summed up this view at the end of his memoir *Inside the Third Reich*, "Dazzled by the possibilities of technology I devoted crucial years of my life to serving it. But in the end my feelings about it are highly skeptical."[3] The counterargument, of course, was that only through technology could one find a way out of the postwar destruction.

In this climate of guilt and determination the Verein Deutscher Ingenieure (VDI, or Society of German Engineers) became an institution where members met to reflect on their special responsibilities. The re-founding of the VDI in 1947 was inaugurated by a conference on the theme "Technik als ethische und kulturelle Aufgabe" (Technology as ethical and cultural task). Subsequent meetings brought together engineers and philosophers to address the special challenges posed to German engineers by World War II and prospective developments in technology. Four of these especially focused engineering-philosophical discussion:

- At Kassel, in 1950, the theme was "Über die Verantwortung des Ingenieurs" (Concerning the responsibility of the engineer).[4]

- At Marburg, 1951, it was "Mensch und Arbeit im technischen Zeitalter" (The human being and work in the technological age).[5]

- At Tübingen, 1953, "Die Wandlungen des Menschen durch die Technik" (Changing humanity through technology).[6]

- At Münster, 1955, "Der Mensch im Kraftfeld der Technik" (Human beings in the force field of technology).[7]

The Kassel meeting drafted the "Engineer's Confession," a kind of moral pledge, thus emphasizing professional ethics. But in later meetings Dessauer—who had returned with honor from an exile first in Turkey and then in Switzerland, forced there by his active opposition to Hitler—was prevailed on to rewrite his *Philosophie der Technik*. Thus in 1956, on the occasion of the VDI centenary, he published his comprehensive *Streit um die Technik*, which built on earlier work and attempted to initiate a general discussion with Heidegger, Jaspers, the Frankfurt school, and others; it immediately became a reference book for questions concerning philosophical aspects of technology.

In the same year engineers and philosophers formed a VDI central committee on "Mensch und Technik" (Humanity and technology), eventually renamed the committee on "Der Ingenieur in Beruf und Gesellschaft" (The engineer in the profession and society). This com-

Background: The Verein Deutscher Ingenieure Haus, Düsseldorf. Foreground: A meeting of the Mensch und Technik Commission of the VDI. Counterclockwise from the center (right of vacant chair), those visible are Ernst Oldemeyer, Alois Huning, Friedrich Rapp, H. H. Holz, Hans Sachsse, Günter Ropohl, and at the head of the table, Karl Landfried, vice president of the Universität Kaiserslautern, where the meeting is taking place. Photos courtesy of Alois Huning. Montage by Emilie Jaffe and Matthew Jaffe, Small Time Press.

mittee was divided into sub-committees on "Pedagogy and Technology," "Religion and Technology," "Language and Technology," and "Sociology and Technology," as well as "Philosophy and Technology." Since its inception members such as Alois Huning, Hans Lenk, Simon Moser, Friedrich Rapp, Günter Ropohl, Hans Sachsse, Klaus Tuchel, and Walther Christoph Zimmerli—some of whom teach in technical institutes or have degrees in both engineering and philosophy—have become the most prominent philosophers of technology in Germany.

Moser, Lenk, and Ropohl, all associated with the technical University of Karlsruhe, have also been the center of what has been called the "Karlsruhe school" of the philosophy of technology.[8]

Moser proclaimed early on that "the ideal 'philosopher of technology' should be a productive philosopher and an active engineer"[9]—a position more than once echoed by Lenk and Ropohl (himself both engineer and philosopher). Indeed, in their own "Toward an Interdisciplinary and Pragmatic Philosophy of Technology: Technology as a Focus for Interdisciplinary Reflection and Systems Research" (1979) Lenk and Ropohl critique both "traditional philosophical [and] technological approaches" before outlining their own "new epistemological and social-philosophical approaches." They conclude that

> just as the multidimensional problems of the technological world cannot be approached with some prospect of success without the participation of social science generalists and philosophical universalists, neither can they be solved adequately and realistically without the corrective input of experts in engineering and the technological sciences, including general technology, of systems analysts and systems planners. A fertile and realistic cooperation spanning antiquated departmental and academic borders, especially between the natural sciences and humanities, the social and technological sciences, is more important today than ever before. (p. 47)

This bridge-building spirit has animated the "Mensch und Technik" committee since its founding.

An initial project of the "Mensch und Technik" committee was the critical evaluation of different interpretations of technology, with articles being published in *VDI-Nachrichten* (the weekly newspaper of the VDI), then collected in annual *Mensch und Technik: Veröffentlichungen*.[10] Throughout the 1960s, "Mensch und Technik" committee work was done largely in committees and occasional reports. But in 1967, in association with the dedication of the new VDI headquarters building in Düsseldorf, the VDI inaugurated a biannual "Day of the Engineer" conference to address general themes. The lectures from the first year on "Technik und Gesellschaft" [Technology and society] were published in 1968 with contributions by philosophers, engineers, economists, and others, in a volume that became a point of reference for other discussions.[11] The same year also witnessed publication of Tuchel's influential *Herausforderung der Technik*, with an eighty-page essay titled "Technological Development and Social Change" followed by an interpretative anthology, "Documents on the Classification and

Interpretation of Technology," plus a dictionary of key terms and a chronology of modern technological development.

In 1970 the VDI organized a three-day public conference in Ludwigshafen on "Wirtschaftliche und gesellschaftliche Auswirkungen des technische Fortschritts" (Economic and social consequences of technical progress), which received extensive media coverage. The proceedings were published in paperback[12] and, as a result of VDI sponsorship, further promoted philosophical, ethical, and political discussions within the engineering community and beyond. The Ludwigshafen conference also for the first time in Germany introduced "technology assessment" as an issue for public discussion.

During the 1970s and 1980s the "Philosophy and Technology" subcommittee focused on relations between technology and values, first on value changes in German public opinion related to the growing awareness of environmental problems, then on issues of technology assessment. A first research project yielded two volumes, both edited by Moser and Huning: *Werte und Wertordnungen in Technik und Gesellschaft* (1975), and *Wertpräferenzen in Technik und Gesellschaft* (1976). A second culminated in draft general guidelines for technology assessment that ranged over technical and economic efficiency, public welfare, safety, health, environmental quality, personal development, and quality of life. These guidelines were formally adopted in 1991 as *VDI-Richtlinien 3780*.[13]

From the 1970s through the early 1990s VDI "Mensch und Technik" members also published a variety of works that moved beyond conference proceedings and research reports into systematic monographs on general themes. As a result, philosophical discussion of technology in Germany has been dominated by the inner circle of the philosophy subcommittee. Indicative of work this inner circle are:

- Hans Lenk, *Philosophie im technologischen Zeitalter* (1971); Lenk, ed., *Technokratie als Ideologie* (1973); Lenk and Moser, eds. *Techne, Technik, Technologie: Philosophische Perspektiven* (1973); Lenk, *Zur Sozialphilosophie der Technik* (1982); Lenk and Ropohl, eds., *Technik und Ethik* (1987); Walter Bungard and Lenk, eds., *Technikbewertung* (1988); and Lenk and Matthias Maring, eds., *Technikverantwortung* (1991).

- Alois Huning, *Das Schaffen des Ingenieurs* (1974; 2d ed., 1978; 3d ed., 1987).

- Friedrich Rapp, ed., *Contributions to a Philosophy of Technology: Studies in the Structure of Thinking in the Technological Sciences* (1974);

Rapp, *Analytische Technikphilosophie* (1978); and Rapp, ed., *Technik und Philosophie* (1990).

- Hans Sachsse, ed., *Technik und Gesellschaft* (3 vols., 1974–1976); and Hans Sachsse, *Anthropologie der Technik* (1978).

- Walther Christoph Zimmerli, ed., *Technik; oder, Wissen wir, was wir tun?* (1976); and Zimmerli, ed., *Herausforderung der Gesellschaft durch den technischen Wandel* (1989).

- Günter Ropohl, *Die unvollkommene Technik* (1985); and Ropohl, *Technologische Aufklärung* (1991).

Rapp's *Analytische Technikphilosophie* (1978; English trans. 1981) exemplifies both the strengths and the weaknesses of this school in the philosophy of technology, and of its efforts to transcend engineering philosophy of technology. Though he distinguishes his approach from those of the engineer, cultural philosopher, social critic, and systems theorist, Rapp's attempt to synthesize such perspectives or to take an alternative point of view remains sparingly descriptive and explanatory. "The aim of this work," he writes, "is to present a philosophical analysis of technology which takes into account the historical and systematic aspects of technological development, provides a thematically ordered overview of the pertinent problems and basic solutions, and, at the same time, makes a contribution of its own to the relevant issues" (1981, p. 21). As he argues in a preface to the English translation, the situation with technology is just so complex that it requires a healthy dose of empirical analysis before any well-founded metaphysical interpretation becomes possible. The primary task of philosophy of technology is thus simply to draw attention to this complexity and to make explicit the precise character of the technological world, how it could have come about, and what consequences follow from it (p. xii).

In 1990 Rapp also edited the first in a ten-volume VDI-published encyclopedia of *Technik und Kultur*, on *Technik und Philosophie*, with major contributions by Huning, Lenk, Rapp, Ropohl, and Zimmerli.[14] Including nine other volumes on technology and religion, science, medicine, education, nature, art, economics, state, and society, this encyclopedia constitutes a worthy successor to Dessauer's *Streit um die Technik*, which had tried to span a comprehensive spectrum of philosophical questions related to technology.

But as Alois Huning has summarized the VDI achievements, it

succeeded in associating almost all leading authors who had written on the problems of technology with the work of the "Mensch und Technik" group. Only Martin Heidegger, Karl Jaspers, and the Frankfurt School . . . remained uninvolved. . . . [Indeed,] prior to 1970 the contributions of these thinkers were scarcely discussed by members of the VDI. Heidegger's language left his work almost inaccessible to engineers, while Jaspers gave the impression of failing to comprehend technology because . . . he criticized the anthropological consequences of technology for both individual and society. The criticism of technological and instrumental reason by the neo-Marxist Frankfurt School was also generally not accepted by engineers, and as a result they remained unaffected by suggestions Horkheimer and Habermas were offering for the mastering of technology.[15]

Thus, despite being able to involve on occasion such leading humanities thinkers as Kurt Hübner and Hans Blumenberg, the "Mensch und Technik" committee has not influenced intellectual life to the degree one might have anticipated. The expansive criteria for technology assessment certainly reach out to include humanities perspectives, and they are to be commended as constituting the broadest established policy for technology assessment in the world. But those addressed by such guidelines remain primarily members of the technical community—and the absences noted by Huning remain significant, raising questions as to whether the VDI school has fully transcended its engineering-philosophy roots.

Pragmatic Phenomenology of Technology

In contrast to the VDI school, the pragmatic phenomenological approach to technology arose within the philosophical community itself and is most vigorously represented by the work of two American philosophers, John Dewey and Don Ihde. Two Dewey interpreters, Paul T. Durbin and Larry Hickman, have made further important contributions to this approach.

The remarkable point about pragmatism in general and Dewey in particular is that despite the clear influence of technology on the American tradition of pragmatism beginning with Charles Peirce in the mid-1800s, it was not until after World War II that this element was given explicit articulation. At the same time, it is also true that in the 1950s the first English-language article to link philosophy and technology was a deft pragmatist analysis by Joseph W. Cohen. Cohen's "Tech-

John Dewey (1859–1952). Drawing by Anne Sharpe.

nology and Philosophy" (1955) begins by stating the ontological priority of technology over science. Human beings were "Homo faber
before [becoming] Homo sapiens. . . . Out of technical processes and
slowly accumulating skills, out of combinations and recombinations of
the tools and expertise of many peoples came the eventual theoretical
organization of technology into science" (p. 409). Having defined
"technology in a sense broad enough to include science" (p. 409), Cohen criticizes the views of technology as evil and as neutral for sunder-
ing intelligence and value. The former is "tradition-bound, anachronis-
tic" romanticism; the latter is a rationalization that lets scientists and

engineers avoid "any disturbing thought that they are also, even as scientists and technicians, concerned with values and as such carry a burden of responsibility for the uses to which their work is put" (p. 413). The truth is that technology is bound up with values, and this insight—which he attributes to Thorstein Veblen, John Dewey, and C. E. Ayers—is "the distinctive contribution of American thought to the philosophy of technology" (p. 413).

According to Dewey, Cohen says, technology is both intelligence and value because "all ideas are intellectual tools employed in experimental operations for the solution of the problems which arise in experience" (p. 416). Dewey's philosophy of technology thus sees technology not as something opposed to value (and hence to democracy, ethics, art, etc.), as antitechnology cultural critics would have it, or as neutral with regard to value, as scientists and engineers think. It is a value, one that must be integrated with other values in culture not by monistic, technocratic management but through "pluralistic planning" (p. 418). However, although Cohen claims that "Dewey's experimental pluralism" (p. 417) provides a better analysis of the role of technology in culture than does Lewis Mumford, and that "instrumentalism is a name for competent reflective thinking in *every* sphere of culture" (p. 416; italics added), he also maintains that "technology and science [provide] the clearest pattern of such thinking" (p. 416). Thus, although originating in philosophy, pragmatic instrumentalism could be interpreted as a sophisticated engineering philosophy of technology that sees other realms of culture as diminished forms of technology.

Durbin (1972, 1978), inspired more by George Herbert Mead than by Dewey, provides both methodological and substantive developments of Cohen's position. For Durbin a philosophy of technology "amounts to a statement as to what one feels a good technological society *ought* to be like, plus some persuasive arguments aimed at getting influential others to agree" (1978, pp. 67–68). In his attempt to carry forward Mead's typically pragmatic argument for "the application of the experimental method to social problems" (p. 71), Durbin calls for members of the technical community to become involved in reform movements related to social problems associated with technological change. In *Social Responsibility in Science, Technology, and Medicine* (1992) he provides detailed descriptions of how progressive social activists in the fields of education, medicine, media, computers, industry, and public interest groups organized around concern about nuclear weapons, nuclear power, and the government can in fact ameliorate problems associated with technology. But it is Hickman's book *John*

Dewey's Pragmatic Technology (1990) that both retrieves the central texts from the Dewey corpus and provides the most extended defense of a pragmatist philosophy of technology. Hickman's study is also undertaken with enlightening and appreciative reference to what a number of other contemporary pragmatists have to say about technology.

Hickman too sees the key to Dewey's philosophy of technology in his instrumentalist epistemology and the priority of practice over theory. As he points out, the pivotal chapter 4 of *The Quest for Certainty* (1929), one of Dewey's most mature and comprehensive works, is a paean to technology. Here Dewey explicitly says that "there is no difference in logical principle between the method of science and the method pursued in technologies" (p. 68). Indeed, in a remarkable acknowledgment that Hickman recovers from a text of over twenty years later, Dewey says, "It is probable that I might have avoided a considerable amount of misunderstanding if I had systematically used 'technology' instead of 'instrumentalism' in connection with the view I put forth regarding the distinctive quality of science as knowledge." [16]

This technological theory of knowledge provides the basis for a contrast between premodern and modern science: "Greek and medieval science formed an art of accepting things as they are enjoyed and suffered. Modern experimental science is an art of control" (p. 80). But such a difference is not just to be noted, it is to be praised:

> The remarkable difference between the attitude which accepts the objects of ordinary perception . . . and that which takes them as starting points . . . is one which . . . marks a revolution in the whole spirit of life, in the entire attitude taken toward whatever is found in existence. . . . [N]ature as it already exists ceases to be something which must be accepted and submitted to, endured or enjoyed, just as it is. It is now something to be modified, to be intentionally controlled. It is material to act upon so as to transform it into new objects which better answer our needs. (pp. 80–81)

Although the "art of accepting things" that he attributes to Greek and medieval science raises questions here, nevertheless, as Hickman rightly sums up Dewey's position, "What Dewey thought significant about inquiry, and what he thought discloses its technological character, is that *every reflective experience is instrumental to further production of meanings, that is, it is technological*" (pp. 40–41; Hickman's emphasis).

But if virtually all knowing, and indeed all human activity, is or ought to be at its core technological, this raises the specter of reductionism. One reviewer of *The Quest for Certainty* even commented on what he saw as Dewey's "reduction of philosophy to technology."[17] As Hickman states the problem, the charge is that "there is a reduction of the function of many tools to the function of one specific type of tool, the extra-organic" (p. 43). His reply is that the criticism presumes "the existence of a sharp line between organism and environment. . . . [But] for purposes of inquiry, the skin is not a very good indicator of where the organism stops and the environment begins" (p. 43). In reality, as Dewey says in *Art as Experience* (1934), "There are things inside the body that are foreign to it, and there are things outside of it that belong to it. . . . On the lower scale, air and food materials are such things; in the higher, tools, whether the pen of the writer or the anvil of the blacksmith, utensils and furnishings, property, friends and institutions—all the supports and sustenances without which a civilized life cannot be."[18] Note the way this passage suggests the idea of tools as extensions of the body. Note too that this reply to one possible formulation of the charge of reductionism does not consider the possibility that if all life is technological then the concept of technology becomes vacuous.

Independent of these questions, Hickman proceeds to develop a comprehensive interpretation of Dewey's philosophy of art and his account of the history of technology, then to use the pragmatic philosophy of technology to reassess the fears of technological determinism articulated by Ellul and others. In an epilogue on "responsible technology" that is in complete harmony with Durbin, Hickman defines technology as "the sum of concrete activities and products of men and women who engage in inquiry in its manifold forms: in the sciences, in the fine and useful arts, in business, in engineering, and in [politics]" (p. 202). Then he adds, "Where technology fails to be responsible, it is not because technology as method has failed, but because inquiry and testing have been misdirected, subsumed to nontechnological ends, or aborted. Ends have been dissociated from means. Fixed religious or political ideologies have taken the place of legitimate, testable inquiry. Economic and class interests have intervened where experimentation would have been appropriate" (p. 202). In short, the problems associated with technologies are caused not by technologies but by nontechnologies—and are to be solved not by less technology, but by more.

As Hickman points out in considering the views of other pragmatists, Webster Hood (1982) has made a case for Dewey as a phenomenologist of technology. The relation between pragmatism and phenom-

Don Ihde, with a background of possible human-instrument-world relations. Photo courtesy of Don Ihde. Montage by Emilie Jaffe and Matthew Jaffe, Small Time Press.

enology is also one that has been the explicit concern of Ihde, one of the leading American phenomenologists and the author (except for one stillborn effort)[19] of the first English-language monograph on philosophy of technology.

Technics and Praxis: A Philosophy of Technology (1979) begins by distinguishing between idealist and materialist attitudes toward technology. The former views technology as applied science, the latter sees science as theoretical technology. Siding with the latter approach, Ihde sketches a phenomenology of human-technology-world relations. He then reflects on some experiential implications of modern technologies such as computers and electronic music and concludes by examining the pioneering phenomenological approaches to technology found in the work of Martin Heidegger, Hans Jonas, and European existentialists. Phenomenologically Ihde is especially good at distinguishing those technologies that extend or embody human experience (the magnifying glass) and those that call for human interpretative or hermeneutic reflection (the thermometer).

Carrying this phenomenological analysis further, Ihde uncovers a basic amplification-reduction structure to all technology-mediated relations. An embodiment technology like the magnifying glass, for instance, amplifies certain microfeatures of the world, but only by reducing our field of vision. "With every amplification, there is a simultaneous and necessary reduction. And ... the amplification tends to stand out, to be dramatic, while the reduction tends to be overlooked. [The result is that] the instrument mediated entity is one which, in comparison with in the flesh relations, appears with a different perspective" (p. 21).

Based on this amplification/reduction, Ihde argues against any view of technology as neutral or pure transparency with utopian possibilities—a perspective that relies on emphasizing amplification while ignoring reduction. At the same time he rejects the idea that technology is a "Frankenstein phenomenon" (p. 40) opposed to the human—a view that emphasizes only reduction while ignoring amplification. Nevertheless, at the same time that he rejects "a hard technological determinism" he admits there are often "latent telic *inclinations*" (p. 42) in technologies that predispose human beings to develop certain life forms over others.

Existential Technics (1983) builds on this suggestion by analyzing how technology becomes involved not just in our interpretation of, or theory construction about, the natural world, but also in how technology influences our understanding of what it is to be human, that is, our self-image or self-interpretation. As he concludes in the introductory, programmatic chapter, "We end up modeling ourselves on the very 'world' we project and interpret ourselves in terms of technology" (p. 22). Part 2 of *Consequences of Phenomenology* (1986), especially the essay "Technology and the Human: From Progress to Ambiguity," carries forward such existential concerns and concludes that "the deeper question of technics and the human remains one about the variable possibilities of our seeing itself" (p. 90)—including the seeing of ourselves.

Technology and the Lifeworld: From Garden to Earth (1990) and *Instrumental Realism: The Interface between Philosophy of Science and Philosophy of Technology* (1991), originally conceived as one book, constitute the most comprehensive presentation of Ihde's philosophy of technology. *Technology and the Lifeworld* rests on the argument that human beings are not able to lead a nontechnological life in some garden state because on the earth they are inherently technological creatures. Having made his special argument to this effect, Ihde reviews the relevant insights of others in the phenomenological tradition (Heidegger, Husserl, and Merleau-Ponty) and describes the pivotal historical importance of technology to the rise of modern science (Galileo and the telescope).

Then, following a reprise of the human-technology-world analysis from earlier works, in the latter half of the book Ihde extends his reflections on the existential import of technology and examines these in cross-cultural perspective. As he emphasizes in *Philosophy of Technology: An Introduction* (1993), he "*celebrates* a certain disappearance of a 'core,' or a 'foundation'" to culture in the technological society, but admits the need to find "post-enlightenment means of securing intercultural . . . modes of tolerance and cultural pluralism" (p. 115). *Instrumental Realism* constitutes an extended dialogue, especially related to the phenomenological interpretation of technoscience, with the work of such authors as Thomas Kuhn, Michel Foucault, Heidegger, Hubert Dreyfus, Patrick Heelan, Ian Hacking, Robert Ackerman, Peter Galison, and Bruno Latour.

Ihde not only wrote the first monograph on philosophy of technology in English, he has also produced the most extensive corpus devoted to the subject and has established a book series devoted to philosophy of technology. Indeed, Hickman's book on Dewey's philosophy of technology was the first contribution to this series. He further emphasizes that his vision of philosophy *of* technology is distinct from other studies in philosophy *and* technology, which tend to be antitechnology. But in light of the importance he gives to technology in human experience, his strong sympathies with pragmatism, and his criticisms of the critics of technology, again it is not clear to what extent his phenomenological philosophy of technology is truly other than a sophisticated and subtle engineering philosophy of technology.

The Question of Marxist Philosophy of Technology

Against this background it might be argued that one of the most important bridges in the philosophy of technology—if not a substantial

The tomb of Karl Marx (1818–1883) in London.
Photo by Victoria Leanse.

philosophy of technology in its own right—is the analysis of technology that constitutes a central theme in the thought of Karl Marx, a contemporary of Kapp, who in fact shares many of Kapp's basic views. But whereas Kapp was influenced in his reform of Hegelianism by the newly emerging discipline of geography, Marx was influenced by the newly emerging discipline of sociology—especially by those sociologists also known as socialists, who pointed up inherent disorders in the relations between technology and society.

The founding but inadequate insight Marxism attributes to those termed "utopian socialists," the first of whom was Henri de Saint-Simon (1760–1825).[20] The thesis of Saint-Simon is that development of modern science and technology must, in order to promote human welfare, be complemented by a reorganization of the society in which they exist. The social basis of modern or technological society is Christianity, which had "put the land of Europe under cultivation, drained the marshes, and made the climate healthy" as well as "caused roads and bridges to be built [and] organized the largest political community which has ever existed." But "having rendered these important ser-vices," the Christian religion, in its laws, judges, ethics, and preachers, had also "become a burden on society"[21] and thus in need of transfor-mation.

> New Christianity is called upon . . . to link together the scien-tists, artists, and industrialists, and to make them the manag-ing directors of the human race [and] to put the arts, experi-mental sciences and industry in the front rank of sacred studies. . . . In the end, New Christianity is called upon to pro-nounce anathema upon theology, and to condemn as unholy any doctrine trying to teach men that there is any other way of obtaining eternal life, except that of working with all their might for the improvement of the conditions of life of their fellow men.[22]

Such management of science and technology by this linkage of scien-tists, artists, and industrialists constitutes an early vision of tech-nocracy.

For Marx, however, rhetoric and technocratic management are not enough. Social reordering of the productive or technological society must be based on structural knowledge of the production process, which is in fact more than just technical. Marx's own analysis of this process begins with the idea that human life is essentially *"sensuous activity, practice"* ("Theses on Feuerbach" 1). This activity "appropriates particular nature-given materials for particular human wants" (*Das Kapital*, 1867, vol. 1, pt. 1, chap. 1, sec. 2 [trans., 1967, p. 42]). Marx's argument is a correction to Hegel, an attempt to extend the Hegelian dialectic of consciousness into the real world. Two decades earlier in the *Economic and Philosophic Manuscripts of 1844* Marx had suggested that although Hegel grasps *"work* (or labor) as the *essence* . . . of human-ity," "the only work Hegel knows and recognizes is the *abstract spiritual* kind."[23] This argument for replacing abstract with real work was fur-ther developed in the *Grundriße* manuscripts of 1857–1858, then first published selectively as *Zur Kritik der politischen Ökonomie* (1859).[24] But Marx's mature effort to take the dialectic, which was left "standing on its head" in Hegel, and turn it "rightside up again"[25] (a project with which Kapp certainly agreed), appropriately enough finds expression in a comprehensive critique of political economy.

When Marx subtitles *Das Kapital* a "critique of political economy," he unites the "critical" tradition of Kant and subsequent German phi-losophy with the practical world of politics and economics. The idea of political (as opposed to household) economy takes its bearings from a reevaluation of technology associated with the Industrial Revolution. Classical political thought was concerned with how to minimize wealth and restrict its pernicious influence. Modern political economy, as defined by Adam Smith and David Ricardo, inquires into the nature and causes of wealth in order to promote a politics— now called "governmental policy"—that maximizes production. Marx's critique of political economy claims to expose the fundamental preconditions of this modern political theory and to correct its shortcomings.

Marx opens his critique with the observation that the capitalist world presents itself not as a world of ideas but as "an enormous collection of commodities, with the individual commodity [*warensammlung*] as its elemental unit." [26] There is thus a fundamental need to analyze this unit, the commodity. The conclusion of *Das Kapital* is that the political economic analysis of commodity rests on two mistakes. Although it recognizes the primacy of making over doing, it fails to appreciate that making is always social; and it takes consumer products to be things independent of making and using processes. The elementary factors of production are labor, the material labored on, and the instruments of labor. But the material labored on is—except with extractive industries such as mining—always the product of some previous production process; the same goes for the instruments of labor. Furthermore, both are simply means within some form of the production process. Thus what is crucial is "not the articles made, but how they are made, and by what instruments" (*Das Kapital*, vol. 1, pt. 3, chap. 7, sec. 1 [trans., 1967, p. 180]). Materials and instruments are only related to production processes; the processes are what is primary.

Because of its uncritical analysis of production primarily in terms of what are now called inputs and outputs—that is, input costs (of materials and labor) and output sales (figured as exchange values of commodities)—political economy failed as a true inquiry into the production process as a whole. Instead, it analyzed production only from the point of view of the members of that particular class, the bourgeois, who were the individual property owners of these processes. Political economy has been tied to limited class interests.

Marx's liberation of political economy from bourgeois class interests entails a new analysis of the production process. It examines "how the instruments of labor are converted from tools into machines" (*Das Kapital*, vol. 1, pt. 4, chap. 13, sec. 1 [trans., 1967, p. 371]) and the way machines themselves tend to become organized into a system in which "the subject of labor goes through a connected series of detailed processes" (*Das Kapital*, vol. 1, pt. 4, chap. 13, sec. 1 [trans., 1967, p. 379]). The basis of this latter transformation is "modern science of technology" [*moderne Wissenschaft der Technologie*] (*Das Kapital*, vol. 1, pt. 4, chap. 13, sec. 9 [trans., 1967, p. 486]), which analyzes the production process into its constituent functions. Marx effectively describes the ways this "new technology" undermines the traditional skills and satisfactions of craft production and places the worker under the specter of an autonomous factory in which labor functions have become equal and interchangeable.

What Marx calls the labor process is at the most basic level "human action with a view to the production of use-values, appropriation of natural substances to human requirements" (*Das Kapital*, vol. 1, pt. 3, chap. 7, sec. 1 [trans., 1967, p. 183]). The instrument-aided creation of artifacts is intended to serve human purposes by transforming materials into useful objects. The specifically human part of this transformation is that it takes place first in the mind, since "at the end of every labor-process, we get a result that already exists in the imagination of the laborer at its commencement" (*Das Kapital*, vol. 1, pt. 3, chap. 7, sec. 1 [trans., 1967, p. 178]). The problem is that the primary use-values that arise from the imaginative engagement of the human being with the world are replaced by exchange-values within capitalist economic life.

This is a problem because, although exchange-value may be rooted in particular, concrete use-values, it is possible to treat artifacts solely as products of material and labor inputs instead of as objects that serve real human needs and purposes. Minimum necessary exchange-value is determined by input costs, primarily of labor, while maximum exchange-value is determined by the market. The difference between the two is profit for the owner of a production process. Since the owner cannot control the market, and since the reduction of labor is possible mainly through technical development, a symbiotic relationship readily develops between the cash nexus and technological progress. To the same degree that use-values recede into the background, there emerges a preoccupation with exchange-values.

One industrial technique to receive an extensive Marxist analysis is the division of labor. Marx analyzes division of labor as it progressed from the "detail work" with hand tools in manufacture (1550–1750) to the mechanized but still "detail" work of machines in the factory system of the Industrial Revolution (ca. 1750–1850). More important, he shows how this "mechanization" of the labor force through the division of labor was a necessary precondition for the mechanization of technology more commonly associated with machinery and the Industrial Revolution. The alienation and physical degradation of the laborer that ensue from this obsession with technology divorced from use-value as a way of enhancing profit are well known.

Marx, however, goes beyond the ontological grounding of technology in human purpose and use-value in a distinctly modern admiration for machines and awe at the power of modern technology to subjugate a hostile nature. Although use-consumption provides production with its immediate purpose, Marx sees the larger purpose as nothing other than production itself, since production "is the real starting point and therefore the predominant factor. Consumption as urgency, as need, is itself an intrinsic factor of productive activity."[27] Absent any idea of human nature as oriented toward something other than itself, Marx declares productivity the "species essence" of humanity. The human being is a producer, an *animal laborans*, whose ultimate fulfillment lies in remaking alien nature into a humanized world. Production in the objective world is the active species life of humanity, and its essential powers are displayed in the history of industry—albeit so far only in alienated form. Although Marx is often thought to subordinate technology to economics, a strong case can be made that he does just the opposite, subordinates economics to technology.

In short, just as technology plays a crucial role in creating the problem of industrial capitalism, it will play an equally crucial role in the creation of socialist society. Not only will a socialized technology eliminate scarcity from the planet, but as the "species life" of humanity it will also eradicate the metaphysical alienation that has challenged and baffled philosophers from Parmenides to Hegel. Technology will accomplish what philosophy and religion have failed to achieve. By making nature the reflection of essential human powers, it will allow humanity finally to affirm itself in and through the world, and not just in the mind or heaven.

Marx also argues that technology further reveals the "fitness of the laborer for varied work, consequently the greatest development of his varied aptitudes" (*Das Kapital*, vol. 1, pt. 4, chap. 13, sec. 9 [trans., 1967, p. 488]). If all jobs are equal, then each worker should be able to do anything he or she wants. The problem is that in a capitalist economy, where individuals own the means of production, workers are wage slaves in the production process. In a communist society, where the means of production are no longer privately owned, a worker will be free to "become accomplished in any branch he wishes"; it will be possible to "do one thing today and another tomorrow." [28] Carried through to its perfection and liberated from the capitalist mode of production, modern technology makes possible true human freedom.

Marx's analysis has had two different, but not wholly unrelated, heritages—one political, the other intellectual. The manifest failure of its political heritage, which was dependent on a necessarily selective interpretation, should not be used to justify any wholesale rejection of its intellectual heritage, which nevertheless reflects another selective interpretation. Indeed, even within the confines of the political heritage there have been intellectual developments that deserve consideration.

The Political Heritage of Marxism

Among those intellectual developments within the political heritage that deserve consideration, the most important is the theory of a Scientific Technological Revolution (STR), or the merging of science and technology into technoscience that dominated discussions in the USSR from the late 1960s until its demise. The term "scientific-technological revolution" was originally suggested in the early 1950s by the Western Marxists J. D. Bernal and Victor Perlo; Marxists in the socialist bloc countries took up the discussion late in the same decade. Communist Party ideologues at first criticized the concept, maintaining it was a deviation from true Marxism-Leninism. Thus it was not until the early 1960s, with a change in the official position of the Communist Party–USSR, that the STR concept became widely accepted in the social sciences or in politics, although as Julian Cooper has argued, STR theory was continuous with the idea that science and technology play a leading role in the revolutionary transformation of society, a central theme in Marxism-Leninism from 1917 on.[29]

The background of opposition to Western European philosophy of technology against which STR theory emerged was clearest in the German Democratic Republic. Not only was engineering philosophy of technology characterized as idealistic bourgeois ideology in the service of reactionary imperialism, but Western European humanities philosophy of technology was also rejected as a pessimistic concentration on the negative aspects of technological progress and for its failure to recognize technology as a productive force influenced by social conditions.[30] Hermann Ley's *Dämon Technik?* for instance, argued that Western European thinking constituted a "new witches' trial of technology." [31] But in 1965 an East Berlin conference on "Marxist-Leninist

Philosophy and the Technological Revolution" undertook to formulate a more constructive Marxist position.[32] Bringing together a number of Central European philosophers, the conference focused on six questions associated with the idea of technological revolution: its essence and history, the role of science, the socialist image of humanity, planning in the technological revolution, methodological problems of modern science, and philosophical problems in industrial science. That same year Erwin Herlitzius's bibliography on "Technik und Philosophie" was published.[33] Both led to genuine philosophical discussions of technology.[34]

Two years later the Czech social philosopher Radovan Richta edited *Civilization at the Crossroads* (1967), a cooperative work in which sixty sociologists, economists, psychologists, historians, engineers, scientists, politicians, and philosophers dealt specifically with the nature of the STR. In contrast to the Industrial Revolution, which was based on mechanical power and factory organization, the STR rests on principles of automation and cybernetic management. The two revolutions thus have different internal structures and social consequences. In the first, science and technology remained relatively independent; in the second, technology becomes a scientific enterprise and science is revealed as having immediate technological implications.

The STR demands highly educated workers, so that human development and creativity become the most effective ways to increase production. While recognizing many of the ills of technological development, Richta and coauthors do not criticize technology but call for its improvement. What they condemn is a one-sided technological progress under social conditions that fail to provide for general human self-realization. This position includes an explicit critique of capitalist social organization and an implicit rejection of Stalinist centralization.[35]

Between 1950 and 1965, Soviet discussions of technology stressed questions related to automation and cybernetics. The key concept in cybernetics is information. Norbert Wiener's description of information as something sui generis, neither matter nor energy, initially led some Marxists to suspect cybernetics of being a new form of idealism. In 1955 this misunderstanding was corrected in two influential articles by S. L. Sobolev, A. I. Kitov, and A. A. Liapunov, and by Ernst Kolman.[36] The resulting interest in cybernetics and its transformation of technology contributed to the Russian understanding of the technical side of the STR.[37]

The most sustained articulation of communist thinking on the STR was contained in the interdisciplinary volume *Man, Science, Technology: A Marxist Analysis of the Scientific and Technological Revolution* (1973).[38] Beginning with analyses of science and technology, their internal relations and influence on production, *Man, Science, Technology* sought to clarify the meaning of the term "revolution" as applied to science and technology. Revolution is defined as involving "radical qualitative changes in social structures in the course of the progressive development of society" (p. 19). In this sense scientific revolutions result either from "the discovery of fundamentally new phenomena of laws" or from of "the utilization of new methods and technical means" (p. 20). In technology the "substitution of new technical media for old ones, by employing quite different principles, signifies revolution" (p. 21). What has now happened is that these two revolutions are merging: technology is a new cognitive method for science; science offers new principles for technology. The STR in a "narrow sense"—that is, as applied to science and technology in isolation from their social and economic conditions—refers to "a convergence of revolutionary changes in science and revolutionary changes in technology into a united process" (p. 24).

The STR as the unification of science-technology as a productive force "places between man and nature, not tools or machines, but self-regulating, self-adjusting processes of production" (p. 369). But technical revolutions do not of themselves lead to production revolutions without a corresponding social revolution. The STR thus can be described, in a restricted sense, as a technical revolution in the instruments of production. But in a broad sense, it will require a basic change in the organization of the production process in order to realize the full potential of the new means.[39]

The failure of the Soviet Union to profit from such analyses does not of itself prove their falsity. Indeed, the results of the failure could even be argued to confirm their truth. Also, because of the difficulties of dealing with the social side of the STR, Russian philosophy of technology during the late 1970s and 1980s especially tended to concentrate on the technical side, emphasizing analyses of the methodological principles of the new engineering sciences. As surveyed by Vitaly Gorokhov (1990), these studies clearly do not extend beyond the engineering tradition.

The Intellectual Heritage of Marxism

The intellectual heritage of Marxism in the West raises other problems. This heritage is most dynamically reflected in the work of such authors as Max Horkheimer and Theodor Adorno, Herbert Marcuse, and Jürgen Habermas. As Patrick Murray (1982) has pointed out, the distinctive character of the founders of this neo-Marxist Frankfurt school is to shift critique from a focus on the inadequacies of political economy to a questioning of the character of the Enlightenment, and to attribute the misuses of technology not simply to economics but to culture.

The achievement of the Enlightenment is its promotion of subjective reason (reason used to satisfy a subject), or what Horkheimer also called "instrumental reason"—without, however, providing a guideline in objective reason (theory) for how the new powers of reason are to be used. It is this lack of theory that leads to the dialectic of turning the Enlightenment against itself and the production of social orders dominated by the military and the "culture industry," that is, brute force and entertainment. As antidote Horkheimer and Adorno argue that "the Enlightenment *must consider itself*, if men are not to be wholly betrayed." [40] This very critique of Enlightenment is a reassertion of theoretical reflection. "The function of theory today is to reflect upon" and to criticize "the socially conditioned tendency toward neo-Positivism or the instrumentalization of thought." [41] Unfortunately neither Horkheimer nor Adorno is able to go beyond critique. Like the Enlightenment they criticize, they ultimately provide no substantive theory.

In the attempt to locate substantive guidelines for technological power, Marcuse, having repeated in popular form basic Horkheimer criticisms of instrumental rationality, picks up and develops suggestions from Horkheimer and Adorno about the possibilities for marginal members of society to reassert substantive rationality. Originally the appeal had been to blacks and Jews. With Marcuse this is transferred to the counterculture, to artists and poets, environmentalists, feminists, and others. In defending the possibilities of a countercultural transformation of technology Marcuse also develops the idea of alternative science and technology, arguing that Western science and technology are inherently flawed by their logic of domination. Habermas (1975), rejecting this view, returns to the more traditional Marxist view that it is not science and technology in themselves but the social conditions in which they exist, especially their ideological interpretation, that constitute the problem. And in place of Marcuse's apotheosis of the counterculture, Habermas develops an extended theory of communicative action as a substantive guide for political and technical development. [42]

Habermas's theory of communicative action provides a basis for linking neo-Marxist thought and modern liberalism as articulated, for instance, in John Stuart Mill's essay "On Liberty" (1859). Habermas, like Mill, argues that the state has an obligation to maximize communicative competence. A new generation of North American neo-Marxists has further extended this link. William Leiss, a former student of Marcuse, after first defending Marcuse's idea of the need for a transformed science and technology (1972), went on to attempt a critique of the expression of human needs in the consumer society (1976) that again echoes Mill's criticisms of a hedonistic utilitarianism. In his most recent work Leiss argues, against the prevalent feeling of living "under technology's thumb" that "government [should] assist citizens in using the rapidly accumulating stock of technical knowledge to inform themselves about the choices that confront them and about the relation between those choices and the fundamental values of free and democratic societies." The antidote to technological fatalism is "an informed citizenry and enlightened public policies" (1990, p. 7).

In a related manner Andrew Feenberg's *Critical Theory of Technology* (1991) projects a general analysis of technology and its relation to culture with the aim of opening opportunities for democratic development. Not unlike Ellul, Feenberg identifies the primary moments of contemporary technical practice as what he calls decontextualization, reductionism, autonomization, and positioning. But contra Ellul, Feenberg offers specific proposals for reconfiguring the diverse possibilities of technology. These include action by which laborers would recontextualize their labor, public recognition of the human significance of vocation, investment of aesthetic value in technological products, and the pursuit of voluntary collegial cooperation in work. This is clearly an attempt, however unrealistic it may appear, to reenclose technology within a humanities perspective.

A Brief for the Primacy of Humanities Philosophy of Technology

The issues adumbrated by the initial imaginary dialogue between engineering and humanities philosophies of technology are not quickly or easily transcended. Attempts to do so arising from within that engineering philosophy associated with the VDI, from pragmatist phenomenologists in the United States, and from Marxists only confirm their persistence. But in order to judge these efforts, and to further their legitimate achievements, it is useful to consider more directly the tension between engineering and humanities philosophies of technology.

With regard to the question of philosophical primacy, one can advance the following thesis. Based on the historicophilosophical studies presented so far, there are at least three possible grounds for proposing the primacy of humanities philosophy of technology (HPT) over engineering philosophy of technology (EPT). These may be described as arguments from historical subservience, from inclusiveness, and from spiritual continuity—each exposing slightly different aspects of the EPT versus HPT tension.

An argument from historical subservience appears at first to point toward the primacy of engineering philosophy of technology over humanities philosophy of technology. At the level of overt or explicit use of the phrase "philosophy of technology," EPT can claim to being prior and thus in some sense primary. But EPT arose explicitly in the process of rejecting a prior, implicit HPT, even by subversive use of selected aspects of humanities philosophy. It was within the rhetorical tradition of humanities philosophy that the establishment of engineering philosophy took place. Recognizing that humanities (and philosophy) conceived technology—and not technology that conceived the humanities—grants HPT priority at the level of covert or implicit existence.

The argument from inclusiveness is itself more inclusive, enlarging this implicit historical priority cognitively, functionally, and anthropologically. Cognitively, HPT includes a knowledge of historical alternatives that EPT lacks. Historicophilosophical studies are characteristic of HPT; for EPT, history tends toward a distinctly Whiggish account of technological progress if not, to put it pungently, toward the belief that "history is bunk."[43]

With regard to functions, making is largely not an end in itself, is not self-justifying. All activity, Aristotle argues at the opening of the *Nicomachean Ethics*, aims at some good. In some cases the goods can be external to the activities that produce them (with the making of such things as pots and houses); in others they are inherent in the undertakings themselves (with doings such as playing music or rational discourse).[44] Although various makings can be subordinated to one another—for example, making pots for cooking food, cooking food for eating—functionally this cannot, the Stagarite maintains, go on forever. There must be a final good in itself, the "master art" of which properly orders and directs all others and is not itself properly ordered or directed by them. Such is politics, the defining or characteristic behavior of the human, that which humans can do only or best (*Nicomachean Ethics* 1.7.1097b11). The postmodern affirmation of a diversity of groundless human projects does not subvert this analysis, since most people still think they have reasons for making what they make—even when nihilists from Nietzsche to Foucault argue that this is not so.

Moreover, the truth or falsity of the nihilist view easily becomes a core issue in humanities philosophy of technology.

In a supplemental approach to this same issue, Aristotle (*Nicomachean Ethics* 1.5) distinguishes three human ways of life: life in pursuit of physical pleasure, in pursuit of virtue and honor, and in pursuit of knowledge and wisdom. The second is more inherently good than the first, the third than the second. At least in part this is due to conceptual subordination: wisdom can bring honor, honor can lead to pleasure, but physical pleasure qua pleasure does not promote honor or honor wisdom. But the life of pleasure is profoundly dependent on making, or what today is called engineering and technology, to produce the richness and luxury in food, shelter, and clothing on which physical pleasures rely. Humanities analyses, by contrast, are open to the broader ways of life characterized by the pursuits of honor and wisdom—from the doings of politics to the nondoings of speculative discourse. Humanities historical research is complemented by, even grounded in, a wider conceptual sense of the human than is engineering. HPT is likewise both historically and anthropologically more inclusive than EPT.

The point at issue here can be further illustrated by contrasting the rhetorical strategies of an engineering text such as Reuleaux's *Kinematics of Machinery* (1875) with some comparably influential and fundamental humanities text such as *The Education of Henry Adams* (1906). Reuleaux's restriction of focus, considering machines only in terms of physical forces and geometrical motions, conceptually articulated in terms of graphics and the symbols of mathematics, enables him to achieve a technical explanation that provides the foundation for mechanical engineering as a systematic pursuit. But Henry Adams's refusal to limit his consideration of machines to the strictly physical, and his attempt to relate them to art, literature, religion, and the general symbols of culture, opens up interpretive understandings denied to but inclusive of Reuleaux's analytic explanations. A major section from a book like Reuleaux's could be included in a narrative such as Adams's, but not vice versa.

The romantic critique of modern science and technology contains further restatement of the anthropological version of inclusiveness. The romantic argument for such inclusiveness is simply that the humanities are more nearly coextensive with human activity than the technologies, or that the humanities include more of human life than the engineering disciplines. To study mechanical engineering, electrical engineering, electronic engineering, and such does not contribute to an understanding of the human to anywhere near the same extent or depth as do studies in art, literature, religion, and so on. The primacy of HPT over EPT is thus a primacy of anthropological understanding. EPT, insofar as it begins with technology and seeks to develop technological explanations of other disciplines, especially those in the humanities, denies or tries to invert the way engineering disciplines are properly allowed to stand on their own within the humanities.

These initial two arguments are both present, if not quite so openly stated, in historicophilosophical investigation of the EPT-HPT relationship. Another argument, only alluded to, is from what may be called spiritual continuity. This is the argument that from its inception philosophy has questioned the technological, and that this questioning remains among its deepest obligations. Recall Socrates' account of his own questioning in response to the words from the Delphic oracle that no one was wiser than he. He found it hard to think of himself as having any knowledge at all. So he decided to test or interpret these words by means of an encounter with those who appeared or claimed to have knowledge. Having found both politicians and poets wanting in this respect,

> Finally [he said] I went to the artisans, because I was conscious of knowing nothing, so to speak, but I knew that I would find that they knew many noble, ingenious things. And in this I was not disappointed; they knew things I did not know and to that extent were wiser than I. But . . . the good craftsmen seemed to me to go wrong as much as the poets: because they practiced their *technai* well, both thought themselves wise in other, most important things, and this error of theirs obscured the wisdom they had, so that I asked myself, on behalf of the words from the oracle, whether I should prefer to be as I am, neither wise in their wisdom nor ignorant in their ignorance, or to be in both ways as they are. The answer I gave myself and the words from the oracle, is that it is better for me to be as I am. (*Apology of Socrates* 22d–e)

Socrates' conclusion is that his wisdom, such as it is, is a kind of learned ignorance. "What is possible," says Socrates, "is that in fact the god is wise, and that the words from the oracle mean to say that human wisdom is worth little or nothing; and it appears that he does not mean to speak to Socrates, but uses my name, making of me an example, as if to say, 'This one among you, human beings, is wisest, who, like Socrates, recognizes that he is in truth worthless in respect to wisdom" (23a–b).

In like manner, other central texts in the humanities can be identified by their tendency to open up new areas of questioning and quandary. Augustine's *Confessions* and *City of God* together adumbrate new mysteries of temporality and "progress," Cervantes' *Don Quixote* presents a paradoxical image of the relation between imagination and action, and Melville's *Moby Dick* narrates the dark forces unleashed by a technically based pursuit of enlightenment. Each in its own way rejects or struggles against a technical delimitation of perspective that might well issue in objective power—Christian righteousness, Cartesian certainty, and industrial determination, respectively—in order to ask questions about human hopes and hindrances in ways that deepen subjectivity.

Like Socrates, we must remain open to the possibility that others do possess wisdom. We must be willing to seek them out and to ask questions. And it may well be that this conversation will at times and of necessity take on a technical tone. A synthesis of the range of issues raised by engineering and humanities philosophy of technology will include

- conceptual distinctions between tools, machines, and cybernetic devices;

- methodological discussions of invention, design, and production;

- epistemological analyses of engineering science;

- speculations on the ontological status of artifacts and works of art versus natural entities;

- the ethical problems engendered by a broad spectrum of specialized technologies; and

- the multifarious political to cultural ramifications of technological pursuits.

But such a comprehensive, systematic, or interdisciplinary analysis must remain finally subordinate to questioning the technical even when we venture to engage its powers.

The implications of this questioning must also be acknowledged. Socrates himself recognized that mathematical knowledge could be employed to create "winds, waters, seasons, and various things," but he argued that one could become involved in such pursuits only if one thought all ethical and political questions were already fully answered (Xenophon, *Memorabilia* 1.1.12 and 15). Often this insistent, sometimes conservative return to questions of justice, virtue, and piety will be perceived as romanticism if not mere churlishness. On occasion the return will degenerate into ritual, not to say mechanical, performance. But were the philosophy of technology to become identified solely with a philosophical extension of technological attitudes, it not only would close itself off to the rich otherness of reality, it would also abandon its claim to be philosophy. Questioning is indeed the ancestral heritage and vital home of thinking. But true questioning must engage what it questions if it is to be more than intellectual gossip.

CHAPTER FOUR

• • • •

The Philosophical Questioning of Technology

Although the use of the term "philosophy of technology" as the name for a special branch of study in the humanities is relatively recent, the name designates investigations continuous with those that have been pursued for centuries under such headings for traditional divisions of philosophy as "logic," "theory of knowledge," "metaphysics," and "moral and social philosophy." Moreover, despite the impression sometimes created by the growing currency of the term in titles given to articles, books, courses of instruction, and learned societies that it denotes a clearly delimited discipline which deals with a group of closely interrelated questions, the philosophy of technology as currently cultivated is not a well-defined area of analysis. On the contrary, contributors to the area often manifest sharply contrasting aims and methods; and the discussions commonly classified as belonging to it collectively range over most of the heterogeneous set of problems that have been the traditional concern of philosophy.

The foregoing sentences, with "philosophy of science" substituted for "philosophy of technology" and a few other minor modifications, constitute the third paragraph of the preface to Ernest Nagel's classic *The Structure of Science*.[1] What Nagel said of the philosophy of science in the early 1960s applies mutatis mutandis to the philosophy of technology today. As we further consider the scope of philosophical questions concerning technology, this passage thus directs our attention first to the relation between the philosophy of science and the philosophy of technology, and from there to the fundamental branches of philosophy.

Science and Ideas

Science is a special kind of knowledge expressed through ideas and theories. Philosophy is likewise engaged with science through ideas and theories—both the theories *of* science (that is, scientific theories) and theories *about* science. Scientific theories include heliocentric astronomy (as variously formulated by Copernicus, Kepler, Galileo, Newton, etc.), circulation of the blood (Harvey), biological evolution (Darwin), relativity (Einstein), and so forth. Such theories and their associated ideas strongly influence our visions of the natural order (cosmology) and ourselves (psychology). They thus constitute implicit philosophies—even scientific philosophies—insofar as philosophy simply connotes a worldview.

It is ideas or theories about science, however, that constitute philosophy of science in the primary sense. Scientists assume well-confirmed theories to be true. When one questions this truth, wonders about the cognitive status or structure of scientific theories, one begins to develop ideas *about* science rather than simply theories *of* science. What is science? Is science true? What constitutes truth in science? What is the logic of scientific argumentation and explanation? What is the reality of scientific entities such as laws, atoms, or quarks? What is the meaning of science—that is, how is science related to other aspects of human life, including ethics and politics? Such questions constitute core issues in the philosophy of science.

Technology and Ideas

Because technology, understood here as the making and using of artifacts, is primarily a practice or activity, the relation between technology and ideas is not nearly as obvious as that between science and ideas. The existence of distinctly technological ideas and theories, for instance, is not as apparent as the existence of scientific theories. When ideas are associated with technology, they often seem to be merely scientific ideas employed in a practical context. Indeed, this very analogy has led many to think of modern technology as applied science and has inhibited development of an independent philosophy of technology.

Nevertheless, there do exist distinctly technological ideas, as revealed in the technological sciences. The idea of the machine (in its many permutations from Aristotle through Vitruvius to Franz Reuleaux and Alan Turing), the concepts of a switch, invention, efficiency, optimization; the theories of hydraulics and aerodynamics, of kinematics and cybernetics, of queuing, information, and network theory—these are all inherently technological. Such ideas are found not in the sciences of physics, chemistry, or biology, but in the disciplines of mechanical, civil, electrical, electronic, and industrial engineering.

Indeed, the use of mechanics in science (as in Newton's "celestial mechanics") can reasonably be argued to derive from early modern technologies (of, especially, clocks), so that science in some senses might accurately be described as applied technology.

Because of their inherently practical character such ideas disclose a *Lebenswelt*, and the general articulation of this lifeworld readily takes on the character of philosophy in the sense of a worldview. Such a worldview or consciousness has been described by Jacques Ellul in *La Technique* (1954)—and in a different but related way by Ernesto Mayz Vallenilla's *Ezboza de una crítica de la razón técnica* (1974). Because of the practical character of the ideas incorporated in this consciousness, the questioning of distinctly technological ideas has a different content than the questioning of scientific ideas. The assumption among technologists is not that technological theories are true but that they work, and that the works to which they give rise are good or useful. When one questions this working and its usefulness, when one raises doubts or wonders about the practical character or moral status of technological actions and their results as well as the ideas on which they are based, then one begins to develop ideas about rather than simply theories of technology. What is technology? Is technology always good or useful? What constitutes goodness within technology? What is the logic of technological thought and action? What kind of reality do technological objects possess? What kind of knowledge do the engineering sciences contain? What is the meaning of technology—that is, how is technology related to other aspects of human life? Such questions constitute core issues in the philosophy of technology.

Because of this difference in the kinds of questions initially raised about science and about technology, the philosophy of science is more closely associated with logic and epistemology, the philosophy of technology with ethics and practical philosophy. But it is a mistake to limit the philosophy of technology to practical issues or to consider it only a form of applied philosophy. Technology is subject to the full range of concerns typical of the traditional divisions of philosophy—a spectrum of issues running from the conceptual and epistemological through the ethical, political, and religious to the metaphysical.

Conceptual Issues

A primary conceptual issue is the already mentioned relation between science and technology. Nagel, for instance, like many professional philosophers of science, appears to equate technology with applied science. Although there may be some prima facie evidence for this, the character of the "application" is not without ambiguities. Working to extend the tradition of logical empiricism, Mario Bunge has done much to explicate the various senses of "applied" in this context.

In different philosophies of science, however, the definition of technology as applied science is not so obvious. The phenomenological tradition of the philosophy of science, which questions science as a whole, is also apt to question technology as a whole, and to look at the relation between the two as much closer than one of independent application. José Ortega y Gasset and Martin Heidegger are clearly the seminal figures here, although American phenomenologists such as Hans Jonas and Don Ihde extend and enrich this tradition of reflection.

Jonas, for instance, in his historicophilosophical essays on the rise of modern science and technology (1974), sees the two as close correlates. Ihde's *Technics and Praxis* (1979) distinguishes between "idealist" and "materialist" attitudes toward technology—the former understanding technology as growing out of science, the latter viewing science as emerging from technology. It is Ihde's contention that the idealist view has dominated in Western philosophy from Plato to Descartes and needs to be corrected by a materialist approach. Indeed, although European phenomenologists have paid limited attention to technology, for American phenomenologists technology has become a general theme. Ihde maintains in *Instrumental Realism* (1991) that reflection on technology can even serve as a foundation for the philosophy of science. In *Existential Technics* (1983) and *Technology and the Lifeworld* (1990), he extends phenomenological analysis by showing how technology influences not just our explanations of the natural world (that is, science) but also our self-understanding and historical reality.

Even in the Anglo-American school of analytic philosophy, however, the undermining of logical empiricism associated especially with historian-philosopher Thomas Kuhn's *The Structure of Scientific Revolutions* (1962) has influenced and altered the conception of the science-technology relationship. For instance, Kuhn observes that "part of our difficulty in seeing the profound differences between science and technology must relate to the fact that progress is an obvious attribute of both fields." [2] Post-Kuhnian philosophy of science is certainly much more sensitive not only to the complexities of the historical record but also to the pragmatic (not to say technological) character of modern science. It may even be that Kuhn's distinction between paradigm-accepting normal science and paradigm-shifting revolutionary science would be better illustrated by changes in technical apparatus than in the conceptual frameworks of science.

Supporting this approach is the work of both philosophers and historians of science. Patrick Heelan's philosophy of science (1983) argues that the modern scientific interpretation of the world is dependent on the prior existence of what he calls a "carpentered environment" that presents the world as a Euclidean space. Peter Galison has argued that twentieth-century physics includes at least two experimental traditions, one based on image-producing devices such as cloud chambers and another on electronic digital devices such as Geiger counters. Within these different technical traditions are further internal continuities and external discontinuities in pedagogy and argumentation.[3]

Studies of the influence of different technical artifacts in science point toward further conceptual distinctions among the types of artifacts themselves—that is, between containers, utilities, structures, objects of art, tools, machines, automata, systems—and their differential engagements with human thought and action in bricolage, craft, art, technique, technology, engineering. At another level, questions readily arise about distinctions between technology as object, as activity, as knowledge, and as volition, and about differences between premodern *techne* and modern technology. In such ways conceptual issues shade into logical and epistemological ones.

Logic and Epistemological Issues

The logic of technology is not the same as that found in either "the primitive mind" or premodern speculative thought, both of which exhibit what Lucien Levy-Bruhl (in regard to the former) terms a "participation mystique" and Aristotle (in reference to the latter) speaks of as the unity between mind and its object. Ernst Kapp in *Grundlinien einer Philosophie der Technik* (1877) initially proposed what might be called a "projection mystique" as the founding logos of *techne*; technics in various forms are conceived as *Organprojectionen*, or extensions of some aspect of the human organism. Although it does not fit as well with premodern technical experience, this approach does provide a logical foundation for the Renaissance aspiration to pursue "the relief of man's estate" (Francis Bacon) through technology as a humanization of the world.

Indeed, modern logic can be interpreted as carrying the Baconian "conquest of nature" into that second nature called language. In modern form, technology seeks to overcome domination by the world; modern logic likewise seeks to extend the demand for freedom into the conceptual and linguistic realm. The founder of modern mathematical logic, Gottlob Frege, after pointing out its necessary role in mechanics, alludes to the modern philosophical aspiration "to break the domination of the word over the human spirit" (*Begriffsschrift*, preface), an end for which his system can serve as "a useful tool." Conceiving the world in terms of functions with arguments and their relations rather than as substances with essences and accidents removes a certain bias against the manipulation of the world, opens up the world to a movement originating with the human rather than subjecting the human to a movement originating with the world. Jean Piaget's "genetic epistemology," which sees formal operational thinking as based in biological evolution and entailing a process of "continuous construction" and "invention," is but another aspect of this logic. Indeed, once the natural world is reconceived in evolutionary terms, modern technology can even recover a measure of the participatory mystique—as it does, for instance, in Oswald Spengler's notion of technology as "tactics of living" (1931, p. 1).

Within such a logical framework, propositions are not properly true or false, but rather more or less useful or appropriate to a context. Propositions that are not strictly true or false are further linked in arguments that are not strictly valid or invalid. This obviously suggests a pragmatic logic, and indeed pragmatist philosophies of science such as John Dewey's have tended to view science as an inherently technological endeavor. Over the past three decades, however, the unique logic of the pursuit of context appropriateness has become a subject of intense investigation beyond the confines of pragmatism. A pioneer in the field, Herbert Simon, in *The Sciences of the Artificial* (1969), makes the general case for an engineering design methodology that employs utility theory, statistical decision theory, algorithms, and heuristics for choosing both *optimal* and *satisfactory* alternatives, imperative logics, factorization and means-end analysis, resource allocation schemes, and so on. This logic of context appropriateness, also called "bounded rationality," is found not only in engineering design, but also in operations research, management science, and artificial intelligence.

The recent development of risk-cost-benefit analysis as another element in this logical framework has given rise in the English-speaking philosophical community to extensive work on identifying and overcoming some of the weaknesses of technical rationality. Within a different tradition, the neo-Marxist Frankfurt school criticism of "instrumental rationality" has tried to place technical rationality within its socioeconomic context. Finally, Heidegger attempts to step outside the modern logical framework by approaching technological knowledge not just in anthropological terms, but as a kind of truth in the sense of a revealing or disclosure—thus reintroducing, albeit in a quite nontraditional way, the notion of epistemology.

Technical rationality as "bounded" or context-dependent rationality is related to technical knowledge as information. The epistemology of the information sciences is closely associated with mathematical information theory and computer science—a science which is no longer of nature. This epistemology focuses on the technical possibilities of signal transmission and reception under diverse conditions as well as on the various ways information when electronically encoded can be sorted and accessed. Much discussion of artificial intelligence and the computer simulation of cognitive processes is related to this topic—and has been criticized by philosophers such as Hubert Dreyfus for failing to distinguish information in a technical sense from true human knowledge. Rafael Capurro's *Hermeneutik der Fachinformation* (1986) also brings the perspective of hermeneutics (as an epistemologically related discipline) to bear on the sorting and accessing of technologically dependent, scientific information.

Ethical Issues

Traditionally, ethics has focused on interpersonal behavior, on how human beings should act toward one another, because this was the area manifesting the most substantive freedom of choice. In analyzing such behavior, the science of ethics has developed at least three general theories for the grounding of particular moral precepts: natural law theory, utilitarian theory, and deontological theory. The first theory focuses on a preexisting framework (law or order versus disorder), the second on consequences (goods versus evils), and the third on the inner character of the action itself (rational or right versus irrational or wrong).

During the past three hundred years, as a result of technological development and the enormous powers it has placed in human hands, traditional theories have been applied in new ways, especially in those professions most intimately involved with modern technology, and the scope of ethics itself has been enlarged to include relations between human beings and the nonhuman world: animals, nature, and even artifacts. This enlarged scope of ethics is especially evident in such new fields as nuclear ethics, environmental ethics, biomedical ethics, professional engineering ethics, and computer ethics—each of which can be thought of as a branch of the ethics of technology.

Nuclear Ethics

Nuclear ethics, the oldest of these new fields, deals with two distinct but related technologies: nuclear weapons and nuclear power. For such philosophers as Albert Einstein, Bertrand Russell, and Karl Jaspers, nuclear weapons intensify basic questions about technological progress and, in words from Jaspers that echo Einstein, call for "a new way of thinking" that will lead to an inner change in human self-understanding (Jaspers 1961, p. 204).

From a more restricted perspective, there is the issue of the moral status of deterrence theory and, with regard to both weapons and nuclear power, the proper apportioning of risk and responsibility for present and future generations. In part because a utilitarian calculus has been the primary justification for the development of nuclear weapons and nuclear power, criticisms have been mounted largely from a deontological and to some extent from a natural law framework. Günther Anders, for example, has argued that not only persons but also artifacts act according to maxims or principles. The maxim of nuclear weapons is total destruction. Anders, after reformulating the Kantian categorical imperative as, "Have and use only those things, the inherent maxims of which could become your own maxims and thus the maxims of a general law," argues the inherent irrationality or wrongness of nuclear weapons (1961, p. 18). To make them is simply self-contradictory.

Criticisms of nuclear power that stress its inherently destructive character (at least over the long term, in relation to the disposal of nuclear waste) often depend on much the same kind of moral argument even when it is not explicitly articulated as such. Recently, however, moral analyses of the risks inherent in nuclear weapons and power-generation technologies have refocused discussion within a utilitarian framework. The claim that nuclear technology contributes less to global climate change than do coal-fired power plants, for instance, is a strongly consequentialist argument.

Environmental Ethics

A natural law critique of nuclear weapons has been put forth primarily by philosophers within the Thomist tradition. In a general sense, however, environmental ethics, and to some extent the "alternative technology" movement, arising in response to various chemical pollutions of the environment and the dangers posed for the terrestrial ecosystem, often makes use of a kind of natural law framework, although it is seldom explicitly called that. But the appeal to a preexisting ecological order with which human technical actions ought to be in harmony and the affinities commonly voiced for certain non-Western natural law traditions (Buddhism, Taoism, Hinduism, etc.) reveal a theoretical disposition similar to some possible interpretations of Thomist natural law ethics.

The fundamental belief of the natural law position that it is immoral to pollute or excessively disturb the natural environment can, of course, be buttressed by appeal to utilitarian self-interest and risk-cost-benefit analysis. Destruction of the natural environment also often harms human beings or places them at unwarranted risk. The further idea that natural species should be preserved can be defended on deontological grounds as well as by postulating certain rights for animals, plants, and perhaps even inorganic nature. The animal rights protest against the experimental use of animals and certain forms of factory farming are common extensions of a deontological environmentalism. In truth, the tension between preservationists who defend wilderness as a good in itself and conservationists who would manage wilderness for its long-range utilitarian benefit is ultimately not just a policy difference but one of substantive moral principle.

Recent recognition of questions related to the overcrowding of satellite orbits and pollution from "space junk" as well as problems with the contamination of the moon and planets from various space exploration vehicles is generating the related field of "space ethics," which extends environmental concerns beyond the terrestrial domain. The unique issues in both terrestrial and planetary contexts concern the obligations of humans as implicated in complex wholes. Debates about the rights of nature focus on the whole in spatial or geographic terms; discussions of responsibilities to future generations consider the whole from a temporal perspective.

Biomedical Ethics

Despite the development of environmental law and the recent creation of government agencies to protect the environment, both of which are associated with the rise of environmental ethics, it is biomedical ethics that is the single most highly developed area of interaction between ethics and technology. Perhaps this could have been expected, given that on average something approaching 10 percent of the gross national product of the nations of Western Europe and North America is devoted to medicine, a field in which technological advances have the most direct impact on the largest number of individuals.

As an academic discipline, bioethics has been pursued both as ethics applied to a particular technology and as the development of a special deontology.[4] As a field, bioethics is conveniently divided into moral issues associated with different stages of human life. Abortion, in vitro fertilization, fetal experimentation, and their associated moral dilemmas are all related to the beginning of life. Physician-patient relationships and questions of confidentiality and informed consent occur for the human adult. The morality of organ transplants, euthanasia, and problems of defining death in the presence of heart-lung machines and other high-tech devices for prolonging life are associated with the end of life. Covering all periods are issues of the allocation of scarce technomedical resources, health care policy, and the protocols of biomedical research, including experimentation on animals. In each of these areas competing claims are based on appeals to utility, rights, and natural law. One obvious example concerns abortion, which tends to be defended on utilitarian and sometimes deontological grounds (greater evils would result if it were illegal; a woman has a right to an abortion) and criticized on natural law and, less commonly, deontological grounds (abortion is not in harmony with a natural moral order; the fetus has a right to life).

In relation to the practice of medical professionals, there has also arisen what might be called applied (or regionalized) deontologism, the ethics of physician responsibilities and patient rights. The concept of a social role as a cluster of constraints guiding behavior was given classic formulation by F. H. Bradley in "My Station and Its Duties" (1876). To accept a role is to accept certain patterns of behavior, and to reject those patterns while continuing to occupy the role is contradictory or irrational. What Kant tried to spell out for any rational being, Bradley (extending Hegel's concept of *Sittlichkeit*) applied to particular social roles; yet when these roles are in fact professions intimately engaged with the powers of modern technology, they take on a new and especially weighty, not to say transformed, character.

Professional Engineering Ethics

This transformation is apparent in recent discussions of the ethics of not just health care professionals, but of professional engineers. In the early 1900s it was often assumed that the primary obligation of

Hans Jonas (1903–1993). Drawing by C. Verdadero.

the medical doctor was to patients, of the engineer to employers. In the 1960s, however, as a result of increased technological powers placed in the hands of both physicians and engineers, such presumptions began to be questioned, and arguments were made that primary responsibilities were not to individual patients or to employers but to society as a whole. In response, in the United States, biomedical research centers have created institutional review boards, hospitals have formed ethics committees with representatives from outside the medical profession, while professional engineering societies have formulated ethics codes that affirm the primacy of the public welfare and have developed mechanisms to support "whistle blowing" in which engineers "go public" on questionable practices by their employers. Role responsibilities have been enlarged beyond the traditional bounds and on occasion supplemented by what John Ladd has termed "the ethics of power."[5]

The most general discussion of this enlarged ethical responsibility in the use of increased technological powers is Hans Jonas's *The Imperative of Responsibility* (1984). For Jonas the problem equally present in nuclear weapons and power, environmental pollution, and biomedical technologies is that long-range consequences of a global nature cannot always be fully known. His thesis is that "the new kinds and dimensions of action require a commensurate ethic of foresight and responsibility which is as novel as the eventualities which it must meet" (p. 18). This new imperative of responsibility in turn calls for "a new kind of humility—a humility owed, not like former humility to the smallness of our power, but to the excessive magnitude of it, which is the excess of our power to act over our power to foresee and our power to evaluate and to judge" (p. 22). To correlate our power to act and our ability to judge, that is, to stimulate the development of this new humility, Jonas proposes the practice of a "heuristics of fear" that would always consider worst-case scenarios before undertaking any technological project.

Jonas's position is closely related to Jacques Ellul's argument for an "ethics of nonpower," which also calls for a voluntary limiting of technical power, but ultimately on theological grounds rather than secular deontological ones. Both thinkers have nevertheless sketched out an agenda for ethical debate about technology that exercises influence within the philosophical community and the public at large. At the same time, although they bring together or synthesize nuclear ethics, environmental pollution, biomedical technologies, and to some extent engineering ethics, both Jonas and Ellul slight the most recent aspect of the encounter between ethics and technology, that is, computer ethics.

Computer Ethics

Computer ethics, in its initial formulations, was restricted to concerns about threats to individual privacy and corporate security—institutional computer monitoring of individual privacy and individual breaches of mainframe computer databases or networks. These are, of course, two sides of the same coin. Related to these issues, as surveyed by Deborah Johnson's *Computer Ethics* (1985), are concerns about the ethics codes of computer professionals, liability for malfunctioning computer programs, intellectual property rights, and the relations between computers and social power (How can we assure fair access to computers?). Still another issue of note focuses on the anthropological implications of artificial intelligence. Claims have even been advanced, paralleling the extension of moral categories in environmental ethics, that computers should be accorded certain rights.

More generally, however, computer ethics raises basic questions about the use and structural character of information. What are the ethical guidelines, for instance, for the creation, dissemination, and utilization of information—not just in and with computers, but in and with all information-processing media, from telephone and radio to television and satellite? Moreover, isn't it possible, as computer scientist Joseph Weizenbaum has argued (1976), that the incomprehensibility of some computer programs places them beyond the realm of proper human responsibility? Isn't it true that certain kinds of information-based technologies are so complex that in principle they cannot be understood or even tested by their designers? In this regard Walther Zimmerli (1986) has argued that responsibility for "data pollution" cannot be effectively dealt with by the general principles of either the utilitarian or the deontological moral framework. What he calls "the paradox of information technology" (that more information leads to less control) requires the development of an information ethics for specific cases (casuistry) perhaps not unlike that often practiced in medicine. In some instances perhaps computers and computer-dependent artifacts, expressly because they escape human control, should not even be created.

Even if the reliability of complex information system artifacts could be ensured by means of advanced techniques such as object-oriented programming, there is a moral question of the proper relation to the virtual realities they create as well as the reified principles or decision-making processes of expert systems. It is generally accepted, for instance, that moral engagement with another entails some recognition of the other as truly other, and that rigid application of even high ethical principles is not conducive to sound moral behavior. But do virtual realities possess deep otherness? And do not expert systems, even the most complex, depend on something like a strict adherence to formulas? Does not charity, as the perfection of justice, ultimately require a suspension of law—but only at the right time, in a way that cannot be determined in advance and is not itself subject to any final formulation?

Supplemental Issues and Comparisons

Adjunct to questions of technology and ethics, and extending axiological reflection, are issues of technology and aesthetics; yet the aesthetics of technology has received almost no truly philosophical attention. There exist cultural studies on the aesthetic impact of modern technology, and Wolhee Choe (1989) has drawn parallels between aesthetic and technological creativity. Is there a concept of technological beauty distinct from beauty as manifested in other realms? Engineers and architects have on occasion argued that there is, although this has yet to be investigated in any comprehensive manner or related to nontechnical conceptions of beauty. Such questions raise the possibility of an ethics of design that might cross or distinguish boundaries between the arts, humanities, architecture, and engineering.

In each of the five new fields of ethics traditional disagreements between natural law, utilitarianism, and deontology emerge in new contexts. Is a certain technological action right because of its inherent character, or should it be judged only by its good or bad consequences? What is the relationship between nature—especially human nature—and technology? What is important is that technology itself seems to favor certain kinds of ethical frameworks. It is perhaps no accident that the technological conquest of nature should have undermined the natural law tradition in ethics, and that a society dominated by technical or instrumental rationality should be strongly utilitarian in its cultural biases. Ethical discussion of technology has also given rise to certain categories in ethics—responsibility, safety, and risk, for instance—that were not as prominent in premodern moral philosophy.

While raising new ethical interests, however, technology also tends to undermine the significance of that kind of individual human action that ethics traditionally deals with. At the individual or personal level the "ethics of nonpower" is not so much a moral option as an imposed reality. In an advanced technological society, the thrust of technology is dependent not so much on individual as on group decisions. It is this realization that contributes to the theoretical attractiveness of socialism in a technological setting and has promoted the rise of what are called science and technology policy studies.

Issues of Political Philosophy

Technology policy studies—that is, studies of politics in technology and of the political guidance of modern technology—transcend narrow technical and economic interests and invoke the question of political life as a whole. The aim of political life has traditionally been construed to be justice, so that central to the pursuit of political philosophy has been an explication and clarification of the essence of justice. The rise of modern technology is in fact correlated with certain transformations in the understanding of justice—the emergence of new metaphors of governance, for instance,[6] not to mention ideas about rights to work or certain types of medical care—so that the investigation of these transformations itself becomes an aspect of the political philosophy of technology.

A subsequent question is how the benefits of modern technology are to be justly or fairly distributed, as in the "social question" that came to the fore in Britain in the wake of the Industrial Revolution. Today, however, questions are advanced as much about how technical costs and risks are to be apportioned. This transformation from a concern for the just distribution of benefits to the just distribution of costs or risks raises anew the question of technical progress—whose evident reality was employed at the beginning of the modern period in arguments that contributed to the original transformations in the conception of justice. Currently it seems that we undertake technological actions less and less for the good of our descendants and more for the benefit they bring us or some group in the present, and that these actions often hold others, including our own progeny, hostage to the risks of our technical deeds.

A different kind of fundamental political issue is the autonomy or neutrality of technological action and institutions. The traditional view has been that social institutions (family, religion, economy, state) tend toward a certain independence in ways that call for an attentive effort to incorporate and subordinate them to any particular vision of justice or the good. Precisely this attentive effort is manifest in classic works of political theory such as Plato's *Laws* and Aristotle's *Politics*. In such works, however, *techne* remains in the background; it seems to be accepted as relatively pliable, readily following the goals embodied in other social institutions. The experience of the nineteenth and twentieth centuries, however, is that this pliability or neutrality can no longer be taken for granted. As Ellul has argued at length, technology in many instances appears to have taken on an institutional character of its own.

As Langdon Winner has provocatively posed the question, we are now forced to ask, "Do artifacts have politics?"[7] Many have argued the affirmative, and that their politics is good. Information dissemination technologies, for instance, are said to promote democracy. Ivan Illich argues, by contrast, that they do just the opposite and urges the articulation of negative design criteria for the political evaluation of technologies that will promote the development of "tools for conviviality" (1973).

But how broad is the concept of artifact or tool in this context? Does it extend beyond physical objects to include social institutions and perhaps even ideas, methods, and systems of thought? Is management, as a technology, also a kind of artifice? Beyond the political-philosophical questions of the just distribution of technological wealth and defenses of political freedom within complex technocratic frameworks, there is the question of the proper extension of complex, large-scale, immaterial artifice. Picking up on questions raised in environmental and computer ethics are inquiries about the economic and psychological implications of a politics of Spaceship Earth. Does planetary management by means of an earth system science constitute a utopian or an anti-utopian achievement?

Religious Issues

According to Rudolf Otto, the fundamental concept of religion is that of the holy or sacred. Mircea Eliade notes further that the sacred is defined primarily in its opposition to the profane or secular. The sacred is characterized by special forms of space and of time, as in church structures and liturgical actions. What, then, is the relation between technological space and time and the holy? Are there techniques peculiar to the sacred? If so, how do sacred technologies differ from secular ones?

The most fundamental opposition is perhaps between a religion of the Earth as sacred and modern technology. This is suggested by feminist historians such as Carolyn Merchant[8] and specifically advanced by apologists for the ways of life of archaic peoples. Mander's *In the Absence of the Sacred* (1991) is particularly forceful in arguing that "the central assumption of technological society [demands] overpowering nature and native peoples" (p. 6) and that "lacking a sense of the sacred we are doomed to a bad result" (p. 191).

What Mander identifies as an absence of the sacred can also be interpreted, however, as its transformation. The implication of studies by Max Weber and others[9] is that modern technology is fostered by a simultaneous contraction of the realm publicly recognized as sacred (secularization) and the expansion of the realm able to be privately understood in spiritual terms (*devotio moderna*). In the premodern world, both politics and *techne* were publicly recognized as having religious significance; this broadly exoteric religion was complemented by a much smaller domain of esoteric practice with which were associated certain techniques for spiritual transformation. The Protestant Reformation, however, in conjunction with the Enlightenment, severely restricted, if it did not effectively sever, the public from the exoteric religious domain while simultaneously opening the public world up to private spiritual commitments that Weber refers to as this-worldly asceticism.

Friedrich Dessauer, Catholic research engineer and explorer of the philosophy of technology as a synthesizing discipline, went even further to view technological invention and the technological transformation of the world as a participation in divine creation. Within this context the question of technological progress becomes one of theodicy. What Dessauer's perspective also tends to overlook, if not abandon, is the traditional notion of spiritual techniques for the transformation of self—techniques that were traditionally incorporated into philosophy as well, validating its claim to be an ascent to wisdom and a means of participating in ultimate reality.

This-worldly asceticism and a mysticism of technology—as Protestant and Catholic forms of a unification between religion and technology—can be contrasted to other theological options. The opposition between sacred Jerusalem and secular Athens (Tertullian) can lead to a fundamental religious critique of technology. An Augustinian theology of conversion can ground an aspiration to transform technology. Thomist views of grace as building on (rather than transforming) nature can take technology as good in itself as well as prepatory to a higher good. And the Lutheran theology of perennial tension between nature and grace can lead to an ongoing paradoxical opposition between technique and spirit.[10]

Metaphysical Issues

Two central issues in the political philosophy of technology—those of autonomy (and determinism) versus neutrality (and freedom) and of progress—are at root metaphysical. Discussions of autonomous technology, for instance, exhibit a structure not unlike discussions of the one and the many, the central question in metaphysics. In an obvious sense reality is one; everything is a thing. In another equally obvious sense it is many; all things are different kinds of things. The crucial issue is which senses are more real and which more illusory, and to what extent. Likewise, at some level of abstraction technology does appear to be one and autonomous; all technology is technology, with a broad historical trajectory that appears to transcend particular times and places. At another level the diversity of technologies belies any strong unity; unity appears no more than nominal. The root issue, a metaphysical one, concerns the different realities present in the different levels of analysis.

The very idea of technological progress depends on some minimal unity in technology over time, one within which change and improve-

Ernesto Mayz Vallenilla

Ernesto Mayz Vallenilla. Drawing by C. Verdadero.

ment can take place. The idea of such a weak continuity that incorporates progressive change is, of course, the common presumption of
modernity. But among those who have challenged this view is Ernesto
Mayz Vallenilla. In his critique of technical reason (1974) Mayz
Vallenilla first outlines a Kantian analysis of the categories of technology—identified as totality, finality, and perfection—which he argues
found of a kind of functional autonomy, and thereby create an alienation
from the human. A subsequent study on the foundations of modern
technology itself argues that this technology is not so much the outcome
of a "gradual evolution" in technics as a new project that "overcomes
traditional anthropomorphic, anthropocentric, and geocentric" limits.
The three sources of what he calls metatechnics—a technics beyond or
outside technics—are instruments that radically alter (rather than extend) human sensory perception, instruments that redesign and recon-
struct the human body as well as the world, and instruments that trans-
mute matter and energy (Mayz Vallenilla 1990, pp. 19 and 22–23).

The very possibility of such a break in the history of technology, not
to mention in history itself, raises the fundamental metaphysical issue

concerning temporality. In words adapted from the closing paragraph of Leo Strauss's debate with Alexandre Kojève on tyranny and modern technique,

> La philosophie au sense strict et classique [suppose] qu'il y a un ordre éternel et inchangeable dans lequel l'Histoire prend place, et qui n'est, en aucune manière, affecté par l'Histoire. [Mais] cette hypothèse n'est pas évidente par elle-même; [ceux qui se livrent exclusivement à la technique] la rejette en faveur de l'idée que l'Etre se crée lui-même au cours de l'Histoire. . . . Sur la base des hypothèses [de la modernité technique], un attachement absolu aux intérêts humains devient la source de la connaissance philosophique: l'homme doit se sentir absolument chez lui sur la terre; it doit être absolument un citoyen de la terre, sinon un citoyen d'une partie de la terre inhabitable. Sur les bases des hypothèses classiques, la philosophie exige un détachement radical des intérêts humains: l'homme ne doit pas être absolument chez lui sur terre, il doit être citoyen de l'ensemble.[11]

The opposition between these two hypotheses may be the ultimate metaphysical issue posed by modern technology.

Not just the central issues of the political philosophy of technology, however, but all previous issues are implicated in metaphysics or first philosophy. With regard to conceptual issues, What are the differences in being that distinguish natural objects from artifacts, objects of art from technological objects, tools from machines and cybernetic devices? With regard to logic and epistemology, What is it about being that technological knowledge grasps? What is it about being that makes technological knowledge possible? With regard to ethics, What is essential and what is accidental about technological existence? That is, what is real and unalterable about technology, and what is accidental and therefore changeable or controllable—and thus subject to ethical reflection? With regard to theology, What is the relation between ultimate reality and technology? In what ways are human and cosmic destiny implicated in technological destiny? In sum, what is the relation between the true, the good, the beautiful, the just, and the transcendent being as disclosed in nontechnical and in technological reality?

Questioning the Questions

The philosophy of technology as currently practiced is not a well-defined area of analysis. In fact, contributors to the philosophy of technology often manifest sharply contrasting aims and methods, and discussions easily classified as belonging to it range over most of the heterogeneous problems that have been the traditional concern of philosophy. Like all previous philosophy, the philosophy of technology raises in a new form perennial questions that are not subject to any straightforward resolution. But this is to say no more than that the philosophy of the making and using of artifice is not the same as the sciences of the artificial. Philosophy is not science, nor is it technology.

In a world of science and technology, however, one may inquire about the usefulness of this philosophical wondering about technology, this investigation of the many philosophical questions it raises. What purpose does this questioning serve? What is its goal?

Consider, for instance, what can be termed the paradox of development. In response to high-technology urban life and its discontents, the developer takes construction into the fields outside the city. The new houses draw inhabitants in large measure because of their openness to nature, despite being unrelated to the landscape where they are placed or to any social context. Yet in short order and as effectively as the city, these suburbs too obscure nature beneath their ever expanding and then decaying structures, roads, and shopping malls.

"Aren't you sorry about what has happened to the world you moved here to enjoy?" one asks the inhabitants. "Yes, a little," they say, "but you can't stop progress"—which, by definition, is good. Questioning the good of technological progress just makes people feel perplexed if not depressed. So why do it?

What is the goal of the questioning of technology? The answer, simply put, is that it is not to serve technology—that in truth it may even on occasion slow down or interrupt technological development. To notice the paradox of development and thus to question technology is to take a first step outside technology. Indeed, to wonder about the development paradox is to deepen philosophical questioning. Although to question technology is to take a first small step beyond technology, to question the questioning of technology is to remain with and to immerse oneself in philosophy.

It was for this very reason that Francis Bacon, at the dawn of the modern age, sought to turn people away from philosophical questioning and toward more practical affairs. Our time, however, having witnessed the technological questioning—not to say destruction—of many things that on other grounds are found to be true, good, beautiful, just, or real perhaps inclines us to exercise greater forbearance toward philosophy, even toward the heterogeneous philosophical questioning of technology.

CHAPTER FIVE

• • • •

Philosophical Questions about *Techne*

The previous chapter makes the case for technology as a significant theme for modern philosophical reflection. In light of this argument, however, a question readily arises with regard to the relation between philosophy and technology in the premodern period. Did premodern philosophers ignore technology or its ancestor, *techne*? Or did they perhaps attend to it in ways that have not always been sufficiently appreciated? Can their nonattending or attending now contribute to the philosophy of technology? In other words, is it appropriate for the philosophy of technology to acknowledge *techne* philosophically? Such questions in their turn invite reflection on history, and especially the history of technology, which must now be called upon to supplement commonsense experience.

Observations on the History of Technology

The history of technology exhibits tensions similar to those present in philosophy of technology—tensions between, for example, internalist (technical) and externalist (social) history. Like engineering philosophy of technology, internalist histories of the making and using of artifacts and of the artifacts themselves arose among engineers and their supporters who became interested in the development of their profession. Like humanities philosophy of technology, externalist studies of the influence of technology on social institutions, and vice versa, initially emerged among those with transtechnical interests, especially social scientists. Indeed, leading historians of technology more or less regularly ally themselves with one or the other of the two traditions in the philosophy of technology.

History of technology as such first makes its appearance in the early modern period with catalogs and chronologies of crafts and inventions. It explicitly aims to promote consciousness and development of the new industrial arts and technologies. Early achievements of this approach are the great French *Encyclopédie, ou Dictionnaire raisonné des sciences, des arts et des métiers* [Encyclopedia, or Rational dictionary of sciences, arts, and crafts] (1751–1772), Johann Beckmann's *Geschichte der Erfindungen* [History of inventions] (1784–1805), and J. H. M. von Poppe's *Geschichte der Technologie* [History of technology] (1807–1811). This phase is prolonged into the twentieth century by Abbot Payson Usher's *A History of Mechanical Inventions* (1929) and historically deepened by André Leroi-Gourhan's *Evolution et techniques* [Evolution and technics] (2 vols., 1943 and 1945). The collaborative, multivolume Charles Singer et al., eds., *A History of Technology* (1954–1958),[1] which calls itself a reference text for "students of technology and applied science," is surely a culmination of this approach. These histories, in diverse ways, focus on technological change as a phenomenon internal to the technical realm, as progressive, and as inherently beneficial.

The idea of a social history of technology can be traced back at least to Arnold Toynbee's argument that more important than the French Revolution in politics was the Industrial Revolution in manufacturing. His *Lectures on the Industrial Revolution in England* (1884) initiated a discussion that has produced a major literature. The same is true for the debate initiated by Max Weber's *The Protestant Ethic and the Spirit of Capitalism* (1905). With *Technics and Civilization* (1934) Lewis Mumford helped integrate such specialized aspects of the social history of technology, so that internalist histories began to be supplemented regularly by critical histories of the social and cultural impacts of technology. Siegfried Gideon's *The Machine Takes Command* (1948) and Roger Burlingame's *Backgrounds of Power* (1949) are further representatives of this second approach.[2] Examples of team-written, multivolume historical overviews that likewise exhibit affinities for externalist, social history are

- Maurice Daumas, ed., *Histoire générale des techniques* (1962–1979);[3]

- Melvin Kranzberg and Carroll Pursell Jr., eds., *Technology in Western Civilization* (1967);[4] and

- Bertrand Gille, ed., *Histoire des techniques* (1978).[5]

Discussions in the historiography of technology regularly remark on the different emphases of internalist and externalist histories. One effort to bridge the two approaches is what is sometimes termed the contextualist approach, which sees technology as a social construction, at the level of both internal developments and external relations. Indicative of this effort is the volume by Wiebe E. Bijker, Thomas P. Hughes, and Trevor Pinch, eds., *The Social Construction of Technological Systems* (1987). What is seldom noticed within the historiographic discussions, however, is the extent to which histories of technology nevertheless fail to address certain key philosophical issues. Whether internalist, externalist, or contextualist, most histories of technology—although they provide required documentation and are useful to promote broad philosophical reflection—are finally limited in what they can contribute to an understanding of premodern ideas about premodern technics. Indeed, from the perspective of philosophy, what is needed is what may, for want of a better phrase, be referred to as a history of ideas about technology—that is, the study of how different periods and individuals have conceived of and evaluated the human making activity, and how ideas have interacted with technologies of various sorts.

Techne and Technology

Virtually all historians (except Mumford, who prefers the term "technics") use the word "technology" to refer to both ancient and modern, primitive and advanced making activities, or knowledge of how to make and use artifacts, or the artifacts themselves. For instance, Singer defines technology as "how things are commonly done or made [and] what things are done or made."[6] While rightly objecting to this definition as wide enough to include even legislation or the making of laws, Kranzberg and Pursell define technology as the human "effort to cope with [the] physical environment . . . and [the] attempts to subdue or control that environment by means of . . . imagination and ingenuity in the use of all available resources."[7] Indeed, they specifically reject any limiting of the term to "those things which characterize the technology of our own time, such as machinery and prime movers."[8] Technology "is nothing more than the area of interaction between ourselves, as individuals, and our environment, whether material or spiritual, natural or man-made"; it is "the most fundamental aspect" of the human condition.[9]

But Kranzberg and Pursell remain at odds with themselves in two respects. In the first instance they try to limit technology to action on the physical environment, while in the second the environment is allowed both material and spiritual dimensions. In neither formulation do they escape their own quite valid objection to Singer. If technology is "nothing more" than they suggest, it is hard to imagine anything that is left out. Certainly not legislation, which, being concerned with "who gets what, when, how" (Harold Lasswell), is in some sense an "effort to cope with [the] physical environment." On their own terms, technology becomes veritably coextensive with human activity and thus fails to exclude the fine arts, spiritual disciplines such as yoga and meditation, or even language (although in practice they clearly do exclude such subjects).

In another attempt to negotiate this issue, Maurice Daumas, editor of a multivolume French history of technology, errs on the other side by being too restrictive. For Daumas, his collaborative work is first a "description of techniques and their development." [10] But then "techniques" are limited to "only those human activities whose object it is to collect, adapt, and transform raw material in order to improve the conditions of human existence." [11] This eliminates the techniques of accounting, banking, the conduct of military operations, and so forth. Furthermore, unlike Singer, who would explicitly include language as a technology, [12] Daumas judiciously deals only with "the methods of transmitting, recording, and writing it—paper, the proliferation of written texts, and so on." Still, by his qualifying phrase "to improve the conditions of human existence," not to mention the problematic character of the ideal of improvement, does Daumas really intend to exclude the inventions of nerve gas, nuclear weapons, and instruments of torture from his history?

With regard to present purposes, however, both conceptual frameworks presume to equate Greek τέχνη (and Latin *ars*) with the English "technology" (and all of these with the German *Technik*, French *technique*, Spanish *técnica*, and so on). A study of the historical origins of the word "technology" (for etymology is a good place to begin the history of ideas) can, however, suggest the questionable character of this identification. It may also help to clarify conceptions of the essence of technology as much as technological or social histories.

Techne *in Greek Usage*

The Greek τέχνη, commonly translated as "art," "craft," or "skill," has behind it the Indo-European stem *tekhn-*, probably meaning "woodwork" or "carpentry," and is akin to the Greek *tekton* and Sanskrit *taksan*, meaning a "carpenter" or "builder," and the Sanskrit *taksati*, "he forms," "constructs," or "builds." One could compare also the Hittite *takkss-*, "to join" or "build," and the Latin *texere*, "to weave," hence

figuratively "to construct," and *tegere,* "to cover," hence "put a roof on."
In nonphilosophical literature *techne* is used to refer to cleverness and
cunning in getting, making, or doing as well as to specific trades,
crafts, and skills of every kind.[13]

In philosophical works, however, *techne* comes to be conceived not
only as an activity of some particular sort or character, but as a kind
of knowledge. In Plato, who is the first to deal at length with this
notion, *techne* and *episteme,* art and systematic or scientific knowledge,
are closely associated. (Note, too, that in nonphilosophical usage *episteme* itself commonly means "acquaintance with," "skill," or "disciplined experience," as in the *episteme* of archery or war.) In the *Gorgias,*
for instance, Socrates argues that every *techne* is involved with *logoi*
(words, speech, reason) bearing upon the specific subject matter of the
art (450b). Moreover, Socrates goes on to distinguish between two
types of *techne,* one that consists mainly of physical work and requires
minimal use of language (such as painting or sculpture) and another
that is more intimately bound up with speech and requires little physical exertion (such as arithmetic, logistic, or astronomy) (450c ff.).

At the same time, those human activities that are devoid of art, that
are nontechnical, *atechnos,* are activities such as cooking and persuading—each being a mere knack or routine way of operating, a *tribe,*
based simply on experience, *empeiria* (501a). (For the association of
atechnos with *tribe* see also *Phaedrus* 260e). Such pursuits are not art
because they have no awareness of the nature, *phusis,* or cause, *aition,*
of what they make or do; they are *alogos* (cf. 465a). To say that such
actions are nonlogical is to say that they are not based on a consciousness of the true nature of the things they deal with; they are simply
means. In modern parlance, they are "pure technique." In the *Ion,* poets
who exercise their craft of making, *poiesis,* by virtue of divine inspiration are also said to be devoid of *techne* or art (cf. 533d); if poets possessed an art, they would be able to explain their creations to others
and to teach (532c). Evidently, then, *techne* in the early Plato refers to
all human activities that can be talked or reasoned about—all activities that are neither spontaneous nor the result of some unconscious
drive or intuitive perception. If such usage seems at first to make technics coextensive with human activity in a way reminiscent of Kranzberg and Pursell, it also stresses the "logical" character of *techne*—not,
of course, in the modern sense of mathematized deduction, but in the
Greek sense of being involved with language and hence with consciousness or knowledge of the inner nature of things. One might even

venture that, in marked contrast to Kranzberg and Pursell, *techne* lays emphasis on the nonutilitarian or transhuman aspects of that activity. This general understanding of *techne*, however, is extended in the direction of modern notions of technology by a classification of knowledge developed in the later dialogue *Philebus*. Here Plato divides knowledge into two classes: that involved with education and upbringing, and that involved with making or producing (55c). Of the second, technical knowledge, there are again two kinds: one sort (such as music, medicine, and agriculture) that proceeds by conjecture and intuition based simply on practice and experience, and one (such as carpentry) that consciously involves the use of numbering, measuring, weighing (55e-56c). The latter possesses greater exactness or precision, *akribeia* (a word that also implies deeper insight), and this is *techne* in the primary sense. Thus *techne* is clearly distinguished from all human activity and knowledge of a political sort (education and, by extension, governing) so as to be associated more closely with the activities of making or producing that operate upon the nonhuman material world. Those activities are most truly *techne* that involve the greatest quantitative precision.

Up to this point in his classification scheme, Plato has used *techne* and *episteme*, art and systematic knowledge, almost interchangeably. Now he proceeds to speak of a "philosophic" arithmetic, which differs from the arithmetic of the carpenter in that it deals not with numbered things, but with numbers alone (56d). But in referring to this latter still more precise or penetrating *episteme*, he no longer employs the word *techne*. Thus *techne* is also conceived to be distinct from what we would call pure theory, or any knowledge that does not bear upon the material world in some practical manner. (Cf. the distinction between pure and applied arts or sciences at *Statesman* 258e.) While tying *techne* into consciousness and defining the primary type of *techne* as that which can use mathematics to express itself, he nevertheless distinguishes *techne* from pure consciousness or consciousness of a nonmaterial reality. This ultimate reality is, of course, grasped only in a provisional or inadequate way through mathematics. The deepest cognition, *gnosis*, of being is to be had only through dialectic, *dialegein* (*Philebus* 58a ff.).

Plato's discussion points toward a conception easily associated, at least intuitively, with modern technology—that of rationalized production, or production made maximally efficient through mathematical analysis. Yet the Greek term τεχνολογία has yet to appear. The first appearance of *technologia* (or one of its cognates) is found in Aristotle, although not in any of his major discussions of what today would commonly be considered *technai*. For Aristotle as for Plato, *techne* is a special knowledge of the world that informs human activity accordingly. As a type of awareness of the world, it lies between unconscious experience and knowledge of first principles; *techne* is part of the continuum that moves from sense impressions and memories through experience to systematic knowledge, *episteme* (*Metaphysics* 1.1.980b25 ff.). "From experience again—that is, from the universal come to rest in its entirety in the soul, the one alongside the many, the unity that is a single identity within them all—originate art [*techne*] and science [*episteme*]: art in the realm of coming to be, science in the realm of being" (*Posterior Analytics* 2.19.100a6–9). Yet while continuing to stress the epistemic character of *techne*, Aristotle does not think of it solely as a kind of knowledge, but reaches back to pick up the commonsense notion of *techne* as activity. *Techne* is not strictly activity, but it is a capacity for action, founded in a special kind of knowledge.

According to Aristotle's formal definition, *techne* is ἕξις μετὰ λόγου ἀληθοῦς ποιτικὴ (*Nicomachean Ethics* 6.4.1140a11). Translated literally, this defines *techne* as a habit (or stable disposition to act in a specific manner) with a true *logos* concerned with (or ordered toward) making (the human production of material objects). To paraphrase, *techne* is an ability to make that depends on correct awareness of or reasoning about the thing to be made. The absence of *techne* in an ability to make involves either the absence of any *logos* (consciousness) or the presence of a false *logos* (false consciousness) (*Nicomachean Ethics* 6.4.1040a20–23). Once again the nonutilitarian character of technical knowledge comes to the fore; insofar as it is true, this *logos* is based on a mental grasping or cognition, *gnosis*, of causes, *aition* (*Metaphysics* 1.1.981b6–7) and speech.

Mention of the connection between *logos* and *aition* recalls Aristotle's distinction of the four causes—a distinction that, incidentally, is consistently illustrated with references to technical products. According to this discussion (cf. *Physics* 2.3) the "why" of a thing is answered only by grasping the "that out of which" it comes (material cause), its *eidos* or archetype (formal cause), the "what makes of what is made and what causes change of what is changed" (efficient cause), and the *telos* or "that for the sake of which" a thing is made (final cause). What is important in such discussions is that Aristotle does not limit the technical, as we might be tempted to do, to efficient causation. The making of artifacts involves all four causes.

Techne, then, is *episteme* in that it involves true consciousness of the world and hence can be taught or communicated (*Metaphysics*

1.1.981b8–10); but it is to be distinguished from *episteme* insofar as it bears upon changing rather than unchanging things (cf. *Nicomachean Ethics* 6.6.1141b 31–36). Aristotle agrees with the later Plato in stressing the "logical" character of *techne* while separating it from knowledge of human affairs, on the one hand, and pure theory, on the other. What is absent from Aristotle's understanding of *techne* is any suggestion that part of its "logic" needs to be the use of quantitative or mathematical concepts; even at the highest levels *logos* is not restricted to mathematical reasoning. Compare, for example, Aristotle's reference at *Politics* 1.11.1258b35–40 to those activities that are most truly *technai* as those in which there is the least element of chance, with *Politics* 8.6.1341a17, where he explicitly refers to flute playing as requiring great *techne*. Unlike Plato, Aristotle is able to think of medicine as *techne* in the primary sense (cf. *Metaphysics* 1.1.981a13 ff.).

Plato and Aristotle agree, then, in stressing the "logical" character of *techne*, even when they disagree on their understandings of the character of the *logos* involved. Yet neither feels drawn to join these two words—to speak of a *logos* of *techne*. *Techne* simply uses *logos*. Here Plato's distinction between numbering, measuring, and weighing in carpentry and a philosophic numbering, measuring, and weighing is suggestive. For Plato, carpentry merely uses a more general or universal arithmetical *episteme*. Although arithmetic is a *logos* to be used by carpentry, it is not *logos* of carpentry in the sense of being derived from or limited to this particular *techne*, nor is it the entire *logos* of carpentry. There are elements of consciousness in carpentry that cannot be expressed through this *logos*, that are not capable of being expressed in the language of arithmetic. Furthermore, there are elements of any *techne* that, because of its involvement with the particulars of the material world, cannot be expressed at all.

To put it simply, what can be grasped or known by *techne* through *logos* is the form or idea, *eidos*, the whatness of the thing to be made. What is not as able to be grasped is the activity, the "how to do it" of the actual making, *poiesis*. Here Plato's example from the *Cratylus* (389a–390b), of the carpenter who repairs a broken shuttle, is instructive. In making repairs a carpenter looks not to some broken shuttle but to the form, to "that which is fitted by nature to act as a shuttle." It is this that the carpenter must reembody, *apodidomi* (literally "give back to," "restore") in the "that out of which he makes" (the material) "not according to his own will, but according to its nature." Again, in the *Timaeus* (29a), Plato says that the *demiourgos* made the world by looking to an eternal and unchanging form or pattern, *paradeigmatos*, which is apprehended by reason, *logos*. As to the how or activity of making, the becoming as opposed to being, this can be grasped only through *pistis*, belief or trust, the mental disposition that in the *Republic* (511d and 534a) Plato associates with the perception of material things. Clearly, then, it is the *eidos* that is grasped by the *logos*—and sometimes by a mathematical *logos*—that is operative in *techne*. But the matter, that out of which a thing is made, and the consequent process of making do not fall within the logical structure of the art.

Aristotle's analysis of carpentry makes the same point. In an artifact, he argues, no material part comes from the carpenter, nor is any part of carpentry (the art) in what is produced. Instead, the shape, *morphe*, and form, *eidos*, are engendered in matter by motion. It is the soul, which possesses a form of knowledge, that moves the hands (or some other part of the carpenter) with a definite motion, one varying with the varying character of the object to be produced, the hands that in turn move the tools, and the tools that move the matter (*On the Generation of Animals* 1.22.730b10–20). Elsewhere, even more pointedly (if more abstractly), Aristotle argues that it is part of *techne* "to know the form and the matter," but the matter, *hyle*, only "up to a point" (*Physics* 2.2.194a23). "Matter is unknowable [*agnosis*] in itself" (*Metaphysics* 7.10.103a9). Only as informed, or related to form, can matter be grasped by mind. Yet relative to every work of *techne* there is a matter and a form, and it is "the *matter* [my emphasis] that governs the making [*poiesis*] and generation of any work of art" (*Metaphysics* 7.9.1034a10–11). "*Techne* imitates nature [*phusis*]" (*Physics* 2.2.194a21; *Meteorology* 4.3.38166; *On the Cosmos* 5.396b12) by uniting form and matter in a particular something (cf. also *On the Generation of Animals* 2.4.740b25–29). The form is the idea in the mind of the artist (*Metaphysics* 7.7.1032a35), but its union with matter is, as it were, at the mercy of matter and its specific receptivity. Form cannot be forced into or imposed upon matter; an artisan must let the matter guide the way it receives form. The ultimate guide for the making activity as activity is not reason but perception, *aisthesis* (*Nicomachean Ethics* 2.9.1109b23; cf. 2.2.1104a1-9). On one occasion Aristotle goes so far as to describe the coming together of form and matter, the becoming of an entity, as dependent on matter's "desire" or "reaching out" for form (*Physics* 1.9. 192a18).

At issue, as Thomas Aquinas notes in his commentary on this last passage, is whether matter, at least any particular matter, is not just privation of form, but a real something in its own right. Although with respect to the object to be made the matter can be spoken of as formless, in reality it is itself something that "seeks form or further form according to its proper nature" (*Commentary on Aristotle's Physics* 1, lec. 15, par. 8). As Thomas says elsewhere (*Summa theologiae* 1, qu. 85, art. 7), "Act and form are received into matter according to the capacity of the matter." Absent an artisan's deep sensitivity to the particular characteristics of this ordering toward form, this "desire" of matter, the result will almost surely be a weak unity, one tending to either rapid physical decomposition or aesthetic disorientation (which is only decomposition of another sort), or both. Premodern artisans were interested in bringing about as perfect a union as possible, while recognizing that they could never completely duplicate the substantial union of form and matter found in nature (cf. *Physics* 2.1.193a12–17). *Techne* in the classical understanding—and this cannot be emphasized enough when comparing ancient and modern making activities—is thus fundamentally oriented toward particulars instead of toward the efficient production of many things of the same kind in order to make money. Mass production would be unthinkable to the classical mind, and not just for technical reasons.

In book 1 of the *Republic* Thrasymachus advances a conception of *techne* as the power that pursues its own interests or the interests of its possessor—as a means by which the stronger dominates the weaker. Thrasymachus's *techne*, like modern technology, is oriented toward the extrinsic end of making money (although not necessarily as efficiently as possible). Socrates rejects such a view, arguing that *techne* as *techne* "does not consider its own advantage . . . but the advantage of that of which it is the art." "There is no kind of knowledge [*episteme*] that considers or commands the advantage of the stronger, but rather of what is weaker and ruled by it" (342c-d). *Techne* and *episteme* are both fundamentally oriented toward some otherness and its good, its "desires," and its "proper nature." When this otherness is material (as in the case of carpentry), and when matter is understood as inherently particular (as it is by Aristotle), then *techne* will be radically limited in its use of *logos*. Because it is matter that gives a particular its particularity (cf. *Metaphysics* 7.8.1033b20–1034a7), individuals themselves cannot be known in their particularity through the logical universal. *Logos* breaks down before particulars.

This inability of a technical *logos* to comprehend particulars in their particularity can be further elucidated by comparison with the inability of law, *nomos* (for the connection of *nomos* with *logos*, see *Politics* 3.15.1286a15–17), to take account of every political circumstance. "Well laid down laws should themselves determine all the things they can," affirms Aristotle, "and leave as few as possible to the decision of the judges" (*Rhetoric* 1.1.1354a32–34). Yet law is always a universal or general statement, and about some things it is not possible to make a universal statement that will always be correct (*Nicomachean Ethics* 5.10.1137b13–15); law is unable to speak with precision, *akribeia*, because of the difficulty any general principle has in embracing all particulars (*Politics* 3.11.1282b5). Or again, "Some things can and others cannot be comprehended under law. . . . For matters of detail about which men deliberate cannot be included in legislation" (*Politics* 3.16.1287b19–23). (For Plato's concurrence on this issue, see *Statesman* 295a ff. and *Laws* 6.769d.)

Indeed, so much is this the case that Aristotle feels compelled to distinguish between legislation, *nomothetikos*, and politics, *politike*, the daily activity of deciding what is right in particular cases (*Nicomachean Ethics* 6.8). The proper operation of both is grounded in φρόνησις or *prudentia*, which, in a definition that exactly parallels one given earlier for *techne*, is described as a habit with a true *logos*, περὶ τὰ ἀνθρώπινα ἀγαθὰ πρακτικήν—that is, concerned with humanly good action, *prakitikos* (*Nicomachean Ethics* 6.5.1140b21). (Cf. the previous distinction between making, *poiesis*, and acting or doing, *praxis*, at *Nicomachean Ethics* 6.4.1140a1–6; plus the use of *poiein* in *Categories* 4 and 9 and *Topics* 1.9 to refer to action in general, including both making and doing.)

Politics as an active involvement in public affairs, by contrast, is concerned with action and deliberation about particulars (*Nicomachean Ethics* 6.8.1141b27–30). Human beings are forced to deliberate, to consider, and to take counsel with others (see *Politics* 3.15.1286a26–30 and 3.11) with regard to events that happen in a definite way for the most part but not always, so that the results are not fully determinate, necessary. In short, deliberation is mostly about means and processes, not about ends or ideals (*Nicomachean Ethics* 3.3). Political wisdom thus focuses on *nomos*, law, and politics on *dike*, justice (*Nicomachean Ethics* 5.1). At the same time, although *nomos* cannot fully determine *dike*, usually only one formed by or grounded in *nomos* will be able to do justice. (Compare the relation between virtue, *arete*, and *logos* described in *Nicomachean Ethics* 6.1.) Judges are educated by the law in order to perfect or complete it (*Politics* 3.16.1287a25–28 and 1287a25–26). They are the functional equivalent of artisans (cf. *Nicomachean Ethics* 10.9). Therefore, "as in relation to the other arts [*technai*], so in relation to the political [art, and its product, the political] organization, it is impossible that everything should be written down with precision" (*Politics*

2.8.1269a10). Besides, "in any art of any kind it is absurd to govern procedure by written rules" (*Politics* 3.15.1286a10). Once again, in making and in doing insofar as it involves making, the logical element breaks down before particulars.[14]

Greek Techne *in Other Terms*

Although the classical Greek term *techne* is not a living part of any language, the essential insight expressed within it is not wholly lacking in later discussions. Jacques Maritain, for instance, has captured something of the traditional understanding of *techne* in his commentaries on the arts:

> In the . . . arts the general end . . . is beauty. But . . . as an individual and original realization of beauty, the work which the artist is about to make is for him an end in itself: not the general end of his art, but the particular end which rules his present activity and in relation to which all the means must be ruled. Now, in order to *judge* suitably concerning this individual end, that is to say, in order to conceive the work-to-be-made, reason alone is not enough, a *good disposition of the appetite* is necessary. . . . The artist has to love, he has to love *what he is making*.[15] And, because in the . . . arts the work-to-be-made is . . . an end in itself, and because this end is something absolutely individual, something entirely unique, each occasion presents to the artist a new and unique way of striving after the end, and therefore of ruling the matter. . . . [Thus] it is by using prudential rules not fixed beforehand but determined according to the contingency of singular cases, it is in an always new and unforeseeable manner that the artist applies the rules of his art.[16]

Note, however, that in the scholastic adaptation of Aristotle's definition of art which Maritain employs—"art is the right rule about things to be made"—"rule" is a translation of the Latin *ratio,* which in turn translates the Greek *logos.* Thus when, in the last sentence, Maritain speaks of "prudential rules," he is evidently using the term in a different sense. By limiting *logos,* which includes both speech and reason, *oratio,* and *ratio* to *ratio* alone, his position is inherently more rationalist than Aristotle's.

Art et scolastique, from which these quotations come, was first published in 1920. Three decades later, in *Creative Intuition in Art and Poetry,* Maritain returns to this theme, with an even more pronounced rationalist tone. After rewording the scholastic definition of art as "the straight intellectual determination of works to be made,"[17] he goes on to discuss once again the character of rules in artistic activity. In this further discussion Maritain nevertheless makes three points relevant to the present issue. First, such rules "are subjected to a law of perpetual renewal." The eternal laws of art are "not to be found at the level of the particular rules of making."[18] Second,

> the work to be made, in the case of the fine arts, is an end in itself, and an end totally singular, absolutely unique. Then, every time and for every single work, there is for the artist a new and unique way to strive after the end, and to impose on matter the form of the mind. As result, the rules of making— which, as concerns art in general, are fixed and determined, as opposed to the rules used by prudence—come in the fine arts to share in the infinite suppleness and adaptability of the rules used by prudence, because they deal every time with the utter singularity of a new case, which is, in actual fact, unprecedented. It is, then, with prudential rules not fixed beforehand but determined according to the contingency of singular cases, it is with the virtues proper to prudence—perspicacity, circumspection, precaution, industry, boldness, shrewdness, and guile—that the craftsmanship of the artist succeeds in engendering beauty.[19]

Third,

> because the work to be made is an end in itself and a certain singular and original, totally unique participation in beauty, reason alone is not enough for the artist to form and conceive this work within himself in an infallible creative judgement. . . . To produce in beauty the artist must be in love with beauty. Such undeviating love is a supra-artistic rule—a precondition, not sufficient as to the ways of making, yet necessary as to the vital animation of art—which is presupposed by all the rules of art.[20]

In contrast to his earlier remarks, Maritain now stresses love not of the object made, but of the ideal of beauty. One may wonder whether, in the case of the practical arts, there is a corresponding love of the ideal of the useful as a practical manifestation of the good.

Recognition of the issue involved is not limited to the scholastic tradition. To cite one philosopher with quite different affinities, James Feibleman, in discussing the place of skill in technological activity, makes the following supplementary observations:

The human individual is also a material object and if in harmony with his tools is capable of depths of understanding of them as material objects when he has used them long enough. Such love for particular kinds of material objects comes only through a prolonged familiarity with their use and is not confined to their form but extends more deeply into the material. (1966, p. 327)

However, it takes

a long time and a great deal of concentration to become deeply acquainted with any material object. The mystical knowledge of matter has long been practiced but seldom recognized. Abstract knowledge is easy to acquire and to identify as such, but concrete knowledge is a different thing. Concrete knowledge uses quite different channels. It is absorbed by means of the sense organs and muscles. It comes through exteroceptors, such as the eye and the ear, and also through proprioceptors in the muscles. (1966, p. 328)

In other words, there is at the heart of technical activity, if not of *techne* itself, an irreducible, nonlogical component.[21] There is an aspect of *techne* that necessarily cannot be brought into consciousness except through the immediacy of a singular, direct encounter, an encounter that takes place through sensorimotor activity and is properly grounded in one of the various forms of love, *storge, philia, eros, agape.* Only love can encompass or grasp the singular.[22]

The point at issue in these citations from Maritain and Feibleman is occasionally expressed by artists themselves. For instance, although in many of his letters and poems Michelangelo likens his work to the heroic imposition of form upon matter—an image of the artist that has been dominant in the West since the Renaissance—in one instance he speaks of merely releasing form imprisoned in a block of marble.[23] He sees his work as an instrument only at the service of matter. This dialectic between active and passive responses to matter needs further elucidation, yet for present purposes it is enough simply to correct an imbalanced picture by noting that even the most "heroic" artists sometimes speak of "following the materials."[24]

Note too that this understanding of *techne* implicitly rejects a contrast, such as that expressed by Mumford, between technics and art. For Mumford,

Art, in the only sense in which one can separate art from technics, is primarily the domain of the person, and the purpose of art, apart from various incidental technical functions that may be associated with it, is to widen the province of personality, so that feelings, emotions, attitudes, and values, in the special individualized form in which they happen in one particular person, in one particular culture, can be transmitted with all their force and meaning to other persons or to other cultures. (1952, p. 16)

But this is to transfer the realm of particularity from the objectivity of the work to the subjectivity of the worker in a manner that is at odds with the Greek notion of *techne* or what Mumford with unconscious irony opposes to technics.

From Techne *to Technology*

Here, then, is the most fundamental difference between Greek *techne* and modern technology. *Techne* involves *logos*, but only in grasping form, not in directing the actual process of production, the activity qua activity. There is no *logos* of this activity. But is this not precisely what modern technology proposes to furnish—a *logos* of the activity, a rationalization of the process of production, independent of, if not actually divorced from, any particular conception of *eidos* or form? Is this not precisely why it can so vigorously claim to be neutral, to be dependent in use on whatever human beings want to do with it, on purely extrinsic ends?

All this can be thrown into relief once more by considering the teachability of *techne*, something both Plato and Aristotle affirm. Although *techne* is involved with language, and is hence teachable, one must be careful about reading into this notion modern conceptions of teachability. It is not teachable in the way modern engineering schools teach technology. What are teachable are the forms of beauty, not the processes of production. With regard to the practice of *techne*—that is, technical action—*logos* is not enough for Aristotle. Like virtue, *techne* is learned primarily through practical imitation: "Human beings become builders by building" (*Nicomachean Ethics* 2.1.1103a35). This explains the absence of any general treatises on *techne* in the Aristotelian corpus—an absence with only one exception. In this exception, in Aristotle's treatise on the *techne* of rhetoric, the words *techne* and *logos* are joined for the first time.

In writing on rhetoric, Aristotle makes four attempts to unite *techne* and *logos* (*Rhetoric* 1.1.1354b17, 1354b27, 1355a19, and 1.2.1356a11). The exact meaning of each occurrence is debatable. That the term does not mean the *techne* of *logos*, the art of words, a synonym for *techne retorike*, is indicated by the parallel use of *techne tou logou*. It is possible that Aristotle means only "words about *techne*" or perhaps "systematic thought concerning an art." Although even this weak sense would be significant, in each instance there is intimation that *logos* of *techne* might mean something stronger, that Aristotle is trying to refer to a *logos* of the activity of the *techne* of persuasion. This is indicated, for example, by the way Aristotle, unlike Plato, argues for the divorce of rhetoric from considerations of truth; rhetoric is a *techne* of the "means" of persuasion (1355b10). The *Rhetoric* is a treatise as much on "how to" as on "what"—even on "how to" never mind "what." Apparently when dealing with the art of persuasion, which operates through the medium of words—a rarefied, not to say "artificial" material—there can be systematic discourse not only about forms or ends, but also about means or processes.

Although Aristotle here argues that this is equally true of other such arts as medicine, one cannot help but suspect that the argument itself is a use of rhetoric. Isn't it enough to note that Aristotle wrote no other such technological treatises? Be that as it may, words alone, divorced from reason, can acquire power simply as a means. Indeed, it is because of this divorce that Plato criticizes rhetoric (cf. *Gorgias* and *Phaedrus*) in a way that can easily be applied to modern technology. So it is not without raising pregnant questions that the Greek term for "technology" comes to mean the study of grammar or rhetoric, and that we find "technologist" used to refer to the grammarian or rhetorician.

A full history of this usage need not be attempted here. Occurrences involving these or closely related meanings are to be found in the works of Dionysius of Halicarnassus and Longinus (first century B.C.E.), Hermogenes (second century C.E.), Porphyry (third century), Sextus Empiricus (third century), Iamblichus (third century), Basil and Gregory of Nazianzus (fourth century), Photius (ninth century), Zonaras (twelfth century), and Eustratius (twelfth century). Saint Basil even complains once (*Epistle* 90) that *technologousi loipon, ou theologousin, hoi anthropoi* ("Human beings practice technology [rhetoric], not theology [prayer]").

Cicero (106–43 B.C.E.) transliterates the Greek term into Latin once (*Letters to Atticus* 4.16), but it is not until fifteen hundred years later that the word really enters the Latin vocabulary—with a quite different meaning. The French Protestant rhetorician Peter Ramus (1515–1572) uses it to refer not to the *logos* of one *techne* (namely rhetoric) but to the *logos* of the relations among all *technai*. Reflecting the passion for method associated with Renaissance humanism, for Ramus *technologia* systematically orders and arranges the arts and sciences. Ramus in fact criticizes Aristotle for his haphazard ordering of the arts. As Walter Ong points out, however, Ramus's concern for ordering shifts from the "things" of the world to "things" in a linguistic sense.

> The "things" which the Greek philosopher [Aristotle] fails to chart correctly turn out to be neither external reality, nor the predicaments, nor even the topics, but the arts. . . . Ramus assumes that the chief business of all classification is the classification of the arts and sciences themselves. . . . This curious shift in Ramus' thought is obviously another manifestation of . . . his tendency to be "objective" not by turning to the outside world but by treating the contents of the mind as a set of objects.[25]

Ramus also coins another term, *technometria* (which does not occur in either Greek or Latin), and uses it as a synonym for *technologia*. After Ramus both terms achieve further currency, especially in the work of Puritan theologian William Ames (1576–1633).[26]

The earliest uses of the English "technology" are close in connotation to Ramus's and Ames's *technologia*. At the conclusion of Sir George Buck's *The Third University of England* (1615), a "breife report of the sciences, arts, and faculties" of schools "within and about the most cittie of London," he writes, for instance, of an "apt close of this general technologie." Yet barely a decade later Tobias Venner, in *The Baths of Bathe* (1628), uses the term with a meaning much closer to the current one when writing that he "cannot but lay open Baths Technologie." By 1706, in John Kersey's edition of Edward Phillips's dictionary *The New World of English Words*, technology is given a version of its modern definition by calling it "a Description of Arts, especially the Mechanical." That this same definition does not occur in the 1658 edition of Phillips's dictionary implies that it was in the second half of the seventeenth century that the term acquired its present English meaning.

Interestingly enough, however, the term also has an obsolete English usage corresponding to the ancient Greek. John Twell, in the preface to his *Grammatica Reformata; or, A General Examination of the Art of Grammar* (1683), writes, "There were not any further Essays made in Technology, for above Fourscore years; but all men acquiesced in the Common Grammar." But this usage is already superseded by the time of Kersey's dictionary.

On the Continent, in French and German, similar developments are taking place, perhaps again under the influence of Ramus. Culminating this development is the concept of technology given by Christian Wolff in his *Preliminary Discourse on Philosophy in General* (1728), which developed out of René Réaumur's posthumous *Descriptions of the Arts and Crafts* (beginning in 1761) and the great French *Encyclopédie* (1751–1772), and the ideas of the German educator Johann Beckmann in his *Anleitung zur Technologie* (1777), the first work to use the term in its title. For Wolff, *Technologie* is "the science of the arts and of the works of art" or the use of physics to give "the reason of things which occur through art." With Beckmann technology is a functional description of the process of production; this "general technology," as he calls it, provides a basis for the sound economic and political regulation of trade. This is the Puritan notion of *technometrica* as the science defining all arts, with its coordinate conception of the individual arts as "eupraxia methodically delineated by universal rules," driven home with a vengeance. It is against this background that Jacob Bigelow says, in the preface to his *Elements of Technology* (1831)—the first English work to use "technology" in its title—that he has adopted a word "found in some of the older dictionaries" in order to refer to "the principles, processes, and nomenclatures of the more conspicuous arts, particularly those which involve applications of science."

Philosophy of Technology versus *Philosophia Technes*

The tension between the words *techne* and "technology" points both toward a hiatus between ancient and modern understandings of the making activities and toward the presence of a kind of making outside the realm of the technological. The paradox is that philosophy of technology both includes and does not include an awareness of such alternative ideas about making. Perhaps it could be said that the engineering tradition of the philosophy of technology, which might also be interpreted as having roots in Ramus's conception of technology, does not include it, but that the humanities tradition of the philosophy of technology implicitly does. What follows, then, is the speculative retrieval of a *philosophia technes* to be incorporated into a broad humanities philosophy of technology.

At the foundation of the difference between ancient *techne* and modern technology—and ultimately it could be argued that the adjectives "ancient" and "modern" are redundant here—is a conception of matter, an ontology or metaphysics of matter. Making qua activity viewed solely as production process rather than as production of some one thing, presumes this ontology. It is the theory about the nature of what one is working with that is a primary determinate of how one works, the structure of the working itself.

The premodern or classical ontology involves looking upon matter as a living reality ordered toward taking on form—in accord with whatever form it already possesses and the potentialities contained therein. There is a hierarchy of form to be articulated in thought and attended to in action. Plato envisions the cosmos itself as an organic unity, a living creature; on one occasion he specifically characterizes it as divine, a god (*Timaeus* 30c–31b and 34a–b). In the thought of his Neoplatonic followers, all multiplicity—including its principle, matter—emanates from and is involved in a continual return to the one source of all being, God (see Proclus, *The Elements of Theology*, prop. 57 and cor., with prop. 72, cor.; cf. also Plato's own myth of cosmic reversal at *Statesman* 268d–274d). This is not to deny that matter or its functional equivalent (Plato's receptacle, *Timaeus* 49a) is often conceived by Neoplatonists as resistant to or opposing form—in the extreme case, as the principle of evil (Plotinus, *Enneads* 2.4.16). Yet even Plotinus admits (*Enneads* 4.8.6) that everything, including matter, participates in the good itself "in the measure that each is capable of doing so." And Augustine, arguing that Christianity is the perfection of Platonism, states unequivocally that "matter participates in something belonging to the ideal world, otherwise it would not be matter" (*De vera religione* 11.21). Thus matter is caught up in a cosmic process and in this sense is living.

For Aristotle and Aristotelians, in their less dramatic manner, something similar is involved. No matter, not even that strictly logical construction prime matter, is a purely neutral or lifeless stuff able to be imposed upon at will; it seeks or is related to form—in any particular case, in some particular way. This is why Aristotle can quite legitimately speak of a "desire" on the part of matter. It is also why, traditionally, moral discipline could not be divorced from the making activity; it is moral rather than intellectual discipline that cultivates human receptivity to the needs and desires of another, that develops the ability to respect another being (whether human or not) for what it is in itself. This, incidentally, is also why alchemy is a sacred rather than a profane endeavor: it is "work" taking place at once in a subject and in materials. The transmutation of some base metal into gold is but the exterior correlate of an interior spiritualization or divinization of soul.

Only in circumstances where matter is cut off from this cosmic process—in a case, that is, such as language interpreted as a means to communication—is the idea of a technology thinkable for the ancients.[27]

In the latter half of the seventeenth century, however, the Western ontology of matter underwent a radical transformation. Under the influence of Galileo (1564–1642), Descartes (1596–1650), Newton (1642–1727), and their followers the material world began commonly to be regarded in much the same way as Aristotle looked upon words. Instead of a potentiality unknowable in itself yet ordered toward something higher, matter began to be conceived of as separated from any cosmic process. This trend is easily exemplified by the Cartesian theory of matter as pure, lifeless extension, in itself ordered toward nothing, something to do with as one pleases. More succinctly, matter ceased to be thought of as in any sense living—as having, as it were, any spiritual aspirations of its own. Consider again the case of alchemy as illustrative of the ancient worldview; for the alchemist, matter is an aspect of God. It is not so much opposed or indifferent to spirit as it is a necessary complement. In modern scientific theory, however, matter does come to be conceived of as wholly inert, totally devoid of spirit. Finally, it was through the modern hiatus that human beings began to imagine the possibility of a *logos* of *techne*. Thus it began to make sense to use a term originally applied to the study of the manipulation of words, then to the organization of systems of words, to name the study of the manipulation of nature. Modern technology may, as Heidegger maintains, be the last stage of metaphysics—but not of ancient metaphysics.

Such is not to suggest, of course, that this transformation of the term "technology" took place consciously. This is just the kind of change that takes place, as it were, behind the back of philosophy, and that must be excavated from its sedimented layers. It is no doubt related, however, to the new emphasis placed on the metaphor of the "Book of Nature." Galileo's description of the natural world as a book whose language needs to be correctly understood, and that is in fact "the language of mathematics," suggests something similar.[28] A full examination of this thesis about the etymology of technology emerging from a new ontology of matter would, of course, entail plunging into a lengthy consideration of the philosophy of nature. Nevertheless, perhaps the present brief study can substantiate a questioning of the facile historical identification of *techne* and technology in a way beneficial to larger issues.[29]

One of these larger issues concerns the existence of an implicit philosophy of technology among the ancients, a philosophy that can itself be of service in the present. An etymophilosophical study of the term "technology" also suggests, despite certain material continuities in the history of technology, the possibility of formal discontinuities of greater significance. The history of technology is not nearly so linear and progressive as technological history implies. To borrow an idea from Thomas Kuhn's *The Structure of Scientific Revolutions* (1962), perhaps the development of technology, like the development of science, should be viewed as proceeding within the framework of "paradigms."[30]

The making of artifacts—what things are made, how they are made and used—is not always the result of some straightforward accumulation of technical knowledge or power; it is conditioned not only by social needs and values (as the social historian of technology would argue), but also, and perhaps more significantly, by philosophical ideas. Indeed, it may well be that technology is more akin to art, in which a history of change and multiplicity yields no simple progress. There are periods of achievements and periods of failure, advance and decline, some periods in which ideals are realized or great things made and others in which possibilities are missed or lost. There are different historical periods, marked off by differing ideals and practicalities across physical and historical horizons. Approaching the world of artifacts from such a pluralistic perspective, through the conditioning of different social and conceptual circumstances, would in turn help shake history loose from the debilitating pressures of progressive historicism.

PART TWO

• • • •

Analytical Issues in the Philosophy
of Technology

Although philosophies suffer a natural and historical genesis, including psychological and sociological dimensions, perhaps an apt symbol for such origins is the furrowed brow of Zeus. Zeus, king of the gods, having swallowed his pregnant wife, Metis of wise counsel, stands confounded, concerned and with wonder, before the complexities of the world—especially the world of his own making. From such concern springs his fair daughter, Pallas Athena. Dutiful daughter and patroness of *technai* of every kind, Athena carries her father's wonder forth into the world of human words and deeds, sent as she is to dwell among them.

Part 1 explores technology as a theme for philosophical reflection—first in modern, then in a premodern history of the philosophy of technology, with a playful interlude on technology across the spectrum of philosophy. Each chapter presumes an intuitive and relatively undifferentiated notion of technology. Such presumptions are enough, however, to furrow the brow. To instantiate the basic philosophical approach we will need to come into closer contact with the real world of technology, or at least that world as it is manifested in technological discourse.

CHAPTER SIX

• • • •

From Philosophy to Technology

The introduction raised the prospect of thinking about technology in a philosophical way and pointed toward different aspects of any such effort. One aspect entails the identification of a basic philosophical stance or attitude; a second involves its instantiation in appropriate conceptual engagements with technology. Philosophy of technology must be both a *philosophy* of technology and a philosophy of *technology*.

Chapters 1 and 2 set out, by means of historicophilosophical investigation, to address the first of these issues. Engineering philosophy of technology was distinguished from humanities philosophy of technology, after which chapter 3 considered mediating approaches but finally argued for primacy of the latter. Chapter 4 sought to articulate in more detail, on the scaffolding of traditional diversions within philosophy, the distinctive stance of humanities philosophy of technology. Chapter 5, by way of supplement, proposed to deepen the discussion by demonstrating the implicit existence of a distinctively premodern humanities philosophy of technology, thereby raising the possibility of a fundamental distinction between ancient and modern technology. Interest in such issues of historical recovery may also be described as typical of the humanities approach to philosophical reflection on technology.

Engineering Objections to Humanities Philosophy of Technology

A proponent of engineering philosophy of technology could, however, raise at least three objections to the philosophical primacy of humanities philosophy of technology (HPT). One is that it is simply not possible to have the humanities without technology. As one prestigious commentator on Benjamin Franklin's definition of the human being as a "tool-making animal"[1] has written, "Inventiveness was the indispensable condition for the survival of the human species. Without fur or feather, carapace or scale, ancestral man stood naked to the elements; and without fang or claw or tusk to fight his predators, without speed to elude them, without camouflage to deceive them or the ability to take to the trees like his cousin, the ape, he was physically at a hopeless disadvantage. What he developed to deal with his deficiencies was [technology]." [2] Second, the defense of HPT as more philosophical "stacks the deck"; HPT is necessarily going to be more weighted with philosophy and philosophical sophistication than engineering philosophy of technology (EPT), since philosophy is one of the traditional humanities and engineering is not. A final objection is that to equate "being more philosophical" with "being primary" begs the question; there are serious weaknesses with so-called humanities philosophy of technology.

The first objection misconstrues if not overindulges itself. Although it has a point, it goes too far. At most it is an argument for the primacy of engineering over the humanities, not of engineering philosophy of technology over humanities philosophy of technology. Furthermore, historical priority does not entail logical primacy. Indeed, the imputed historical priority itself is questionable, since it is in no way clear that premodern and modern technology are not the same—a point repeatedly overlooked in many otherwise sophisticated philosophies of technology.

Nevertheless, there remains a readily appreciable truth that must be acknowledged. To the extent that there is any continuity, however attenuated, between premodern and modern technology, and insofar as the humanities are dependent on technology, then to that degree a philosophy of technology that takes its bearings solely from the humanities rather than from technology must be deficient.

There is also some truth to the second objection, that the defense of HPT as "more philosophical" stacks the deck. Of course it is more philosophical, since philosophy is one of the humanities. Consider the situation if the tables were turned. Were one to distinguish between "engineering technology" and "humanities technology," engineering technology would necessarily be more technological than humanities technology, simply because engineering is a technology whereas the humanities are not. At the same time, if there were humanities scholars making claims for works of literary criticism or even philosophical texts as being instances of technology equal in engineering significance to large bridges or skyscrapers, then surely such scholars would deserve criticism as being insufficiently appreciative of the technological character of technology.

An engineer might still respond with the third objection, Why argue about words? The real issue is not which tradition can make a stronger claim to the word "philosophy," but which is really more philosophical in the sense of doing what needs to be done by way of conceptual analysis and reflective clarification of the lifeworld as it has been influenced or transformed by modern technology. Is EPT not in fact making the more important contributions to this task?

Here the answer is both yes and no. It is important to admit that EPT is doing something that needs to be done. A defense of the philosophical primacy of HPT need not imply that EPT should cease to be practiced—or that HPT is perfect as it stands. In the EPT emphasis on paying closer attention to the real world of engineering experience and discourse, it reveals legitimate analytic work to be done. As American engineer-philosopher Billy Vaughn Koen puts it:

> The study of engineering method is important to understand the world we have. The environment of man is a collage of engineering problem solutions. Political alliances and economic structures have changed dramatically as a result of the telephone, the computer, the atomic bomb and space exploration—all undeniably products of the engineering method. Look around the room in which you are now sitting. What do you find that was not developed, produced or delivered by the engineer? What could be more important than to understand the strategy for change whose results surround us now and, some think, threaten to suffocate, to pollute and to bomb us out of existence?
>
> Yet, although we speak freely of technology, it is unlikely that we have the vaguest notion philosophically of what it is or what is befalling us as it soaks deeper into our lives. . . . Now, as we sit immersed in the products of the engineer's labor, we must ask: What is the *engineering* method?
>
> The lack of a ready answer is not surprising. Unlike the extensive analysis of the scientific method, little significant research to date has sought the philosophical foundations of engineering. Library shelves groan under the weight of books by the most scholarly, most respected people of history analyzing the human activity called science. No equivalent reading list treats the engineering method. (1985, pp. 1–2)

The British engineer-philosopher G. F. C. Rogers readily agrees. As he says, although "no one can hope to understand the work of more than a few" of the disciplinary specialists contributing to the complex world of contemporary knowledge, we can and should try to understand the framework of ideas within which each broad group operates. This is especially necessary when the specialists are engineers or technologists because their power to influence the way in which we live has reached an awesome level. . . . Finding ways of harnessing the power of technology for the greater benefit of mankind, and of moderating the social stresses arising from the ever-increasing rate of technological change, poses unparalleled problems for humanity. There is little hope of accomplishing either of these things unless both the public and government understand the nature of engineering and the ways in which technologies are born and develop. (1983, p. 1)

But insofar as the engineering-philosophical analysis of engineering is taken as the basis for a general explanation of the human world, or even of technology, EPT fails to recognize its own limitations and its place within a larger framework. Something is left out, and it is not clear how EPT can be expanded to include the missing element. Argument to this effect no longer is merely a plea for verbal distinctions, but seeks real ones.

One of these real distinctions is between different ends or criteria of judgment. Historically, the rise of EPT entailed an explicit rejection of HPT in the form of what might be called premodern philosophy of technology—not for being less philosophical, but because it was less *technological*. Traditional philosophy has done less to change the world than have gunpowder, printing, and the compass, argued Francis Bacon; therefore philosophy itself (that is, especially natural philosophy) should be changed, should become allied with the making of artifacts. But practical efficacy in changing the world is not the highest or most inclusive criterion of judgment. When someone wants to bring about practical change, it always makes sense to ask why or for what?

The argument here can be made in a collateral way by pointing out that humanities philosophy of technology is inherently more inclusive than engineering philosophy of technology. Inadequate examples notwithstanding, because of the humanities commitment to a plurality of perspectives, humanities philosophy of technology must in principle remain open to the engineering perspective. It is not obvious that engineering qua engineering has a similar principled openness. For instance, questioning the world according to engineering criteria such as practical efficacy or efficiency is only one kind of questioning and can itself be questioned. To defend or argue for the primacy of efficacy or efficiency, one has to make use of other criteria. Even to criticize some particular HPT tendency to slight efficacy and efficiency in defense of tradition or beauty almost necessarily calls for invoking other nonengineering criteria such as democratic principles or economic constraints. When engineering philosophers of technology initiate discussions with humanities philosophers of technology they become more like them than happens when the situation is reversed. Humanities philosophy of technology is more capable of including engineering philosophy than engineering philosophy of technology is of including humanities philosophy.

But why should this kind of inclusiveness be a defining characteristic of philosophy? Isn't engineering or technology inclusive of the humanities, in the practical or material sense that without some technology there would not be any human life, much less any humanities—which also generally use tools of many sorts, from pencils to computers, to perform their distinctive tasks? But the point is that even if engineering includes the humanities on the practical level, once engineers start proposing theories about the nature and meaning of technology they are no longer doing technology but are engaging in a kind of philosophy. Once one starts talking rather than making, then criteria of talk or discourse such as comprehensiveness and inclusiveness properly become factors of judgment, not solely those of practical effectiveness.

Philosophical Objections to Humanities Philosophy of Technology

Yet it is crucial to remember that the defense of humanities philosophy of technology over engineering philosophy of technology is not without its own criticisms of the typical humanities engagements with technology. Humanities philosophy of technology often does fail to pay sufficient attention to engineering experience and technological reality—presuming that it is possible to think on the cheap. It is remarkable, for instance, how little José Ortega y Gasset, Martin Heidegger, and Jacques Ellul seem to know about the real world of engineering. This is not quite so true with regard to Lewis Mumford, but even Mumford, especially in his late works, relies more than one might like on large metaphors that sometimes lose contact with technical experience. There is something going on in EPT that HPT must be altered to include.

There are any number of examples of humanities scholars, especially philosophers, talking about technology in shallow ways. Consider, for instance, Bernard Dauenhauer's phenomenological study *Silence*.[3]

Dauenhauer undertakes to describe various kinds of silence in relation to different kinds of discourse, one of which is technological discourse. But in comparison with his descriptions of scientific, political, moral, religious, and artistic discourse, his characterization of technological discourse is exceptionally thin. The description of political discourse makes references to Aristotle, Montesquieu, Rousseau, the Third Reich, Napoleon; that on artistic discourse to actual works of art such as Picasso's *Guernica* and T. S. Eliot's *The Waste Land*. But in talking about technological discourse Dauenhauer relies on the most general kinds of statements from Heidegger and Marcel. He never appeals to the works and words of engineers themselves.

Although artists and perhaps even politicians might be able to recognize their languages in Dauenhauer's descriptions, it is doubtful whether any engineers could recognize themselves. Indeed, of the four key representatives of the humanities philosophy of technology tradition—Mumford, Ortega, Heidegger, and Ellul—the two professional philosophers exhibit exactly this same weakness in the strongest sense. Although HPT is in principle more inclusive than EPT, it has clearly not exercised or realized this inclusiveness. It is only more technically minded philosophers such as Mario Bunge or more recent contributors to the philosophy of technology such as Don Ihde who begin to rectify this oversight. HPT may be able to be inclusive—but it is not yet nearly inclusive enough. To become inclusive, to realize its full potential, HPT needs to turn *from* philosophy *to* technology, or at least to technological discourse.

The movement at issue is to some extent the opposite of that enunciated by Samuel Florman, a ready representative of engineering philosophy in one of its more expansive forms. On the one hand, in *The Existential Pleasures of Engineering* (1976) Florman is a withering critic of humanities philosophers of technology such as Mumford and Ellul; he defines human beings as inherently technological and defends engineering as itself a liberal art. In *Blaming Technology* (1981) he goes on to defend nuclear power, to argue (contra E. F. Schumacher) that "small is dubious," and to reject the idea of engineering ethics.

On the other hand, Florman admits, and even argues, that engineers cannot be fully civilized by engineering alone; their education should be complemented and enlarged by the liberal arts and the humanities. Indeed, his first book, *Engineering and the Liberal Arts* (1968), was written "to advocate the cause of liberal education for engineers" (p. vii). The first chapter of that book, "The Civilized Engineer," became the title of a fourth volume, *The Civilized Engineer* (1987), in which he pleads "the cause of a humanistic professionalism, an ennobled engineering that will rise out of the ashes of vocational training" (p. 173).

But the weakness of engineering philosophy of technology is also revealed here in one of its representatives who is most open to the humanities. Florman does not want to reduce the humanities to simple utilitarian value. Yet for Florman the humanities remain fundamentally dependent on technology and thus are at best a kind of desirable epiphenomenon that should be granted some reflective influence on the primary phenomenon of technology.

> The "roots" of a civilized society are the technical accomplishments that relieve people of brute effort and make humanity possible. When we speak of the "fruits" of our efforts, of the "flowering" of civilization, we refer to art, philosophy, and science. If the fruits and the blossoms are not returned to nourish the soil, then life loses strength and its flowering becomes less radiant. . . . [I]f technology is not enriched by new beauty and insight, then the growth that follows is less luxuriant and all of humanity is the loser. (1987, p. 181)

Florman fails to give an adequate account of the humanities task, to see that the humanities are themselves a root of civilization.[4]

Given Florman's genuine if failed attempt to take account of the humanities, it is especially appropriate that humanities philosophy of technology make a genuine effort to engage technology on its own terms and not, like Dauenhauer, remain at a superficial distance. With this in mind the focus of attention in part 2 properly shifts from the humanities *philosophy* of technology to humanities philosophy of *technology*—or what may be termed, without qualification, philosophy of technology simpliciter.

Two Usages of the Term "Technology"

In the spirit of this shift it is appropriate to begin with a consideration of the very term "technology." The word "technology" has, in current discourse, narrow and broad meanings, which roughly correspond to the ways it is used by two major professional groups—engineers and social scientists. The latter usage also indicates the way humanities scholars most often employ the term. It is important to recognize such distinctions at the outset, because tension between these two usages, which stretch across a spectrum of conceptual references, easily results in analytic confusion.

Engineering Usage

The engineering usage is more restrictive. To begin with, the word "engineer" itself has etymological and sociological connotations that cast shadows over any engineering concept of "technology."

Etymologically the word "engineer," rooted in the classical Latin *ingenero*, meaning "to implant," "generate," or "produce," readily connotes producing or making, but not only of an artificial sort; the Latin (as in *ingeneratus*, "innate" or "natural") is associated with *natura* as well as with *ars* or *techne*. Yet today engineers often distinguish not just between bringing into being by nature and by technique, but also between engineers and technicians. The engineer works with nature and its laws as revealed by science, whereas the technologist focuses more on the actual construction. The engineer makes with the mind, the technician with the hands; the former is a white-collar worker, the latter a blue-collar worker. Such a difference is exemplified, for instance, in professional distinctions between a bachelor of science degree in engineering and the bachelor (or associate) of technology degree in the applied or industrial arts.

Historically, however, this usage can be contrasted with the original meaning of the term "engineer" and its cognates, which first appeared in the Middle Ages (Latin *ingeniator*) to designate builders and operators of battering rams, catapults, and other "engines of war."[5] Later, somewhat independently, in the eighteenth century the term was used to designate the operators of steam engines. Indeed, attenuation of both references is quite recent.

Reflecting this background, Samuel Johnson's *Dictionary of the English Language* (1755) defines the engineer as "one who directs the artillery of an army," and Noah Webster's *American Dictionary of the English Language* (1828) describes him as "a person skilled in mathematics and mechanics, who forms plans of works for offense or defense, and marks out the ground for fortifications." There is, however, a shift in emphasis between Johnson and Webster; the latter begins to identify the engineer as the one who "forms plans" or thinks things out—albeit with regard to military fortifications. This picks up on a supplementary connotation from the Latin, one that enters English by way of the French. Because natural objects exhibit cohesion within themselves (they "work") and with their environment (they "fit"), an artifact that exhibits either set of properties can be described as *ingeniosus*. Thus the Old French *engignier* is one who contrives or schemes to make things fit—with an implication, perhaps, that they might not otherwise do so. Because of the vaguely impious character of competition with nature, the fourteenth-century English "engynour" who plots and lays snares, even though he may well work with his mind and not with his hands, has certain unsavory connotations.[6]

Originally, then, the distinction between the white-collar engineer and blue-collar technician did not exist—all engineers were khaki-collar soldiers—or it existed in other forms. Outside the realm of military affairs, for example, the general name for one who designs and directs the construction of large-scale artificial structures was "architect"—Latin *architectus*, Greek *architekton*, from *archi-* (primary or master) plus *tekton* (carpenter or builder). Here there is an implied distinction between the designer who exercises a superior or more inclusive view and the technician or worker. Thus Vitruvius's *De architectura*, a work in ten books published at Rome in the first century C.E., deals primarily with urban planning, options in building materials, aesthetic principles, general construction strategies, hydraulics, geometry, mechanics, and so forth.

John Smeaton (1724–1792) was the first person to call himself a "civil engineer." Having initially gone up to London in 1742 to study law, he joined the Royal Society and became involved in scientific works. After serving as architect for rebuilding of the Eddystone Lighthouse in the late 1750s, he began in 1768 to refer to himself as a "civil engineer" to distinguish both his professional origins and his works, although certainly in peacetime many military engineers were employed in tasks similar to his own. While retaining a broad nonmilitary connotation on the Continent, "civil engineering" has come to refer in the English-speaking world more narrowly to the designing, constructing, and maintaining of roads, bridges, water supply and sanitation systems, railroads, and such—that is, publicly funded and utilized projects that are conceived more from the point of view of utility and efficiency than in terms of aesthetic form or symbolic meaning.[7]

The eighteenth century thus witnessed a lateral separation of civil engineering from military engineering among the upper classes. At the same time there occurred a vertical distinguishing of mechanical engineering from mechanics among the lower classes. As an offshoot of the multiple inventions and utilizations of the steam engine, the term "engineer" was used to designate a person, usually of lower-class origins, who operated the same. Closely associated were the "mechanics" who constructed, maintained, and operated these machines.[8] James Watt (1736–1819), for example, was said to be a "practical engineer," to dinstinguish him from the slightly more theoretically based representatives of military and civil engineering.[9]

With the development of "mechanical engineering" as a profession distinct from but allied with artisans, inventors, operators, and scientists—that is, engineers as persons with technological engagements and scientific-mathematical training—the 1800s promoted the Enlightenment vision of a union between science and the practical arts in which science would provide a method for solving practical problems and thus serve as a foundation for systematic progress. Since then engineering has expanded its method to consider a broad range of materials, energies, or products, as in chemical engineering, electrical engineering, radio engineering, electronic engineering, aeronautical engineering, nuclear engineering, and computer engineering.

Thus practiced, engineering has come to be defined, in the words of *Webster's New International Dictionary* (1959) and the *McGraw-Hill Dictionary of Scientific and Technical Terms* (3d ed., 1984)[10] as "the science by which properties of matter and the sources of energy [Webster]/power [McGraw-Hill] in nature are made useful to man in structures, machines, and products." Ralph J. Smith, an authoritative engineering educator, commenting on his own version of this definition—"engineering is the art of applying science to the optimum conversion of natural resources to the benefit of man"[11]—has proceeded to conclude that "the conception and design of a structure, device, or system to meet specified conditions in an optimum manner is engineering."[12] Furthermore, "it is the desire for efficiency and economy that differentiates ceramic engineering from the work of the potter, textile engineering from weaving, and agricultural engineering from farming."[13] "In a broad sense," Smith writes later, "the essence of engineering is design, planning in the mind a device or process or system that will effectively solve a problem or meet a need."[14]

The engineer, then, is not so much one who actually makes or constructs as one who directs, plans, or designs, as is reflected in such metaphorical usages as "the general engineered a coup," meaning he planned or organized it—thought it all out—not that he picked up a gun. "Engineer" continues to be able to refer, in a more restricted sense, to one who operates engines, as in the expression "railroad engineer." Yet in the latter case there is no "engineering" to learn, only the skill of how to control a train. Engineering as a profession is identified with the systematic knowledge of how to design useful artifacts or processes, a discipline that (as the standard engineering educational curriculum illustrates) includes some pure science and mathematics, the "applied" or "engineering sciences" (e.g., strength of materials, thermodynamics, electronics), and is directed toward some social need or desire. But while engineering involves a relationship to these other elements, artifact design is what constitutes the essence of engineering, because it is design that establishes and orders the unique engineering framework that integrates other elements. The term "technology" with its cognates is largely reserved by engineers for more direct involvement with material construction and the manipulation of artifacts.

In fact, engineers (reflecting and influencing the culture at large) tend to take the two cognate chains, technics-technical-technician and technology-technological-technologist (two cases of abstract noun-adjective-practitioner), and conflate them to form the grammatical hybrid technology-technical-technician. This explains how the terms "technical" and "technician" can be in greater currency when qualifying practices or naming practitioners of specific making or manipulating activities, while aspects of these same pursuits can be referred to abstractly as "technology."[15]

This "materialist" or practice-oriented usage is also the foundation of the term "technological sciences" (= systematic *knowledge* of making, or sciences of the industrial arts), which is meant to include traditional military and civil engineering, agricultural engineering, and the new disciplines related to space, computers, and automation.[16] This is the meaning implicitly adopted for "technology" (as a kind of condensed form of "technological sciences") when it is defined by the *McGraw-Hill Dictionary of Scientific and Technical Terms* (1984) as "systematic knowledge of and its application to industrial processes."

In light of such a definition the technician, as someone directly involved with acquiring and using technical knowledge (technology), is naturally less sophisticated than the engineer. The engineering researcher establishes protocols and methods that the technician employs to collect data; the technician likewise uses such data to carry out designs formulated by the engineer. It is this understanding that lies behind, for instance, Smith's distinctions between engineer, scientist, technician, and craftsman:

> The engineer is a man of ideas and a man of action. . . . He develops mental skills but seldom has the opportunity to de-

velop manual skills. In concentrating on the application of science he can obtain only a limited knowledge of science itself. . . . The primary objective of the *scientist* is "to know," to discover new facts, develop new theories, and learn new truths about the *natural* world without concern for the practical application of new knowledge. . . . The engineer is concerned with the *man-made* world. He has primary responsibility for designing and planning research programs, development projects, industrial plants, production procedures, construction methods, sales programs, operation and maintenance procedures and structures, machines, circuits, and processes. . . . The *technician* usually specializes in one aspect of engineering, becoming a draftsman, a cost estimator, a time-study specialist, an equipment salesman, a trouble shooter on industrial controls, an inspector on technical apparatus, or an operator of complex test equipment. . . . [The] technician occupies a position intermediate between the engineer and the skilled *craftsman*. The craftsman, such as the electrician, machinist, welder, patternmaker, instrument-maker, and modelmaker, uses his hands more than his head, tools more than instruments, and mathematics and science rarely.[17]

Without rejecting such a formulation, some engineers nevertheless further distinguish "technologist" and "technician." Philip Sporn, for instance, in his classic little volume *Foundations of Engineering*, distinguishes technician, technologist, and engineer by the comprehensiveness of their abilities. Technicians make particular devices (motors), technologists have mastered some whole field (electric power production), whereas engineers are concerned with a system including the socioeconomic context (electric power systems).[18]

Variations on this view are reflected in such philosophical papers as James K. Feibleman's "Pure Science, Applied Science, and Technology: An Attempt at Definitions" (1961) and C. David Gruender's "On Distinguishing Science and Technology" (1971). For Gruender, the chief distinction between applied science and technology "is in the scope or generality of the problem assigned. Those of broader scope we are inclined to think of as problems of 'applied' science [= engineering?]; those that are closer to being specific and particular we think of as 'technology'" (p. 461).

Thus, just as the adjective "technical" connotes a limited or restricted viewpoint, so the engineering technician works from a more limited standpoint than the engineer. The technician or technologist might, for instance, know how to perform a test, operate a machine, assemble a device (and even be involved in directing others who have a less comprehensive view of some particular operation or construction project), but not necessarily how to conceive, design, or think out such a test or artifact. Consider, for example, such terms as "lab technician," "medical technician" or "medical technologist," and "drafting technician."[19] In each case the person referred to is designated as proficient at performing some operation or construction, but not at fully organizing or understanding the procedures involved. The engineer has a superior or more inclusive view of a material construction than the technical assistant.

Social Science Usage

For social scientists, however, the term "technology" has a much broader meaning. To begin with, it includes all of what the engineer calls technology, along with engineering itself. Such usage has some basis in engineering parlance, as when an engineering school is termed an "institute of technology." Yet this continues to limit technology to those making activities influenced by modern science. Engineering schools are quite recent additions to the academic arena and focus on special kinds of making; making pots, for instance, is not a conspicuous feature of the curriculum at MIT.

In light of their disciplinary origins, one might expect the social sciences to have adopted precisely this restricted usage. Jay Weinstein, for example, has argued at length that both "technology and social science are the specific products of Europe's industrial revolution" and that in each of its three independent beginnings during the middle to late eighteenth century in England (Adam Smith and others), in late eighteenth- and early nineteenth-century France (Henri Saint-Simon and Auguste Comte), and in mid- to late nineteenth-century Germany (Karl Marx) social science arose to remedy defects in technology and extend its aims and methods into society.

> In light of the development concept, technology was seen as knowledge to transform humanity and nature for the better, to free man from the limitations on his powers that were once accepted as inevitable. Social science was to be an adjunct to technology because it is required to help understand these objects: humanity and nature, knowledge, and freedom. In addition, it became clear . . . that social science must be used in understanding the interest and behavior of participants in technological activity: owners, technicians, workers, etc., that

these too are consequential and scientifically comprehensible parts of the innovation process. From these observations it followed that social science and technology are mutually dependent means to achieve their common end: development, progress through the application of scientific principles to human affairs.[20]

Yet social science usage, stimulated by recognition of the social significance of making activities allied with modern natural science—vide the sociocultural reaction to, and now the sociology of, the Industrial Revolution—has extended the term even further to refer to all making of material artifacts, the objects made, their use, and to some extent their intellectual and social contexts. Even crafts such as potting become technologies in this loose sense, because there are certain modern technologies (e.g., industrial ceramics) of which potting is a remote precursor, and because the ways potting affected premodern society are presumed continuous with the impact modern technology has had on the social fabric. Indeed, in the history of technology, which is the primary social science study of technology, technology has sometimes been defined so as to include even the making of nonmaterial things such as laws and languages—although the implications of such definitions have not been widely thought through or adopted.

Compare, for instance, the understandings of "technology" found in the *McGraw-Hill Encyclopedia of Science and Technology* and in *A Dictionary of the Social Sciences*. In the former technology is defined as "systematic knowledge and action, usually of industrial processes but applicable to any recurrent activity" and "closely related to science and to engineering."[21] In the latter the term is defined, first, in regard to primitive societies, as denoting "the body of *knowledge* available for the fashioning of implements and artifacts of all kinds," and second, in regard to industrial societies, as denoting "the body of *knowledge* about (a) scientific principles and discoveries and (b) existing and previous industrial processes, resources of power and materials, and methods of transmission and communication, which are thought to be relevant to the production or improvement of goods and services.[22] Although this definition overemphasizes the cognitive component in technology, both ancient and modern, it nevertheless indicates the much wider range of the social science concept.

Other social science definitions have, however, gone even further. According to the *International Encyclopedia of the Social Sciences*, for instance, "Technology in its broad meaning connotes the practical arts. These arts range from hunting, fishing, gathering, agriculture, animal husbandry, and mining through manufacturing, construction, transportation, provision of food, power, heat, light, etc., to means of communication, medicine, and military technology. Technologies are bodies of skills, knowledge, and procedures for making, using and doing useful things. They are techniques, means for accomplishing recognized purposes."[23]

Some social scientists, it is true, prefer to limit "technology" to modern industry[24] or to distinguish between "technics" and "technology," letting the former stand for primitive arts and crafts and the latter for more sophisticated engineering.[25] Both approaches nevertheless remain minority usages. More characteristic is the view of Peter F. Drucker, who maintains that the subject matter of technology is not so much "how things are done or made" as "how man does or makes."[26] For Drucker, technology includes not only successful but also failed making and all human undertakings insofar as they are (intentionally or unintentionally) oriented toward making and using—so that the history of technology includes a history of work, invention, economics, politics, science, and so forth. Economist Nathan Rosenberg, likewise, prefers to write not about technology so much as "technological phenomena," taking "diversity and complexity" among such phenomena "as axiomatic."[27]

The Extension of "Technology"

Distinctions in the usage of the term "technology" could, of course, be expanded. Michael Fores, for example, in an analysis complementary to that just given, appeals to British usage to distinguish four senses of "technology": (1) that of science policy studies, in which technology encompasses all scientific *and* engineering activities; (2) that of government statistics, in which labor activities in the technology category include all workers up through and including engineers as *opposed* to scientific workers; (3) that of engineers, who would limit technology to craft techniques; and (4) the common dictionary or etymologically correct definition of technology as the "science of the industrial arts."[28] But (1), (2), and (4) are simply aspects of the broad social science usage, whereas (3) is the narrow engineering usage.

This tension between the narrow engineering usage and the broad social science usage of the word "technology" cannot be neatly resolved; it can only be accommodated. One such accommodation would attempt to stipulate around the problem ("We will define technology as . . ."); another might use subscripts to distinguish engineering usage (technology₁) from social science usage (technology₂). Still another could provisionally adopt the more extensive meaning, with the intention of gradually formulating distinctions within it by whatever means become available and appear appropriate during the course of a deeper analysis. This third approach is preferable as both less arbitrary or artificial and more open to whatever distinctions naturally emerge.

Without anticipating subsequent analyses, then, one can suggest that together both the engineering and social science usages point, first, toward the conceptual primacy of the making of material artifacts then, second, toward a large number of elements and influences that go into and arise out of this primary activity, influenced by and influencing its different forms. The thesis is that "technology" is not a univocal term; it does not mean exactly the same thing in all contexts. It is often, and in significant ways, context dependent—both in speech and in the world.[29] But neither is it a pure equivocal such as "date," which can refer to wholly unrelated things on a calendar or a palm tree. There is a primacy of reference to the making of material artifacts, especially since this making has been modified and influenced by modern science, and from this is derived a loose, analogous set of other references. An initial need in the philosophy of technology is for some mapping out or clarification of this conceptual one and many, a conceptual one and many that can be assumed to reflect a real diversity of types of technologies with various interrelations and levels of unity.

Becoming aware of this spectrum of conceptual references is philosophically important on two counts. First, in discussions of the social and ethical consequences of technology debates inevitably arise about whether technology can be limited or even eliminated. But much of the disagreement rests on a failure to clarify differences in assumed definitions. On the one hand, if by technology one means the making activity in general and the using of material artifacts, then obviously technology can never be abandoned and is in fact coeval with if not prior to the emergence of human life (since animals also make and use artifacts such as spiderwebs and bird's nests). On the other hand, if by technology one means some particular form or social embodiment of this general human endeavor, then clearly technology is expendable; technologies have been abandoned repeatedly throughout history, under both peaceful and violent circumstances. Indeed, the history and sociology of technology depend on this interpretation when cultures are analyzed in terms of technological change.

Second, in the formative philosophical discussions a large number of apparently incompatible definitions have been offered for technology. Technology has been variously conceived

- as sensorimotor skills (Feibleman)

- as applied science (Bunge)

- as rational efficient action (Ellul 1954) or the pursuit of technical efficiency (Skolimowski 1966)[30]

- as "tactics for living" (Spengler), means for molding the environment (Jaspers 1949), or control of the environment to meet human needs (Carpenter 1974)

- as means for socially set purposes (Jarvie)

- as pursuit of power (Mumford 1967)

- as "systematic application of scientific or other organized knowledge to practical tasks" (John Kenneth Galbraith) or "knowledge of techniques" (Nathan Rosenberg)[31]

- as means for the realization of "the gestalt of the worker" (Jünger) or any supernatural self-conception (Ortega)

- as self-initiated salvation (Brinkmann 1946)

- as invention and the material realization of transcendent forms (Dessauer 1927 and 1956)

- as a "provoking, setting-up disclosure of nature" (Heidegger 1954)

Some conceptions evidently differ only in words. Yet even after this is taken into account, there remains a variety of definitions, each of which—it seems reasonable to suggest—highlights some real aspect of technology, guided by a tacit restrictive focus. Argument over the truth or falsity of such definitions thus too often hinges on the exclusiveness of a limited perspective. The disagreements at issue call for a more open description of technology that delineates its different types and their interrelationships. As one perceptive observer has argued, what is needed is "not definitional but characterological" framework.[32] Only such an analysis can provide a foundation for assessing the relative truth and significance of each prospecive definition.

Initiating such an open characterization, technology can be described as the making and using of artifacts. Human making, in turn, can be broadly distinguished from human doing—for example, political, moral, religious, and related activities. Admittedly, this does not reflect the etymology of the word "technology" (which became current in the nineteenth century to refer to the industrial arts), nor does it always accord with various feelings and intuitions entrenched in the

English language. Nevertheless, it does serve to demarcate what should be the full scope of a philosophical concern with technology and to draw out what is unique to this study.

Modern philosophy of human action has concentrated almost exclusively on doing—the province of ethics, political philosophy—at the expense of making. The only exception is some limited discussion of making in the philosophy of art and aesthetics. Under the stress of contemporary problems and needs, however, human beings are called to reflect on making in a more comprehensive and fundamental manner—and in ways that find echoes in premodern thought. Similarities and differences in the many aspects of technology await disclosure through an analysis of its various constitutive elements. Where analysis warrants typological relations, these will be denoted by some qualifying adjective (as with the expression "scientific technology") or by distinct words properly defined (as with "technique"). Such constitutes an initial conceptual program in the philosophy of technology.

A Framework for Philosophical Analysis

In undertaking an analysis of diverse types of technology, however, one cannot just dive in. The rich complexity of the subject forces one to adopt at least a provisional classifying or categorizing scheme. Numerous frameworks or preliminary typologies have been proposed and used—although these have often been more for technical, historical, encyclopedic, or educational and heuristic than philosophical purposes.

With regard to technical purposes, there are typological frameworks utilizing distinctions between the various branches of engineering (civil, mechanical, chemical, electrical, etc.), as well as those grounded in differentiations of engineering functions or operations (designing, developing, production, etc.). The former are often also the basis for divisions of labor in social science studies such as Singer et al.'s *History of Technology* (1955–1984) and Kranzberg and Pursell's *Technology in Western Civilization* (1967). The latter can influence economic as well as technical studies.

With regard to uniquely historical purposes, there are the standard periodizations (Greek, Roman, medieval, seventeenth century, etc.) and modifications thereof. Bertrand Gille's *History of Techniques* (1978) relies on such standard divisions of history, in order to write narratives that synthesize, say, civil and mechanical engineering in "The Modern Technical System." Mumford (1934) modifies the standard divisions to distinguish what he terms eotechnic, paleotechnic, and neotechnic phases in the history of technical activity. But as even historians admit, neither approach is completely satisfactory.[33]

With regard to encyclopedic concerns, no classification scheme is so highly articulated as that developed under the tutelage of philosopher Mortimer Adler and found in the fifteenth edition of the *Encyclopaedia Britannica*.[34] In this scheme all knowledge is divided into ten subject areas, beginning with that which bears on (1) matter and energy, that is, physics, and (2) the earth, moving on through (3) the sciences of nonhuman and then (4) human life to (5) human society, (6) art, (7) technology, (8) religion, (9) history, and (10) the branches of knowledge itself, including both science and philosophy. Although its proximity to art is revealing, technology is easily the most anomalous of these major categories; it is, for instance, the only one that does not appear at all in the first edition of the *Britannica* (1771) and is not accorded an entry in the classic eleventh edition (1911), or indeed even in the immediately preceding fourteenth edition (1974).[35] When technology arrives on stage in the *Britannica*, it comes as a star.

As part 7, technology is approached from three main perspectives: its historical development and social impact (particularly on work), its internal divisions (energy conversion, tools, measurement and control, extraction of raw materials, industrial production), and its major fields of application (agriculture, industrial production, construction, transportation, information processing, the military, the city, earth and space exploration). In a kind of echo, "The Technological Sciences" are considered the seventh and last subdivision under science in part 10, with a four-part analysis in terms of history, professional branches (civil, aeronautical, chemical, electrical, mechanical, etc., engineering), agricultural sciences, and interdisciplinary technological sciences (bionics, systems engineering, cybernetics). Although gratifyingly inclusive, any attempt to conceptualize this plethora of divisions quickly produces as much confusion as insight.

Turning to educational or heuristic purposes, options continue to proliferate. Not only are there all the possibilities already mentioned—each appropriate to different pedagogies—but a host of others emerge. Just to mention a few examples: There is the medieval division of the seven mechanical arts in Hugh of St. Victor;[36] Jacob Bigelow's division first by materials used and then by human uses;[37] André Leroi-Gourhan's anthropological classification of techniques into those that do not go beyond the direct action of the hand in grasping, striking, and such and those that extend into fabrication, acquisition, transportation, and consumption;[38] Leo Marx's literary contrasts between technology as consciousness and as machine;[39] Donald W. Shriver Jr.'s axiological perspectives on technology as means, ends, politics, and evolutionary development;[40] Daniel Callahan's Freudian distinctions between preservation, improvement, implementation, destruction, and compensatory technologies;[41] John G. Burke's "typology of technology" as physical, chemical, biological, and social;[42] and so forth.

The inadequacy of such typologies is witnessed by the exclusiveness of their diversity. Each serves as a vehicle for a more or less special argument but proves mostly unable to carry on any sustained dialogue with the others.

A well-considered definition that moves in the right direction is proposed by Frederick Ferré in the Prentice-Hall Foundations of Philosophy series volume *Philosophy of Technology* (1988). Ferré briefly notes many of the ambiguities considered at greater length here and inventories debates about whether technologies are essentially material, science based, possessed by animals, natural or unnatural. He further observes that definitions prescribe as well as describe, and as such must steer a careful course between excessive breadth and restrictive narrowness. His own definition of technology as *"practical implementations of intelligence"* (p. 26), although not developed against any explicit background references to alternative proposals, is nevertheless a judicious advance on previous efforts. Because of their practicality, technologies are not ends in themselves (like the arts and other doings); because they are implementations they are material (thus excluding language per se); and their intelligence is broadly construed to include both the tradition based and the theory based.

Ferré's argument, however, is more concerned to bring technology within the purview of a focused philosophical discussion—to justify philosophy of technology as a subfield of philosophy comparable to philosophy of science, of religion, of language—than to throw light on technology itself. It reaches out to philosophers and near philosophers but not to engineers or technologists. Engineer-philosophers such as Samuel Florman are conspicuous by their absence. Although it is not nearly as parochial as Dauenhauer's discussion, it is nevertheless doubtful that engineers and technologists would find its analyses at many points confirmed by technical experience. It also fails to carry its definition forward into a disciplined consideration of the modes and manifestations of technology, but concentrates instead simply on mapping out existing philosophical discussions surrounding technology.

Explicitly philosophical typologies that do reach out to engineering experience and discourse have been proposed, although usually with some restrictions. Egbert Schuurman (1972), for instance, in his "philosophical analysis of modern technology," distinguishes technological objects and the twin activities that contribute to their fabrication: technological forming and technological designing.[43] Dieter Teichmann (in Rapp, ed. 1974) considers five bases for the "'internal' classification of the technological sciences"—historical development, types of science or laws of nature used, kinds of production supported, functional place in the general productive process, and structural characteristics of the objects produced. He concludes, however, that "the only meaningful classification . . . will be one which takes into account both the objective structure of technology, the classification of the natural sciences and the teaching structures."[44] Mario Bunge (1979b) discerns four branches of technology: material (the traditional forms of engineering), social (psychology, sociology, economics, military science), conceptual (computer science), and general (automata theory, information theory, optimization theory). Stanley Carpenter (1974) differentiates among technology as object, as knowledge, and as process.

Schuurman and Teichmann limit their analyses to modern technology, and Bunge merely "discerns" his, revealing no inherent rationale in the one and the many so discerned. Carpenter both uncritically assumes the primacy of cognition over affectivity and introduces the term "process," connoting a system of repetitive operations, to cover all human technical activities. As is even more evident in Carpenter (1978), his framework is biased toward epistemological issues and routine performances as against metaphysical questions and varieties of technological activity (inventing, designing, etc.).

It remains, then, to propose and develop a typology that can encourage an active dialogue with such previous attempts, protecting and ordering the insights they contain. While disclosing similarities and differences where necessary and appropriate, this typology should also reflect on ancient and modern making and using as is encouraged by social science studies.

The path toward such a philosophical framework is pointed out by one of the most general philosophical analyses to date, Robert McGinn's attempt to answer the question, "What is Technology?" (1978), especially as developed in two later publications. McGinn treats "technology as a *form of human activity* [comparable to] science, art, religion, and sport" (p. 180). (Note the echoes of the *Britannica* scheme.) The key characteristics of this activity are that it (1) has material outcomes, (2) fabricates or is constitutive of those outcomes, (3) is purposive, (4) is resource based and resource expending, (5) utilizes or generates knowledge, (6) is methodological, (7) takes place in a sociocultural-environmental context, and (8) is influenced by individual practitioners' mental sets. As N. Bruce Hannay and McGinn (1980) summarize matters in a subsequent paper, "technology can be characterized as that form of cultural activity devoted to the production or transformation of material objects, or the creation of procedural systems, in order to expand the realm of practical human possibility" (p. 27).

Unlike the Schuurman, Teichmann, Bunge, and Carpenter typologies, McGinn's characterology encompasses both modern and premodern technology and leaves room for most of their insights. The most obvious weakness of McGinn's descriptive analysis of technology as human creative activity is that it seems to imply a restrictive typology in which both artifacts and their use fail to qualify as primary aspects of technology. An artifact, for instance, is the outcome of technology but not itself technology; and McGinn explicitly rejects cloud seeding and agriculture as technology in any strong sense (1978, p. 182).

McGinn's analysis is complemented by his colleague Stephen Kline's response to the same question, "What is Technology?" (1985).[45] In a slight but pointed proposal, Kline recognizes four definitions of technology as artifacts or hardware, as sociotechnical systems of production, as technique or methodology, and as sociotechnical systems of use. Unlike McGinn, Kline recognizes both making and using as technological activities and grants that artifacts can be termed technology.

McGinn goes some way toward adopting Kline's enlarged framework in his book *Science, Technology, and Society* (1991), although he continues to resist according artifacts full status as technology. Moreover, McGinn's later synthesis of the elements in his characterology is skewed toward technological conceptions. To McGinn's mind, many human activities "can be analyzed in terms of six key aspects or components: their inputs, outputs, functions, transformative resources, practitioners, and processes" (p. 16). But to suggest that such categories are equally adequate or revealing for "*art, law, medicine, sport,* and *religion*" (p. 15, his italics) tends to reduce all such human pursuits to technological form. It was in engineering, not the humanities, that the language of inputs and outputs, resources, and processes, was first developed and is most appropriate. This is a language that engineer-philosophers are wont to use in translating the humanities into engineering terms.

McGinn, Kline, and their philosophical predecessors can nevertheless be brought together by a simple observation implicit in their analyses, although its potential has not been fully explored. Technology is pivotally engaged with the human. As such it is to be considered in relation to the essential aspects of a philosophical anthropology—with differences drawn between its manifestations in the mind, through bodily activities, and as independent objects that take their place in the physical and social world.[46] On such a basis distinctions can readily be articulated between technology as knowledge, technology as activity, and technology as object—three fundamental modes for the manifestation of technology.

In this conceptual framework, however, there is one arguable oversimplification. The anthropological interior need not, and in truth should not, be restricted to cognition. The will is an equally real if subtle aspect of the human. McGinn suggests the same by calling attention to the fact that when "material outcomes possess properties resulting from the operation of [chemical, biological, or physical] laws [they] may in a sense be said to be due to the volition of the practitioner" (1978, p. 182). Technology as volition must thus be added as a fourth mode of the manifestation of technology.

The resulting framework can be summarized by means of the diagram in figure 1. A full defense of this framework would require a comprehensive metaphysics, epistemology, philosophical psychology, and philosophical anthropology. For present purposes, however, it is sufficient to note that the framework proposed is not meant to be final or ultimate. It is enough that it be more comprehensive than previous ones, capable of adaptation to alternative positions on major issues in the philosophy of technology, a means to take philosophy more deeply into the realm of the technological, and open to further criticism or modification in response to future considerations.

The present framework is, then, provisional in character. Like the informal procedure of Aristotle's *Categories*, it is put forward by intuitive appeal to a commonsense metaphysics and anthropology. There is much to be said for beginning with commonsense notions, although one must be conscious that this is a beginning only, and that there is a rich and varied tradition of philosophical interpretation of humanity and the world against which commonsense hypotheses should be tested. Indeed, in the course of the analysis a number of efforts will

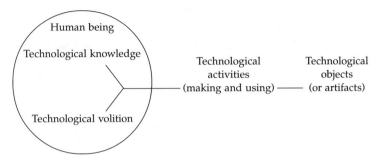

Figure 1. Modes of the manifestation of technology.

be made to do precisely this—that is, to bring into the discussion philosophical ideas, both traditional and modern, that bear on the adequacy of the proposed framework.

At the beginning, then, a framework should be both definite enough to provide some guidance and open enough to allow for adjustments and the possibility of winding up with new ideas. If it is to be philosophical, it should raise philosophical questions while remaining hospitable to different responses to those questions. Thus it is relevant to observe that the framework at hand can support either a technological determinism (in which objects or ideas exercise a controlling influence on human activity) or a theory of human freedom (in which individual volition or creative knowledge plays a dominant role). Technology as knowledge is further interpretable in terms of instrumental reduction or cognitive transcendence. The precise metaphysical status of technological objects is not fixed in advance, nor the structural features of technological activity. All these remain open as diverse paths for a deeper understanding of technology in each of the modes of its manifold forms. Recognizing that one aspect of its adequacy will be the degree to which this fourfold framework functions to orchestrate technological phenomena and philosophical questions, let us turn to more specific analyses.

CHAPTER SEVEN

• • • •

Types of Technology as Object

Artifacts—material objects such as tools, machines, and consumer products—are what most readily come to mind when the word "technology" is mentioned. As engineer David Billington has said, "When people talk about technology today, they usually mean the *products* of modern engineering: computers, power plants, automobiles, nuclear weapons" (1986, p. 87).

Technology as object is the most immediate, not to say the simplest, mode in which technology is found manifest, and it can include all humanly fabricated material artifacts whose function depends on a specific materiality as such. Depending on one's understanding of the terms, there may be some redundancy in this definition. Artifacts might be taken as by definition human fabrications. But in order to exclude nonhuman animal fabrications, which have also been argued to be artifacts, a little redundancy may be useful. Specifying material-ity excludes sociotechnical systems from being technological objects in a primary sense—although these may well be derivative manifesta-tions of technology. The qualification of effective functional depen-dence on some particular kind of material further excludes writing insofar as it can, unlike the hammer, effectively perform a function whether it consists of three-inch-high wooden block letters or of light patterns projected on a movie screen. This last qualification, like that relating technology as object to human beings, should nevertheless remain open for consideration of how far it is true even in a case such as writing. As Ivan Illich (1993) has shown, for instance, differences in the physical characteristics of writing as a physical object can subtly transform its function. For present purposes, however, technology as object will include such artworks as paintings and sculpture, but not poems or novels—only the physical books in which literature may be printed.

The Spectrum of Artifacts

The attempt to formulate a definition already entails the question of the extent of the spectrum of artifacts. Lewis Mumford has divided technological objects, or what he terms technics, into utensils, apparatus, utilities, tools, and machines—without insisting that these are mutually exclusive or complete (1934, pp. 9–12).[1] Slightly modified and enlarged, the following derivative list catalogs some basic types of technology as object:

- *Clothes*—artifacts for covering the human body, both utilitarian and decorative.

- *Utensils*—e.g., baskets, pots, dishes, spoons; storage containers and instruments of the hearth and home.

- *Structures*—e.g., houses and other stationary artifacts within which human activities take place.

- *Apparatus*—e.g., dye vats, brick kilns, containers for some physical or chemical process initiated and controlled by humans.

- *Utilities*—e.g., paths, roads, reservoirs, electric power networks.

- *Tools*—instruments operated manually that act to move or transform the material world, usually outside the home (contrast household utensils); typically, implements a worker uses to perform work, although there are certainly tools of communication and scholarship (paper and pen) and such.

- *Machines*—tools that do not require human energy input because they have an external source of power (wind, water, steam, electricity, etc.) but do require human direction; devices that operate, under human direction, to perform work.[2]

- *Automata* or automated/cybernetic machines—machines that require neither human energy input nor immediate human direction. Automated devices take part of their energy output and recycle it back into the device itself as a form of control. (One common example is a thermostatically controlled heater, where some small fraction of the heat output is used to operate a thermocouple that in turn regulates the heat level.)

Each of these artifacts is meant in some way to be lived with, used, lived within, operated, or set in motion. Each further calls forth its own unique history and analysis: the history of clothes and their different types (bags with cutouts, wrappings, tailored garments); the history of different types of architectural structures (private dwellings, public buildings). Philogist Ludwig Noiré, developing a suggestion by Lazarus Geiger, proposes a basic distinction between tools, utensils, and weapons that parallels the Hindu trinity of Brahma the creator, Vishnu the preserver, and Siva the destroyer: "The tool corresponds to the creative principle. The utensil serves the preservation of life. . . . Thus we understand why utensils almost always are regarded as passive and named from the way in which they are produced, while tools are conceived as active" and named from the actions they perform.[3]

Given a broad definition of technology as object, there are still at least three other possible types:

- Those already hinted at under tools, that is, *tools of doing* or performing (letters, numbers, musical instruments)

- Those explicitly mentioned in the introductory paragraph, artifacts that are not meant to be used in the normal sense but are only contemplated or worshiped (or more accurately, used as a means of worship), that is, *objects of art or religion*

- *Toys,* or artifacts of play and games

The three can easily overlap. Tools of doing can also be tools of religion insofar as religious action can be a kind of doing or performance. The games in which toys function can constitute both artistic and religious performances. Toys too come in types that mimic those in the primary list: toy clothes (for dolls, for dress-up play), toy utensils (for dollhouses and games), toy structures (dollhouses themselves), and so on. Some static toys such as special kinds of dolls may also function like objects of art, as things to be cherished and admired rather than used. The history of moving toys often anticipates developments in utilitarian technology (Hiero's steam engine, fireworks) and itself reflects those developments (from dolls to robots).

Another problem with the distinction between tools of doing and tools of making is that whether an object is one or the other appears to be highly context dependent and not always clearly discernible in the tool itself. If the difference is only one of context, can it be claimed as distinctive of the object? For example, numbers can sometimes be used for doing mathematics, sometimes for making money or even buildings.

But is this latter not really a use of mathematics for purposes of money making or house building? Numbers can be artifacts when written down in some form, but numbers themselves fail to be artifacts on two counts. They function effectively independent of any particular material embodiment. It seems reasonable to argue that they realize their full potential or make the most sense only when they are used for doing. Although it is possible—and indeed sometimes necessary— to subordinate doing to making, so that the doing at issue becomes provisionally a means to making, one must distinguish between intrinsic and extrinsic uses. Language may be used to conduct business, yet it finds its full realization in poetry. This is so because a thing is defined, according to Aristotle, not by its innumerable possible uses but by what it can do uniquely or best (*Nicomachean Ethics* I.7.1097b24–1098a17). A further example: The carpenter's hammer is not in itself an instrument either for knocking out a thief or for pulling weeds in the garden, although in a pinch it can certainly function both ways. But the former operation, if proposed as a definitive description, fails to make sense of the claw, while the latter is unable to give meaning to the flatness of the head. Only when understood as an instrument for fabricating with nails (driving and pulling them, as necessary) can all of its qualities be recognized for what they are and fitted together conceptually.[4]

There are some artifacts that in themselves appear to exhibit features unambiguously distinctive of one of these new categories—musical instruments, stage props, liturgical vestments, paintings, sculpture, icons. Furthermore, insofar as a cup comes to be (not just comes to be used as) a chalice for celebrating the Eucharist, or some machine truly becomes a work of art, each is likely to assume features unnecessary for simply being a cup or a machine: certain ornamentation, perhaps. Nevertheless, there are also clothes and utensils, for instance, that can become tools for the performance of a drama or a liturgical celebration without any alteration.

In considering the broad spectrum of artifacts, it may also be useful to allow for a distinction between primary and secondary types of technology as object. Something could be a work of art in a primary sense and also, in a secondary sense, a member of the superordinate class of technological objects. In like manner, tools of performance, tools of religion, and toys might also be thought of as species of the genus technology as object. Billington, for instance, argues that the multiple manifestations of technology as object really constitute only two basic types of entities: structures and machines (1974, pp. 275–288). The former are static artifacts such as roads, bridges, dams, power plants, buildings (and clothes?); the latter are dynamic artifacts such as cars, ships, television sets, computers (and utensils, tools?). Later he adds that both types of objects are parts of larger systems, which also come in two kinds: networks (streets and electric power grids) and processes (assembly lines and oil refineries). Are systems also types of technology as object? If so, surely they are technology as object in a derivative sense, since they are composed of objects in the primary sense.

A stronger and more consistent argument for a hierarchy of types of technology as object is found in the work of the French mechanologists Jacques Lafitte and Gilbert Simondon. Lafitte, in *Réflexions sur la science des machines* (1932), calls all artifacts machines and then distinguishes between three different kinds: passive machines (clothes, houses, roads, within or on which human use takes place), active machines (artifacts that transform energy, such as the lens, the plane, and tools in general, as well as more complex operating machines), and reflexive machines (the self-governing engine, thermocouple heater, computer). Simondon, in *Du mode d'existence des objets techniques* (1958), is even more monist than Lafitte and describes all technical objects in terms of simple elements (or parts), multielement objects, and ensembles of interacting objects. However, any such expansion of the idea of machine necessarily requires, as it does for Lafitte and Simondon themselves, the introduction of subsidiary and somewhat contrived distinctions between kinds of machines. Yet even independent of such "mechanistic monisms," there are necessarily distinctions to be made.

Types of Machines

Machines pose complex conceptual issues, partly because the term "machine" has shifted its meaning from the antique hand-operated instrument of work to the modern nonmanually operated instrument. The noun "machine" and hence the adjective "mechanical" come from the Greek *mechane* (Latin *machina*), meaning "instrument for lifting heavy weights," with the cognate verb meaning "to make by art, to construct, to contrive by skill or cunning." Going back even further, the word appears related to the hypothetical Indo-European roots *mogh-* and *megh-*, and thus to the German root *maxan* (from which come *Macht* and *machen*), all meaning "to have power," hence the English "may." Since premodern mechanical power originates in the human body and is distributed by the hand—even when it is into a structure cunningly contrived to lift heavy weights—the adjective came in later Latin and in English up until the seventeenth century to have strong associations with manual work. Thus John Donne in his sermons could speak of writing, carving, and acting—indeed, anything that "belongs to the hand"—as "mechanical offices."[5] And only such an understanding of "mechanical" as referring to manual or bodily activity explains the unity underlying Hugh of St. Victor's scheme of the seven mechanical arts of weaving, weapons forging, navigation, agriculture, hunting, medicine, and acting.

Yet with the development of nonmanual sources of power in the modern period the noun changes its meaning, and so does the adjective. Thus historically "machine" has meant at least three different things:

- First, "machine" can refer to the simple machines of classical antiquity—lever, wedge, wheel and axle (or winch), pulley (or block and tackle), screw, and inclined plane (to give one traditional list)—or some combination thereof.[6]

Actually, since (as the science of mechanics has shown) the wedge = an adapted inclined plane; the wheel and axle = a lever pivoting about a fulcrum at its midpoint; the pulley = a form of wheel and axle; and the screw = an inclined plane wrapped in a spiral—this traditional list of six can be reduced to two: the lever and the inclined plane. Levers come in three classes, depending on the relation between the fulcrum and the acting force. Inclined planes are distinguished by slope and thus form a continuum.

- Second, a "machine" can be any implement or large-scale simple machine that requires more than one human being to operate it because of its energy requirements. This is the definition found, for instance, in Vitruvius and applied to "catapults and wine presses."[7]

- Third, the "machine" can be an implement that does not depend on human energy—although it still requires some human monitoring or directing, "driving" in the sense that one "drives a car."[8]

The most general characterization of machines covering all three senses is as "instruments for transmitting force or modifying its application" (to quote a common dictionary definition).

With regard to machines in the first sense, or tools, anthropological analysis has distinguished percusive tools (hammers, axes); cutting, drilling, and abrading tools (knives, augers, saws); tool auxiliaries (workbench, vise); screw-manipulating tools (screwdrivers, wrenches); and measuring and defining tools (rulers, dividers, levels). Notice that among the first four types, adaptations of the lever are more pervasive than adaptations of the inclined plane. With regard to the fifth, perhaps it is useful to distinguish between "instrument" in a restricted sense, as a measuring, recording, or observing device—hand operated or otherwise—and "tool" as a manually operated machine "for transmitting force or modifying its application." There is even a sense in which measuring instruments can be interpreted not as human tools but as tools given to nature to use in speaking to humans.[9]

Admittedly, as one author has observed, "the difference between the tool and the machine has never been clearly defined."[10] But the common notion is that the "tool" is a hand-operated machine or at least that element of direct contact between a machine and the world that in principle can be humanly manipulated, whereas "machine" denotes an instrument in its independence, or that aspect of an instrument that is not dependent on the human.[11] The independence implicit in the idea of the machine corresponds to Hegel's definition of machine as a "self-reliant tool," tool being understood as any instrument of work.[12] Another commentator notes that by making use of appropriately shaped natural objects, human beings actually used tools before making them; the first making was a "making us."[13]

Machines in the third sense can readily be distinguished into four classes: those that depend on human or animal power (horse-drawn plow), those that employ direct mechanical energy from nature (windmill, water wheel), those that create their own mechanical energy from heat (the heat engines: steam engine, internal combustion engine), and those that use some form of abstract energy (electrical, chemical). Of the last two categories there are two further types: those that generate or transform energy, and those that transmit power and perform useful work. The former type (as in the electric dynamo or solar cell) exhibits the strongest tendency toward independence of the human and is sometimes thought to be uniquely modern.[14] Normally the latter (cars, airplanes) will depend on the former and involve some immediate human direction.

Power tools and machine tools are not always carefully distinguished. Power tools are power-driven hand tools such as the electric drill and saw or the air-driven jackhammer or even kitchen and household appliances. A rotary electric handsaw is also different from the electric table or radial-arm saw, which are stationary shop power tools. Machine tools, by contrast, are power tools for metal cutting— machines used to make machines. Some basic classes of machine tools, moving from the early and general to more recent and specialized, are lathes, shapers and planers, drilling machines, milling machines, grinding machines, power saws, presses, turret lathes, production millers, gear-cutting machines, and broaching machines. Recent innovations in machine tools include automatic control by tracing or record-playback techniques and numerical control, as well as chipless methods of removing metal by plasma arc and laser-beam machining.

Finally, one should observe the resistance exhibited in our uneasiness about calling an automatic device, which is designed and fabricated or at least assembled but neither energized nor directly operated by a human being, a machine. At most we seem willing to refer to the automaton as an "automated" or "cybernetic machine." Nevertheless, such a machine is an extension of previous conceptual developments and could be argued to be a fourth type of machine, the machine as cybernetic or self-regulating device.

The Machine (and Object) as Process

As the machine becomes increasingly independent of direct human energy input, its character as object undergoes an important transformation; it becomes not just a static object but the bearer and initiator of operations or of special physical, chemical, or electrical processes. The key shift is no doubt that from tool to machine. The steam engine and internal combustion engine are no longer simply objects after the manner of hammers or saws; they have become containers for processes—and in a much more specific way than a cooking pot or dye vat, in which quite different processes can take place depending on what is placed in them and on external conditions (temperature, etc.). Gasoline placed in an internal combustion engine does something quite specific, and it does this only under conditions established by the engine.

The design and construction of such process engendering and process enclosing machines thus entail the fabrication not merely of a physical object but of a process. What takes place in an electric alternator or internal combustion engine does not take place outside it. The process of electric generation or heat engine operation takes place only under humanly contrived conditions and is to that extent artificial. As machines become more and more independent of human energy input—expanding from mechanical to chemical and electrical processes—and then linked together into systems, machines also become increasingly characterizable as objectified processes.

Historically, the machine as process has evolved from water-powered energy transformers to batch processing production (first of textiles but then of chemicals), machining and assembling operations (using machine tools), and industrial assembly lines. There is a difference, for instance, between using tools or even machines to manufacture a number of separate and discrete artifacts, even when such artifacts are identical or very nearly so, and the bulk manufacture of some product that is homogeneous and uniform throughout and made in separate batches or even continuously. Indeed, the very term "product" tends to connote the latter character. Examples of bulk production include municipal water purification and waste disposal, whiskey distilling and petroleum refining, the forming of paper from wood pulp, and the chemical manufacturing processes that produce paint, plastic, and the like. In each case the "machine" or "object" that makes such processes possible must be thought of as itself a kind of process. Integrated into a technical system, such an object becomes less and less a substantial entity that can enter into an indefinite number of different and distinct relationships and more and more a moment in a system of preestablished relationships. Computer systems or networks are perhaps good examples, but so are assembly lines, chemical processing plants, all of which are dependent on industrial or process engineering.

This notion of the technological object as process can be related to a suggestive philosophical discussion by Norris Clarke that contrasts two ways of looking at the universe: the Aristotelian or substantialist and the Whiteheadian or relational.[15] For Aristotelians the primary categories of being are substance and accident, with all accidents inhering in some particular substance. "There is no such thing in Aristotelian metaphysics," Clarke writes, "as a *single* relation or set of relations linking, or immanent in, several substances at once" (p. 6). But for a relational metaphysics, relation or process becomes the primary reality in which substances function as moments. Clearly the world of advanced technological machines is to some extent more amenable to such a relational metaphysics.

The Engineering Analysis of Machines

Classical mechanics is the branch of physics that deals with the motions of material bodies and the forces acting upon them.[16] So defined, mechanics is subdivided into statics (dealing with bodies at rest) and dynamics (bodies in motion). Before the development of vector analysis by Simon Stevin (1548–1620), mechanics consisted almost exclusively of formulas for the equilibrium of simple levers derived from Archimedes (third century B.C.E.). The work of Galileo (1564–1642) on falling bodies laid the foundation for the modern science of dynamics[17] with its two main branches—kinematics (dealing with the motion of rigid bodies without regard to the forces involved) and kinetics (dealing with the relations between forces and motions).

In mechanical engineering, machines are analyzed and described by means of the science of dynamics. Machines are closed systems that can be analyzed in terms of motions (kinematics) and forces (kinetics). As such, a machine is defined as "a combination of rigid or resistant bodies having definite motions and capable of performing useful work."[18] If a machinelike device does not perform useful work in the (somewhat loose) mechanical engineering sense, it is termed a mechanism. A clock or speedometer, for example, is a mechanism but not a machine.[19] Thus, since the science of dynamics includes kinematics and kinetics, the primary function of a mechanical device can be either the modification of motion (direction) or the modification of motion and force (amplification or reduction). If the former, it is a mechanism; if the latter, a machine. Furthermore, the way the parts of a machine are interconnected to produce a required output motion from a given input, even when the purpose is force directed to useful work, is known as the mechanism of the machine. One can, for example, speak of the mechanism by which a steam engine, as a machine, works.[20]

According to standard engineering analyses, the components (which themselves can be machines) of some machine complex or system are the prime mover, generator, motor, and operator. A machine system typically receives an energy input from some external source (motion from moving air or water, heat from burning coal or gasoline, etc.) and either captures it (as in the mechanical motion from wind) or transforms it into mechanical energy (as in a coal-fired power plant), usually in the form of a rotating shaft, in what is called the prime mover. This mechanical energy is then itself subject to transformation by some generator into electrical, hydraulic, or pneumatic power, which is in turn (no pun intended) used to drive a motor, which then drives what is called an operator. An operator can have a variety of outputs such as materials processing, packaging, conveying, sewing, and washing. This sequence can be represented schematically as in figure 2. On occasion a prime mover can directly drive a motor or operator, as shown by the dotted line. Operators also include direct, manually operated implements such as typewriters and calculating machines, as shown by the double-dotted lines. (The operator obviously resembles, and in many instances is, a tool.)

This conception of a machine as "a closed kinematic chain" or "a combination of resistant bodies so arranged that by their means the

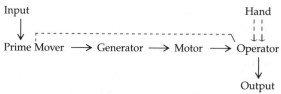

Figure 2. The engineering analysis of a machine system.

mechanical forces of nature can be compelled to do work accompanied by certain determinate motions" goes back to Franz Reuleaux's *Kinematics of Machinery* (1875; trans. 1876, pp. 502 and 503). Reuleaux's definition was however, formulated before the advent of electrical, chemical, and nuclear energies, so that it does not apply to electronic "mechanisms" such as radios, car batteries, and computers—unless "kinematic chain" is interpreted very broadly to include the movement of electrons along a wire or molecules within some chemical compound. In such circumstances the "simple machines" would no longer be limited to the lever, inclined plane, and so on, but would also include the capacitor, resistor, rectifier, amplifier, and such. Thus the possibility arises of reformulating Reuleaux's definition to make it suitably general, and within this genus to distinguish the various species of machines.

In fact, with the engineering of dynamic artifacts beyond what can easily be referred to by the term "machine," the term "device" has acquired special significance as denoting any "mechanism, tool, or other piece of equipment designed for specific uses" (*McGraw-Hill Dictionary of Scientific and Technical Terms*). As the technical dictionary further indicates, the term "device" is especially common in electrical engineering, where it indicates "an electronic element that cannot be divided without destroying its stated function" and is commonly applied to such active elements as transistors and transducers. In ordinary language, the *Oxford English Dictionary*, having noted other possible meanings, defines "device" for the present context as "the result of contriving; something devised or framed by art or inventive power; an invention, contrivance; *esp.* a mechanical contrivance (usually of a simple character) for some particular purpose."

Device is thus not synonymous with artifact, since it denotes instrumentality and in most cases even dependence on some (internal or external) operation. A device accepts some input and uniquely modifies it to produce a desired output. The fundamental aim of engineering analysis is to describe the input-output transformation and to be able to compare input and output in both quantitative (amounts of energy) and qualitative (types of motion) terms. Thus devices are to be distinguished, for instance, from frames or structures and processes; one would not normally speak of a car or radio chassis or a house as a device. Only in some derivative sense would such objects be subject to mechanical engineering analysis. The concept of the device includes tools, machines, and automata, but it is questionable whether it could refer to utensils, apparatus, and utilities—not to mention tools of doing and objects of art—without becoming a metaphor.

Physical, Chemical, and Biological Artifacts

Two basic philosophical questions about artifacts are, What are they, both conceptually and ontologically? and How are their distinctions and operations to be understood? The unifying whatness of artifacts, traditionally conceived, is that they are physical objects made by human beings. This extrinsic fact is, however, reflected in an intrinsic structure. One way of describing the intrinsic structure is to say that the form and matter in an artifact are less unified, less perfectly or profoundly integrated, than in natural objects. If a bed were to sprout, says Aristotle, what would come up is not another bed, but a tree (*Physics* 2.1.193b10; cf. also 193a13). The crucial distinction between nature and artifice is that nature has its source of motion or rest within itself, whereas for artifacts the source of motion or rest is in another (*Physics* 2.1.192b14–19). A machine, for instance, exhibits an inherent tendency to deteriorate because its form and matter are not really one; to function as a machine over extended periods requires regular upkeep from some human source. This is in contrast to a plant (and all living things), which generate and regenerate themselves unassisted by human beings.

Michael Losonsky (1990), having noted Aristotle's rejection of the idea that artifacts have essences in the proper sense (i.e., true unities of form and matter), contrasts Hilary Putnam's theory of artifacts as natural kinds. Losonsky then defends Putnam against his critics by developing a Marxist theory of artifacts as having their own special kind of intrinsic (and socially embedded) structures.

With the advent of chemical and biotechnological or genetic engineering, however, it is no longer clear that even Aristotle could defend the rejection of natural kinds among things made by humans. Chemically engineered polymers unify form and matter at the level of atomic structure and can exhibit stability equivalent to or greater than that of natural compounds. Chemists themselves commonly speak of "synthetic" rather than "artificial" products, which they "process" rather than "make," thus implicitly indicating a close affinity to natural processes. Although chemical transformations have existed under the direction of human beings since the first cooking of foods, this kind of making has traditionally remained in the background of philosophical reflection on the essence of artifice. The modern chemical industry forces it into the foreground.

Chemical transformation, perhaps more than physical, also raises questions concerning the extent to which human alteration of an object turns it into an artifact. Cooking wheat has not commonly been understood to transform wheat from a natural into a technological object; it simply processes the wheat. On the one hand, cooked wheat is transformed by human action, even more profoundly than a rock and stick that have been tied together with a leather thong to make a hammer. On the other, wheat could conceivably be cooked by nature (if fire and rain at once swept through a wheat field), whereas a rock could never become attached to a stick with a leather thong through the action of nature alone. Indeed, perhaps it is precisely because the union of matter remains at such a superficial level that the hammer is considered an artifact whereas the loaf of bread usually is not.

It may also be that because bread making has been identified with feminine and domestic activity, whereas hammer making and the hammered construction of those structures that externally dominate urban life have been conventionally associated with masculinity, that reflection on the artifice found in products of the hearth has been so improperly neglected. The issue here may also be one of degree. Perhaps it is because the rock-laced-to-stick has a function distinct from rock, stick, or thong alone, whereas bread and wheat both function as nourishment, that the hammer is described as an artifact whereas the loaf of bread is not. But then what about a stick or even a tree with a symbol carved into it? Does this constitute sufficient human making to create an artifact? If so, do cosmetic decorations, pierced ears, skin scarring, or plastic surgery turn the human body into an artifact?

The metaphysical issues raised by chemical artifacts are intensified by the possibilities of biological artifacts, which can transform what might otherwise have been taken as aberrations or exceptions into revealing harbingers. In medicine, for instance, to what extent does medical treatment create an artifact? Does a prosthetic device such as the hook that replaces a severed hand or the wooden leg turn the human body into a technological object? Traditionally, such instances were taken to be merely provocative exceptions. But with the advent of organ transplantation (which is dependent on control of the immune system), the implantation of heart pacemakers, and even artificial organs, the exceptional character of such undertakings is undone.

Likewise with the selective breeding of plants and animals, which was once conceived as helping nature to do what could in principle happen of itself—the guiding of nature. Its results only appear to be artifacts but really are not. Only when crossbreeding produces infertile hybrids (e.g., the mule) does one have something approaching a "living artifact." But with the biotechnological, genetic engineering of plants and animals it is possible to combine the extrinsic fact of human making with the absence of an intrinsically weak unity of form and matter that brings about the need for ongoing external maintenance or care.

The possibility of a living artifact is indicated again by that apotheosis of the artificial organ or prosthetic device known as a cyborg—from cyb(ernetic) and org(anism)—and the science of its construction, bionics—from bio(logy) and (electr)onics. This integrated human-machine system, by realizing the fantasies of android, golem, and robot, has, Donna Haraway writes, "made thoroughly ambiguous the difference between natural and artificial, mind and body, self-developing and externally designed, and many other distinctions that used to apply to organisms and machines.[21]

In one sense technological objects or artifacts, which were originally distinct from natural objects, seem to have become natural objects. In another sense, as issues of ecology and environmental contamination suggest, these new "natural artifacts" may not fit in with the larger unities of nature. The ontology of artifacts ultimately may not be able to be divorced from the philosophy of nature.

Animal Artifacts, Social Artifacts, the Planet as Artifact

Reflection on chemical and biological transformations in the structures of things made revives the issue of whether things made by animals can be said to be artifacts. Consider the examples of birds' nests, spiderwebs, and beaver dams. Although not made by human beings, are they not physical objects that have more in common with artifacts—precisely by exhibiting that intrinsic character described as a weak unity of form and matter so that they do not persist without maintenance?

What, too, about social institutions, what Emile Durkheim called "social facts"? Although not material, aren't they made by human beings? As made, don't they also exhibit that paradoxical status of becoming to some degree independent of the maker, as standing over against and apart from the maker and able to exert influence on the world independent of the intentions of that maker, and yet requiring the conscious or unconscious attentions of the maker in order to persist in existence? Marshall McLuhan goes even further. For him,

> It makes no difference whatever whether one considers as artifacts . . . things of a tangible "hardware" nature such as bowls and clubs or forks and spoons, or tools and devices and engines, railways, spacecraft, radios, computers, and so on; or things of a "software" nature such as theories or laws of science, philosophical systems, . . . forms or styles in painting or poetry or drama or music, and so on. All are equally artifacts, all equally human. (McLuhan and McLuhan 1988, p. 3)

Finally, what about the planet Earth? Now clearly influenced not just in part but as a whole by human activity, studied by earth system science and on the verge of being managed by a planetary technology, hasn't it too become a kind of artifact? Bill McKibben (1989), argues that because of the transformation brought about in the global ecosystem as a result of large-scale environmental pollution, and because of the massive high-affluent recreational use of nature, the idea of nature as "the separate and wild province, the world apart from man to which he adapted, under whose rules he was born and died," has itself died (p. 48).

> But now the basis of that faith is lost. The idea of nature will not survive the new global pollution—the carbon dioxide and the CFCs and the like. This new rupture with nature is different not only in scope but also in kind from salmon tins in an English stream. We have changed the atmosphere, and thus we are changing the weather. By changing the weather, we make every spot on earth man-made and artificial. We have deprived nature of its independence, and that is fatal to its meaning. Nature's independence *is* its meaning; without it there is nothing but us. (p. 58)

"By domesticating the earth," McKibben argues, "even though we've done it badly, we've domesticated all that live on it" (p. 84). What used to be wild animals are now just creatures in those enlarged zoos known as wilderness areas.

Increasing recognition that humans have done the domesticating badly has led to proposals for doing it better. The heritage of such proposals runs from R. Buckminster Fuller's "externalized metabolic regenerating organism" of *Operating Manual for Spaceship Earth* (1969, p. 115) and the world computer modeling of *The Limits to Growth* (1972)

to Worldwatch Institute's annual "State of the World" reports and the
"green pragmatism" of volumes such as *Gaia: An Atlas for Planet Man-*
agement and *The Global Ecology Handbook*, not to mention the idea of
restoration ecology.[22] On the one hand, such proposals appear to enhance planetary artificiality in the sense emphasized by McLuhan.
And contra the enthusiasms of Fuller and followers, it can be argued
that such amplified artificiality is leading toward what might be
termed a "green technocracy."[23]

On the other hand, one can question any quick identification of nature with wilderness like that assumed by McKibben and others, with
the resulting opposition between nature and technification. One must
ask whether influencing or modifying nature is the same as making it
into an artifact.[24] Kant, for instance, in the *Critique of Judgment* (1790),
actually argues that nature, if it is to be anything more than mere
mechanism, must be thought of as possessing its own inherent technics. This idea is also present in Félix Duque's *Filosofía de la tecnología*
de la naturaleza (1986). Although neither Kant's nor Duque's analyses
address exactly those questions at issue here, they do suggest a need
for more conceptual clarifications.

On the Human Experience of Tools and Machines

Technological objects, however, are not just objects, energy transforming tools and machines, artifacts, with distinctive internal structures, or things made by human beings; they are also objects that influence human experience. The exploration of this influence leads to
different kinds of distinctions among artifacts.

The most common cultural interpretation of machines (= an early
philosophy of technology) views tools and machines as physical extensions of the human body, or what are sometimes also termed organ
projections. This notion about the human meaning of technology as
object has an immediate intuitive plausibility. Although hinted at by
Aristotle (*De anima* 3.8.431a1–3), it flowers into a full-blown theory only
in such late nineteenth- and early twentieth-century studies as those
of Kapp (1877), Bergson (1907), and Lafitte (1932). More recently it has
been expanded in anthropological theory by Gehlen (1957) and then
by McLuhan (1964) and Paul Levinson (1988) to include even electronic
media as extensions of the human nervous system.

But not to mention other issues, one can legitimately inquire about
the meaning of "extensions" here. There are at least two kinds of extensions: a hammer, for instance, extends by way of enlargement the

power of the arm muscle, while it extends by way of abstraction and magnification the form and hardness of the fist. A comparison might be made with the distinction between the enlargement that takes place through a telescope, where light from a star is collected and concentrated—thus bringing one, as it were, closer to it—and the magnification that takes place in a microscope, which doesn't just bring one close to an object but optically abstracts and transforms its visual features. The former provides artificial experience of a physical possibility, the latter an artificial means for rendering an imaginary possibility.

A pile driver, however, not only abstracts and magnifies form, it also abstracts and magnifies power by placing at human disposal energies that human beings do not otherwise possess. If tools (or machines in the classical sense) increased human power, it was only by enlarging inherent human energies. If machines in the second sense (as tools operated by more than a single individual) did the same, it was only by uniting the energies inherent in a human group. Modern power machines achieve this effect in a different way by placing nonhuman energies at the disposal of personal human direction. Thus, whereas tools are single-function instruments that separate, specify, distribute, or concentrate the total power resident in the human body, machines incorporate the hand as an instrumental link in a person's multifaceted directing or governing of nonmanual energies.[25]

Furthermore, "the technical advance that characterizes specifically the modern age is that from reciprocating motions to rotary motions."[26] Modern machines, unlike tools, typically achieve their effect by means of rotary rather than reciprocating motion. While this is most obvious in the rotary power saw as contrasted with the reciprocating handsaw, it is equally true of the pile driver, which develops its reciprocating power in a pneumatic operator dependent on the rotating shaft of some prime mover. As historian Lynn White Jr. has argued in reference to the discovery of the crank,

> Continuous rotary motion is typical of inorganic matter, whereas reciprocating motion is the sole form of movement found in living things. The crank connects these two kinds of motion; therefore we who are organic find that crank motion does not come easily to us. The great physicist and philosopher Ernst Mach noticed that infants find crank motion hard to learn. Despite the rotary grindstone, even today razors are whetted rather than ground: we find rotary motion an impediment to the greatest sensitivity. The hurdy-gurdy soon went out of use as an instrument for serious music, leaving the re-

ciprocating fiddle-bow ... to become the foundation of modern European musical development. To use a crank, our tendons and muscles must relate themselves to the motion of galaxies and electrons. From this inhuman adventure our race long recoiled.[27]

Both kinetically and kinematically, modern machines, as contrasted with traditional tools, involve a qualitatively distinct separation of human beings from their bodies and primordial bodily awareness. Not all extensions are the same.[28]

This is confirmed by two different connotations of the adjective "mechanical." In the premodern sense a "mechanical task" is one performed manually and thus dependent on human energy. It is in this special sense *bavavoikos*, base or ignoble, because it focuses attention on one's own physical powers, which are extremely limited (cf., e.g., Xenophon *Oeconomicus* 4.2–3). It does not connect with higher, transhuman or spiritual powers but remains on the strictly natural plane. In modern usage, by contrast, a "mechanical task" is one done without attention, repetitively, routinely, or even ritualistically (in the bad sense of that word). Modern machines are base or ignoble in a new way because they alienate us from the sensorimotor, mind-body complex. Consequently, attention is not focused at all, anywhere, and must be entertained by some extraneous sensations—music, colors, and the like, as devised by industrial psychologists. This is why, from the contemporary perspective, a return to mechanical operations in the primitive sense can be seen as a desirable thing, reuniting mind and body; and this desirableness in turn is the source of a difficulty we experience in appreciating the ancient critique of manual work.

Along this line, Mumford and others have argued that "the skilled tool-user becomes more accurate and automatic, in short, more mechanical, as his originally voluntary motions settle down into reflexes" (1934, p. 10). Such generalizations rest on a thin experience with tools. When an operation becomes mechanical, as in a machine, one loses control of it. With a power-driven saw, for example, an artisan cannot respond as sensitively to a piece of wood—a knot, say, or stringy grain that splinters easily and damages a particular work—as with a handsaw. Admittedly, accuracy in the sense of following a superimposed, geometric line, becomes more fully realized with power tools, but only at the expense of a certain responsiveness to materials. As Sōetsu Yanagi summarizes the experience of craftsmen, "No machine can compare with a man's hands. Machinery gives speed, power, complete uniformity, and precision, but it cannot give creativity, adaptability, freedom, heterogeneity. These the machine is incapable of, hence the superiority of the hand, which no amount of rationalism can negate."[29]

The point is confirmed by psychological studies of the modularization of skills and the occasions for positive and negative transfer of skill modules.[30] An individual who develops a tendency to respond to some general task in a predetermined way acquires a modular skill. Although obviously useful, such a skill can also be an occasion for what is called "functional fixedness"—that is, an inability to respond to a variation in the task in an appropriate manner. Power tools may be said to suffer from a kind of reified functional fixedness. As carpenters are well aware, a power saw easily gets "out of hand," and an injury from a handsaw is usually a lot less serious than one from a power saw. It is no accident that finishing work or fine cabinetmaking continues to be done primarily by hand. But surely the artist does not become less sensitive or even necessarily more accurate in the geometric sense as brushstroke techniques become internalized in sensorimotor reflexes. The loss of awareness at the level of technique can actually increase artistic control at the level of the work itself. Thus there seem to be important differences between so-called automatic operations with tools and with machines.

The Social Dimension of Artifacts

Just as the presence or absence of some types of technological objects can reflect differences in technologies in general, so the relative proportions of the various types of technological objects available in some particular society will affect that society in numerous ways.

For instance, machines run by abstract power (electricity, etc.) dominate modern society but are not found at all in traditional or primitive societies. This point is relevant to the discussion about "intermediate," "alternative," or "soft" technologies as opposed to "hard" technologies, and their social and ecological impact in underdeveloped (and developed) countries.[31] But it is also important to remember that types will not always appear in unqualified formal distinctions; usually there will be an admixture of various elements. The real question will be one of proportion and degree, not simple presence or absence—as is the case with machines driven by nonliving sources of power.

These differences in the phenomenology of use are related to differences in the characters of the objects produced, although such differences are not easily conceptualized. Consider, for instance, the distinctions that Mexican philosopher Octavio Paz draws among objects of fine art, industrial technology, and handcraft:

> The industrial object tends to disappear as a form and to become indistinguishable from its function. Its being is its meaning and its meaning is to be useful. It is the diametrical opposite of the work of art [in which the meaning is to be useless but beautiful]. Craftwork is a mediation between these two poles: its forms are not governed by the principle of efficiency but of pleasure, which is always wasteful, and for which no rules exist. The industrial object allows the superfluous no place; craftwork delights in decoration. Its predilection for ornamentation is a violation of the principle of efficiency. The decorative patterns of the handcrafted object generally have no function whatsoever; hence they are ruthlessly eliminated by the industrial designer. The persistence and the proliferation of purely decorative motifs in craftwork reveal to us an intermediate zone between usefulness and aesthetic contemplation. In the work of handicraftsmen there is a constant shifting back and forth between usefulness and beauty. This continual interchange has a name: pleasure. Things are pleasing because they are useful and beautiful. This copulative conjunction defines craftwork, just as the disjunctive conjunction defines art and technology: usefulness or beauty.[32]

There also exist what may be called material distinctions among technological objects, as opposed to the formal distinctions mentioned so far. To take an example that has been the object of a well-known historical study, the stirrup (part of a tool for horseback riding) is capable of an indefinite number of stylistic variations while retaining its basic formal (or technical, functional) properties. Consider, for instance, the similarities and differences between the block, bell-shaped wooden platform stirrup of the fourteenth-century Arabs; the heavily ornamented metal Spanish vaquero stirrup; the even more heavily ornamented Spanish stirrups of the Dutch shoe and pointed boot design; and the thin metal, starkly functional stirrups of the English. Just as the concept of a triangle is capable of being imagined and drawn as red or green, isosceles or scalene, so is the idea of a stirrup as footrest attached to a saddle capable of an indefinite number of reifications.

Thus one can postulate an argument between Platonic and Aristotelian views of technology as object. The former would maintain that a stirrup is a stirrup on the basis of form or function alone, never mind its material embodiment. The latter would see a simplified English riding stirrup as substantially different from an ornate Spanish vaquero stirrup. The point is that in some abstract sense these objects are the same, but at the level of physical realization there are significant typological (not just individual) distinctions based on materials and ornamentation, which reveal differences in use contexts and cultural attitudes. Technological objects are subtly diversified by physical and formal "styles." To substantiate this suggestion, however, calls for both sociohistorical studies and a phenomenology of artifacts.[33]

Toward a Phenomenology of Artifacts

A pivotal initiation of this project can be found in Ivan Illich's *Tools for Conviviality* (1973). There are, of course, anticipations in the Western intellectual tradition long before Illich, especially in the complex of cultural interpretations of the Industrial Revolution. To the societal problems associated with the creation of industrial tools and artifacts there have been two basic responses. One argues that the problems are caused *not* by material objects, but by the social context in which these objects exist. The second argues that the problem is caused by the objects themselves. The first has been termed a socialist response, the second the Luddite—or, more fairly, an artifactist— response.[34]

The tradition of artifactology or artifactist thought before Illich includes, besides Lafitte and Simondon, at least the following eclectic mélange:

- Jacques Ellul's presentation (1954) of a "characterology of technique" as exhibiting automatism of technical choice, self-development, unity (or indivisibility), the linking together of techniques, technical universalism, and technical autonomy.

- Günther Anders's argument (1961) that artifacts can have maxims, so that the Kantian categorical imperative must be extended to read: "Have and use only those *things*, the inherent maxims of which could become your own maxims and thus the maxims of a general law."

- Lewis Mumford's distinction (1964) between authoritarian and democratic technics.

- Marshall McLuhan's thesis (1964) that independent of content, a particular communications medium is its own message.

- Jean Baudrillard's description (1968) of the postmodern "system of objects" as constituting a linguistic-like phenomenon liberated from economies of production.

- Richard Weaver's analysis (1970) of machines as constituting, as is said of military forces before utilization, their own "forces in being" or influence.[35]

It is crucial to note—as my earlier references to Lafitte and Simondon already indicate and the inclusion of Baudrillard here reinforces—that artifactist thought is in no way inherently antitechnology. Artifactology simply subscribes to the thesis that artifacts have consequences; there is room for considerable disagreement about the character of those consequences and whether they are to be promoted or restrained.

In none of the cases listed, however, do the authors provide extended and detailed analysis of the inner structures of artifacts and how such structures give artifacts inherent tendencies toward specific kinds of human engagement and use. Their focus remains largely at the macro and in one case symbolic level, stressing external relations.

Although Ellul makes some observations about the personal and societal effects of machines qua machines—and is commonly misconstrued as opposed to the artificiality qua artificiality of artifacts[36]—his central interest is technical action. As a result his characterology applies to technology and tools more as social institutions than as material objects. Anders limits himself to considering a particular kind of artifact—namely, nuclear weapons—and McLuhan at least in this first book does the same by focusing on communications technologies or media. McLuhan, as well, increasingly clothes analysis in an oracular rhetoric,[37] as does Baudrillard, for whom the "objectlessness" of distinctly contemporary artifacts turns them into signs, a kind of immaterial artifact. Weaver's ideas provide no more than a suggestive analogy for how collections of artifacts influence individual decision and social behavior.

Mumford provides a broader perspective on artifice—one that takes note of differences between machines and tools as well as the distinctive identities exhibited by clothes, containers, structures, apparatus, utensils, and utilities. One perceptive observation concerns how "in the series of objects from utensils to utilities there is the same relation between the workman and the process that one notes in the series between tools and automatic machines: differences in the degree of specialization, the degree of impersonality" (1934, p. 11). His arguments nevertheless remain impressionistic, and as much analogic as analytic. It is also true that even when Mumford analyzes the influence of machines on human affairs, as with the case of the mechanical clock, he does not relate this influence to structurally distinct properties in the artifacts. Indeed, he even argues that these arose first in social organization: the megamachine of rigid hierarchical social organization anticipates the mechanical machine.

Illich, by contrast, puts forth an analysis of the inner structures of tools with concrete implications for the explanation of distinctive human-artifact engagements that can be summarized as in table 2. Although Illich fails to make useful reference to Mumford's broader spectrum of distinctions, he nevertheless provides a pointed analysis of two types of tools and the ways their different structures constrain human engagements, independent of particular intentions. For Illich, tools not only embody or express the intentions of individual human makers and users, but also and as significantly embody what may paradoxically be termed "unintended intentions"—which, for that very reason, must be investigated. There is the need for a phenomenology of the artificial related to but not limited by concerns for the effective manipulation and management of artifacts.

As operational or functional entities, tools can be analyzed into material and formal elements: energy constitutes a kind of prime matter of motion, providing the raw or unformed impulse of operating, while guidance (operating of course through the tool itself) gives the functioning of any tool its formal definition.[38] Because they are dependent on human users for both the material and formal elements of their functioning, hand tools exhibit a unique dependency on and qualitatively distinct engagement with human beings. Insofar as the energy to operate power tools becomes independent of human users, such tools begin to exhibit a certain autonomy. Moreover, because power tools concentrate increasingly greater quanta of energy in the hands of users, they necessarily introduce into the social order inequalities that would otherwise not be present. The person who owns a power tool has more power than one who does not. Power grows out of the structure of the tool rather than from social organization.

This initiating sketch of a phenomenology of artifacts begins to reveal a straightforward sense in which technology can become autonomous in relation to human users (if not makers), and it shows how different kinds of tools can have inherent features that ground distinc-

Table 2. Tools and Their Human Engagements: Preliminary Analysis

	Analytic Elements	
Kinds of Tools	Immediate Source of Energy (matter)	Immediate Source of Guidance (form)
Hand tools	Human beings	Human beings
Power tools	Nonhuman realities	Human beings

tive impacts on societal orders. This is true independent of particular social contexts within which some particular tool might be embedded or the particular social processes it is associated with. It is also relatively simple to see the meaning of Illich's repeated call for new kinds of tools for human beings to work with (tools employing human energy and guidance) instead of more tools to work for humans (tools requiring less and less direct human energy or guidance). The latter less and less allow end-users to introduce their personal intentions into the world, to leave traces of themselves in those rich constructs of traditional artifice that have served for millenia as the dwelling places of humanity. Users now become consumers and leave traces of themselves only in their wastes.

Moreover, with hand tools the general bodily engagement and the dependency on human energy provide the basis for direct, intuitive judgments about the efficacy of a particular tool in its context. If a hand tool does not work, the user knows it immediately and through direct experience. To swing a dull ax and feel in the hands and arms the rebound of momentum that fails to penetrate the grain of the wood, while hearing a thud rather than a sharp crack, is all the evidence the woodsman needs that a blade requires the whetstone. As tools are replaced by machines that become vehicles for utilizing energy originating outside the human body, the user is reduced to operator or manipulator, deprived of many of the direct or immediate indicators of efficacy. To compensate, to provide a new basis for judgment, human users develop a science of mechanics, with its quantified measures and gauges of efficiency. The quantification of efficacy by the input-output calculus of efficiency in turn gives birth to new constructions of artifice, the world of machines. Such an analysis constitutes the suggestive initiation of a comprehensive phenomenology of artifacts and their human engagements. Drawing on the mechanological analyses of Lafitte and Simondon, one can summarize possibilities as in table 3.

Table 3. Tools and Their Engagements: Extended Analysis

	Analytic Elements	
Kinds of Tools	Immediate Source of Energy (matter)	Immediate Source of Guidance (form)
Hand tools	Individual human beings	Individual human beings
Premodern machines	Groups of human beings or animals	Individual human beings
Modern machines	Inanimate nature (wind or water) and technologically controlled nature (heat engine)	Individual human beings or groups of human beings assisted by mechanical controls
Power tools	Technologically controlled and abstracted nature (electricity)	Individual human beings and mechanical or electrical controls
Cybernetic devices	Technologically controlled and abstracted nature (electricity)	Electronic controls

Illich's hand tool/power tool distinction simplifies a conceptual gradient from tools to cybernetic devices. Tools are first of all hand tools, then machines that require energy input from groups of laborers (as with galley slaves rowing a ship) or animals (a team of oxen pulling a moldboard plow) or the readily accessible motions of nature (wind caught by the sail). External input undergoes further transmutation with the development of, first, the heat engine (steam engine, internal combustion engine), then electricity, to drive a mechanical prime mover. The power of the steam engine exponentially exceeds any previous energy source; electricity takes such powers into realms of scientific and conceptual abstraction.

Transmutations in guidance and formal functioning follow. Note, for instance, how coordinate with harnessing power from the heat engine there developed internal technical requirements for technological controls that were initially realized in the mechanical governor—thus introducing a formal decoupling of human operators from machine operation. Such formal decoupling at the level of operation is also coordinate both with emergence of the mathematicized engineering analysis of control and with an expanded external coupling through mass consumption of mass-produced products. Electrical and electronic power tools such as kitchen appliances and personal computers, for their part, reintroduce a level of personal use not possible with large-scale steam-powered industrial machines.

When the computerization of control in turn becomes the primary means for operating large-scale mechanical systems, placing workers behind video screens and touchpad keyboards to monitor information flows, this introduces what may be called the techno-lifeworld of the screen. For example, Shoshana Zuboff, *In the Age of the Smart Machine*, analyzes how "as information technology is used to reproduce, extend, and improve upon the process of substituting machines for human agency, it simultaneously accomplishes something quite different. The devices that automate by translating information into action also register data about those automated activities, thus generating new streams of information." [39] To automate with advanced computers is also to informate. Moreover, "The intrinsic power of [this] informating capacity can change the basis upon which knowledge is developed and applied in the industrial production process by lifting knowledge entirely out of the body's domain. The new technology signals the transposition of work activities to the abstract domain of information. Toil no longer implies physical depletion. 'Work' becomes the manipulation of symbols, and when this occurs, the nature of skill is redefined" (p. 23). The new skill is to sit before multiscreen or window-enhanced monitors and correctly interpret visual displays of information, responding not with the body but merely with digital movements that engage no more than a touchpad or touchscreen. [40]

Despite the dematerialization implicit in this stage of informated engagements, the general analysis suggests the need to be wary of dropping the distinction, a proposal that Illich himself encourages, although in qualified form, by calling physical artifacts and social institutions first- and second-order tools, respectively. For Illich, as for John Dewey, for instance, it sometimes appears that hammers, schools, and logic are all equally tools. Although no doubt partially true, the idea obscures the need for different kinds of analyses when dealing with material objects and social institutions, not to mention ideas. With social institutions, for example, it is quantity of individual interactions and bureaucratic line-and-staff structures that are central, not energy input-output ratios and technical control mechanisms.

Against such a background one can appreciate certain necessary refinements, as well as trajectories for future research, presented by such works as the following:

- Langdon Winner, *Autonomous Technology: Technics-out-of-Control as a Theme in Political Thought* (1977) and *The Whale and the Reactor: A Search for Limits in an Age of High Technology* (1986).

- Don Ihde, *Technics and Praxis: A Philosophy of Technology* (1979) and *Existential Technics* (1983).

- Mihaly Csikszentmihalyi and Eugene Rochberg-Halton, *The Meaning of Things: Domestic Symbols and the Self* (1981).[41]

- Albert Borgmann, *Technology and the Character of Contemporary Life: A Philosophical Inquiry* (1984).

Langdon Winner's *Autonomous Technology* is an extended defense of a thesis found most fully articulated in Ellul's *The Technological Society*, the idea that the rise of modern technology parallels creation of a new form of political life, what Winner calls "technological politics." But unlike Ellul and like Illich, as Anthony Weston has noted, Winner identifies specific design criteria for tools. In the last chapter of his book Winner puts forth three guidelines for an "epistemological Luddism" that would question and thus reintroduce into technological politics some of the character of traditional political life. These would examine technologies in terms of their (1) intelligibility to nonexperts, (2) degree of flexibility, and (3) tendency to foster dependency (1977, pp. 326–327). For Illich a tool is convivial if it (1) can be freely chosen, (2) is an active expression of personal life, and (3) is not monopolized by some professional elite. As Weston observes, Winner's (1) corresponds to Illich's (3), Winner's (2) to Illich's (1), and Winner's (3) to Illich's (2).[42]

Winner's second book, *The Whale and the Reactor*, carries this argument forward. In "Do Artifacts Have Politics?"—the central chapter of its first and governing section—he considers two ways artifacts can embody political implications. In the first, human beings specifically make technologies solve political problems. Examples are Robert Moses' Long Island parkway overpasses, designed to restrict use by buses and thus access by the urban poor; Cyrus McCormick's molding machines, used to break shopfloor labor organization; and the mechanical tomato harvester, which turned truck farming into agribusiness.

> The things we call "technologies" are ways of building order in our world. . . . Consciously or unconsciously, deliberately or inadvertently, societies choose structures for technologies that influence how people are going to work, communicate, travel, consume, and so forth over a very long time. In the processes by which structuring decisions are made, different people are

situated differently and possess unequal degrees of power as well as unequal levels of awareness. . . . For that reason the same careful attention one would give to the rules, roles, and relationships of politics must also be given to such things as the building of highways, the creation of television networks, and the tailoring of seemingly insignificant features on new machines. (pp. 28–29)

In comparison with Illich's argument and its urgency, this simply calls for more carefulness in tool making and using.

In the second case, there are technologies that, independent of any human intention, embody certain inherent political implications. Here Winner cites the arguments of Engels, Plato, and Marx (in that order) and then distinguishes strong and weak versions of the thesis. In the strong version, a certain technology is said to require or necessitate some specific social relations. In the weak version, a technology is said not to require but to be strongly compatible with specific social relations. "My belief that we ought to attend more closely to technical objects themselves is not to say that we can ignore the contexts in which those objects are situated" (p. 39). But in neither version does Winner analyze the inner structure of modern tools.

Consider, by contrast, Don Ihde's *Technics and Praxis*, which examines in detail how tools or instruments can extend human capability and, in the very same process, also restrict access to the world through a simultaneous amplification/reduction structure. Ihde uses the example of a dentist's probe—a small metal rod with a pointed tip—which is able to detect irregularities in a tooth that a finger could not sense. "But at the same time that the probe extends and amplifies, it *reduces* another dimension of the tooth experience. With my finger I sensed the warmth of the tooth, its wetness, etc., aspects which I did not get through the probe at all. The probe, precisely in giving me a finer discrimination related to the micro-features, 'forgot' or reduced the full range of other features sensed with my finger's touch" (p. 21). What Ihde calls the amplification/reduction structure is very close to what McLuhan terms the two laws of enhancement and obsolescence. For McLuhan, "Any new technique . . . or tool, while enabling a new range of activities by the user, pushes aside the older ways of doing things (McLuhan and McLuhan 1988, p. 99). (Others might describe this simply in terms of benefit/cost tradeoffs.)

For McLuhan the laws of enhancement and obsolescence are grounded in the way all artifacts "are extensions of the physical human body or the mind" (p. 93). For Ihde, too, the probe embodies or extends the finger or hand. But instruments not only enter into what Ihde terms embodiment relations, they also can take on hermeneutic or interpretative relations. That is, in diagrammatic description of the relation

Human - Instrument - World,

the instrument can be assimilated to a human-instrument combination so that the user and instrument together confront or interpret the world thus:

[Human-Instrument] ⟶ World.

But human users can also view themselves as apart from the instrument, which is now seen as part of the world, and thus enter into a hermeneutic or interpretative relationship directly with the instrument-world:

Human ⟶ [Instrument-World].

Eyeglasses are engaged through embodiment relations, electron microscopes through hermeneutic ones.

Ihde's consideration of how consideration of how concrete things such as dental probes, telephones, magnifying glasses, optical and electron microscopes, telescopes, electronic musical instruments, and computers exhibit such relationships can be compared with Illich's concerns for the ways power amplification entails freedom reduction. The move in *Existential Technics* toward consideration of how technical engagements influence human self-understandings introduces a recursive or reflective dimension into the human-instrument-world relation that can be diagrammatically indicated as follows:

Human ⟷ Instrument ⟷ World

Concern for how technologies reflect back upon their creators and users, influencing their self-images and self-interpretations, can also be correlated with later works by Illich, such as *ABC* (1988) and *In the Vineyard of the Text* (1993), that reflects on how the technologies of writing have influenced Western European culture.

Mihaly Csikszentmihalyi and Eugene Rochberg-Halton are likewise concerned with the relation between things and self-understanding. In their words,

men and women make order in their selves . . . by first creating and then interacting with the material world. The nature of that transaction will determine, to a great extent, the kind of person that emerges. Thus the things that surround us are inseparable from who we are. The material objects we use are not just tools we can pick up and discard at our convenience; they constitute the framework of experience that gives order to our otherwise shapeless selves.[43]

Their focus, however, is on household things and their symbolic import. It is nevertheless remarkable that in a comprehensive survey of previous approaches to an understanding of things that considers psychological, anthropological, and sociological studies there is no mention of the philosophical approach represented by either Illich or Ihde.

At the same time, by raising the question of the symbolic import of things, Csikszentmihalyi and Rochberg-Halton present again the challenge of immaterialism associated with Baudrillard. This challenge concerns the relation between the inner structure, the functional, and the symbolic characters of artifacts, and it is crucial to Illich's argument for self-learned self-limitation in the making and using of technology. Any attempt to focus ethical-political reflection on material artifacts— especially one arguing for the experiential learning of self-limitations—must address the counterthesis of Baudrillard and others regarding the immaterial sign character of contemporary objects. For Baudrillard, for instance, *"there are no limits to consumption"* [44] because modern things are more like words than physical objects. Just as conversation is inherently limitless, so is modern consumption:

> We want to consume more and more. [Read: "We want to talk more and more."] This compulsion to consume [to talk] is not the consequence of some psychological determinant . . . nor is it simply the power of emulation. It is a total idealist practice which has no longer anything to do (beyond a certain point) with the satisfaction of needs, nor with the reality principle; it becomes energized in the . . . object-signs of consumption. . . . Hence, the desire to "moderate" consumption or to establish a normalizing network of needs is naive and absurd moralism.[45]

Albert Borgmann's explication of contemporary artifacts in terms of what he calls the device paradigm provides the beginning of an analytic response. Borgmann also, alone among serious philosophers of artifice writing in the wake of Illich's *Tools for Conviviality* (1973), grants it a measure of recognition—even while taking issue with at least one thesis of the text.[46]

Borgmann contrasts traditional things with modern devices. "A thing . . . is inseparable from its context, namely, its world, and from our commerce with the thing and its world, namely, engagement. The experience of a thing is always and also a bodily and social engagement with the thing's world" (p. 41). A device, by contrast, seeks to realize the promise of technology "to bring the forces of nature and culture under control, to liberate us from misery and toil, and to enrich our lives" (p. 41), but in a material object cut loose from all bodily and social engagement. In contrast to a fireplace, for example, "a central heating plant procures mere warmth and disburdens us of all other elements" (p. 42). In its very disburdenment, the device takes on a disembodied or immaterial character, like a word or a sign.

But human beings are not just the users of words and signs, they are embodied beings whose lives are realized through what Borgmann calls focal things and practices. While recognizing, with Baudrillard, the presence and influence of devices, he nevertheless, like Illich, calls for "*the recognition and restraint* of the [device] paradigm. To restrain the paradigm is to restrict it to its proper sphere. Its proper sphere is the background or periphery of focal things and practices. Technology so reformed is no longer the characteristic and dominant way in which we take up with reality; rather it is a way of proceeding that we follow at certain times and up to a point, one that is left behind when we reach the threshold of our focal and final concerns" (p. 220).

According to Borgmann, such a reform will take place not so much out of crisis as out of focal concern. It is not the stick of necessity so much as the carrot of "the significance of things and the dignity of humans" (p. 220) that can lead from the nonconvivial to a convivial world. Whether this is as true in the world of the screen as in a world of tools is perhaps another issue to be addressed by artifactist thought.

CHAPTER EIGHT

• • • •

Types of Technology as Knowledge

Etymologically the very word "techno-logy"—after the manner of "biology" or "sociology"—is often thought to connote knowledge. Technology as knowledge is also the manifestation of technology that has received the most sustained analytic scrutiny. This no doubt reflects the epistemological proclivities of modern philosophy. As intellectual historians have often noted, although the problem of knowledge plays an important role in philosophy from Plato on, it is with Descartes, Locke, Hume, and Kant that issues of the essence and limits of knowledge have taken center stage. The historicizing of knowledge that follows Kant—from Hegel through Nietzsche to Cassirer and Heidegger—simply extends the modern attempt to ground philosophy in the knowing subject rather than in wonder at the things that are.

Prescinding from the resultant manifold definitions and theories of knowledge itself, however, technological knowledge can simply be contrasted with knowledge of nature. The latter bears on natural objects, the former on artifacts—which could thus be differentiated according to the kinds of technological objects known. Just as natural science is distinguished into physics (dealing with nonliving nature) and biology (dealing with living nature), so technological knowledge may be thought of as composed of architecture (dealing with structures) and mechanics (dealing with machines), not to mention civil, mechanical, chemical, electrical, and other types of engineering.

Although the wealth of information or data about technological objects can be classified as technological knowledge and divided up into such categories, the shortcomings of such an initial contrast and resultant classification are a failure to appreciate the ways there can also be technological knowledge of nontechnological objects. Furthermore, the classification scheme reveals nothing about the unique epistemological structure of technology as knowledge.

More analytic epistemological scrutiny of technology has argued for the following distinctions, working from the least to the most conceptual:

- *Sensorimotor skills* or *technemes* (Gille 1986, p. 1144). The sensorimotor skills of making and using are preconscious "knowhow" more than "know that," acquired by intuitive as well as trial and error learning or imitative apprenticeship to some master craftsman, and thus do not qualify as knowledge in the strict sense. They have nevertheless been accorded considerable attention by phenomenologically inclined philosophers.

- *Technical maxims* (Carpenter 1974, p. 164), *rules of thumb* of prescientific work (Bunge 1967, p. 132), or *recipes* (Gille 1986, p. 1146). These constitute an initial attempt to articulate successful making or using skills. Example: "To cook rice, bring water to a boil, add one-half volume of rice, and simmer for twenty minutes." Indeed, most cookbook recipes are technical maxims, as are many heuristic strategies for problem solving.

- *Descriptive laws* (Carpenter) or *technological rules* (Bunge). These take the form "If A then B," with concrete reference to experience. As Carpenter says, descriptive laws "are like scientific laws in being explicitly descriptive and only implicitly prescriptive of action, but they are not yet scientific in that the theoretical framework which could explain the law is not yet explicit" (Carpenter 1974, p. 165). Because they are usually generalizations derived directly from experience without systematic integration, engineers often refer to such formulas as "empirical laws." Example: Couloumb's sliding wedge analysis for determining the stability of earthwork structures (essentially a modified parallelogram of forces laid out on an embankment cross section), formulated not with the use of engineering geology and physics, but simply by means of traditional design tables and his own observations about which sizes and shapes of fortifications held up well under such and such conditions.[1] (Note that there are also many empirical rules of using, e.g., those developed by Frederick W. Taylor from his time-and-motion studies at the Watertown arsenal.)

- *Technological theories*. Technological theories, according to Bunge, are of two types: substantive and operative. "Substantive technological theories are essentially applications, to nearly real situations, of scientific theories" (1967, p. 122). Examples: aerodynamics or the theory of flight as an application of fluid dynamics; thermodynamics; electronics. Substantive theories constitute the so-called engineering sciences and are applied science in the strict sense.[2] Operative technological theories "are from the start concerned with the operations of men and man-machine complexes in nearly real situations" (p. 123). Examples: decision theory, operations research. Substantive theory employs both the content and method of science; operative theory applies only the method of science to problems of action to develop "scientific theories concerning action" (p. 122). (The former are thus more tied up with making, the latter with using.)

The inherently epistemological character of these distinctions may be suggested as follows. According to a widely accepted analytic definition (which can be traced back to Plato), knowledge is justified true belief. True beliefs concerning the making and using of artifacts can be justified by appeal to skills, maxims, laws, rules, or theories, thus yielding different kinds of technology as knowledge. Different epistemologies of technology and epistemologies of different technologies debate the interaction and relative weights of these various types of technology as knowledge. These are further subject to realist, instrumentalist, pragmatic, and other interpretations, although engineers, like scientists, readily assume the realist stance.

Cognitive Development and Myth in Technology

The cognitive psychologist Jean Piaget (1896–1981) distinguishes four stages in the development of intelligence: sensorimotor cognition or the development of the psychomotor complex (birth to age two); pre-operational or imaginative thinking (ages two to seven); concrete operational thinking or the interiorization of objective functional relationships (ages seven to eleven); and formal operational or abstract thinking (from puberty on). Although these stages bear some resemblance to the four types of technology as knowledge, there is no simple one-to-one correspondence. Nevertheless, skills are characteristic of the earliest stage, technological theories of the latest. Furthermore, much of Piaget's reflection on child development is based on observations of the mastery of technical skills and technical operations, so his studies can be used to help examine various aspects of technology as knowledge.

For instance, according to Piaget the stages of cognitive development form both a necessary and a cumulative sequence. Sensorimotor skills and preoperational imagination are necessary preludes to concrete and formal operational thinking; they are also stages that are never completely abandoned but are retained as live cognitive options. This implies a certain primacy for psychophysical development, but also perhaps the existence of another type of technology as knowledge: imaginative or mythopoeic knowledge.

Indeed, before Piaget anthropologist James Frazer (1854–1941) and ethnologist-philosopher Lucien Levy-Bruhl (1857–1930) distinguish, respectively, between magical, religious, and scientific thinking and between prelogical and logical thinking, arguing in each case that the former are not just embryonic or failed versions of the latter. For Levy-Bruhl, prelogical intelligence is what he terms "corporate" and is oriented toward mystical participation in the life of the natural world that is quite unlike the analytic habits of scientific or logical thinking. It has, Levy-Bruhl and others argue, its own integrity, its own structures and achievements, which are different from and not necessarily inferior or merely preparatory to those of logical thinking.[3]

Mircea Eliade, in his study of the mythologies of mining and metallurgy and of alchemy, provides a ready appreciation of this kind of knowledge. For instance, in relation to the mythologies and rituals of the Iron Age, Eliade concludes that "the image, the symbol and the rite anticipate—sometimes even make possible—the practical applications of a discovery" (1971, p. 24). Meteorites, the first sources of iron, fall from the sky as gifts from the gods; they call attention to the solar disk, inspiring its imitation in the wheeled chariot. "The hammer, successor to the axe of the Stone Age, becomes the emblem of powerful gods, the gods of the storm" (p. 30). Mythopoeic technology as knowledge views metals as divine and fashions them into ornaments, amulets, and statuettes. Variations of such symbolization surely continue in our own time to influence making practices.

The Phenomenology of Technical Skill

Another fruitful suggestion from Piaget's work is his argument in defense of the cognitive character of skills, a position elaborated upon by the work of phenomenologists. Such studies also provide another, slightly different model of cognitive development. Hubert Dreyfus, for instance, utilizing the studies of Patricia Benner,[4] argues that "as human beings acquire a skill through instruction and experience, they do not appear to leap suddenly from rule-guided "knowing that" to experience-based know-how. A careful study of the skill-acquisition process shows that a person usually passes through at least five stages of qualitatively different perceptions of his task and/or mode of decision-making as his skill improves" (Dreyfus and Dreyfus 1986, p. 19). The five stages of skill development are those of the novice, the advanced beginner, competency, proficiency, and expertise.

It is important to note that all of these are stages within the domain of skill as such; there is no transformation, even at the level of expertise, to abstract or formal and therefore conceptually teachable knowledge. The teaching of skills remains confined to the methods of apprenticeship and imitation. At the same time—as the very terms themselves indicate—the first two stages are tied down to the particularity of the learner, whereas the latter three are truly emancipated from individuality, although they continue to exhibit the character of what Michael Polanyi calls unspecifiability.

The work of Polanyi as well as that of Bertrand Gille deserves mention in the present context because of their careful distinctions between different kinds of skill. For Polanyi one basic distinction is between skill and connoisseurship, two kinds of unspecifiable or nondiscursive knowledge. The sensorimotor expertise of an automobile mechanic includes both the ability to hear or perceive the differences between the most subtle sounds of an engine (connoisseurship) and also the ability to adjust nuts, bolts, and valves with just the right degree of tightness or looseness that conditions may require (skill). There are, it seems, both practical and cognitive dimensions to skill as a type of technology as knowledge.

For Gille what is transmitted from master to apprentice by gesture and the spoken word in the context of the workshop are what he calls "technemes" bearing on three aspects of craft making. First is the choice of materials to be used. Second comes the appropriateness of certain actions and tools. Third, even the object to be made is circumscribed by "more or less precise norms" (1986, p. 1145). "At this stage, the acquisition of knowledge is imitation" (p. 1146).

In a more general manner Donald Schön (1983) argues against the common conception of technical rationality he finds imposed on the professions by the Enlightenment view of reason and science. Against the model of rationality that requires the rigorous application of scientific theory, Schön outlines a process of tacit or intuitive reflection-in-action that moves through ascending cycles of problem recognition and problem transformation toward greater or more expansive mastery. He finds this skill development process exhibited by professionals as diverse as the engineer, psychotherapist, business manager, and town planner.

Maxims, Laws, Rules, and Theories

Attempts to analyze the explicitly cognitive dimensions of technology have largely focused on explicating the relation between maxims, laws, rules, and theories, and have centered on a discussion of how far tech-nology is properly described as applied science. In general those repre-senting the positivist tradition in the philosophy of science have ar-gued that technology is applied science and have sought to create an epistemology of technology based on the covering law model of scien-tific explanation.

The idea of technology as applied science is clearly the received view among both engineers and scientists. Its strongest philosophical representative has no doubt been Mario Bunge, an avowed proponent of "scientific realism" logically articulated. For Bunge technology is to be strongly distinguished from technics (craft skills of the artisan), from technical practice (engineering practice, medical therapy, etc.), and from pseudotechnology (astrology, alchemy, homeopathy, psycho-analysis, etc.).[5] Technics and engineering practice are kinds of activity, not kinds of knowledge. Unlike technics, however, engineering prac-tice is based on knowledge, indeed, on technology—the use of science to guide human making. Astrology is a pseudotechnology precisely because it applies pseudoscientific rather than scientific knowledge to the guidance of human affairs. For Bunge, technology is *"the scientific study of the artificial* . . . or the field of knowledge concerned with de-signing artifacts and planning their realization, operation, adjustment, maintenance and monitoring in the light of scientific knowledge" (1985, p. 231). Thus the key issue in the epistemology of technology is "the process whereby the crafts are given a technological basis, and . . . converted into applied science" (1967, p. 129).

Full understanding of this process requires a further distinction be-tween scientific laws and technological rules, the central or pivotal con-cepts in science and in technology. Scientific knowledge is composed of an articulated set of observations, laws, and theories. The corre-sponding elements of technology are actions, rules, and theories. This comparison is summarized in table 4. Scientific laws (which can be integrated into general theories) describe objective patterns of empiri-cal phenomena or facts in nature and can be more or less true; techno-logical rules prescribe courses of action and can only be more or less effective.

What happens when the maxims or rules of thumb of prescientific craft (which are often justified by mythological as opposed to scientific laws) are transformed into technological rules is that they are "grounded" in scientific law. "A rule is *grounded* if and only if it is based on a set of law formulas capable of accounting for its effective-ness" (1967, p. 129). The precise structure or logic of this being "based on" or "accounting for"—the grounding of rules in nomological state-

Table 4. A Comparison of Cognitive Elements in Science and in Technology

	Cognitive Elements		
Bodies of Knowledge	Empirical or direct worldly Engagements	Conceptual Entities (lower level)	Conceptual Entities (higher level)
Science	Scientific observations	Scientific laws	Scientific theories
Technology	Technological actions	Technological rules	Technological theories

ments through nomopragmatic statements—is complex and deserves further epistemological treatment. In a brief illustration, the nomological statement "Water boils at 100°C" ("nomological" because it simply describes an empirical conjunction between boiling and temperature) grounds the nomopragmatic statement "If water is heated to 100°C, then it boils" ("nomopragmatic" because it introduces the operative action of heating). This in turn can be the basis for any number of technological rules: "To boil water, heat it to 100°C," "To keep water from boiling, keep its temperature below 100°C," and so on. The essential point is that "whereas technics can make do with rules of thumb, or recipes, technology requires *rules based on law statements* (i.e., nomopragmatic statements)" (1985, p. 242). Besides, "the blind application of rules of thumb has never paid in the long run: the best policy is, first, to try to ground our rules, and second, to try to transform some law formulas into effective technological rules" (1967, p. 133). In either case the result is technological theories—and technology as applied science.

Technology as knowledge, on this view, is thus to be internally distinguished from science by the presence of rules and nomopragmatic statements and externally based on problems and goals. The problems of science are cognitive ones solved by observations that result in the accumulation of information about the world. Science aims at understanding, and its central element is a scientific law that purports to describe the way the world is. The problems of technology, by contrast, are practical. Technology aims at control, and its central element is a rule (sometimes called a law) that purports to prescribe the way the world can be manipulated. Science and technology are thus externally distinguished by ends or intentions: scientific knowledge aims at knowing the world, technological knowledge at controlling or manipulating it.[6] This explains the difference, for Bunge again, between scientific prediction (which is a means of confirming theory) and technological forecast (which, by suggesting how to influence circumstances, is a means of control). Such a difference in aims also accounts for differences between scientific and technological experiments: the former test the truth of some theory, the latter assess theory effectiveness. Such different aims, when prolonged into action, produce different experimental structures.[7]

Grounded technological rules can be integrated into technological theories of two different kinds—what Bunge calls substantive and operative theories:

> Substantive technological theories are essentially applications, to nearly real situations, of scientific theories. . . . Operative technological theories . . . are from the start concerned with the operations of men and man-machine complexes in nearly real situations. . . . Substantive technological theories are always preceded by scientific theories, whereas operative theories are born in applied research and may have little if anything to do with substantive theories. (1967, p. 122)

Substantive technological theories utilize already accepted scientific theories, and they almost always simplify them—in ways that have, for instance, been informatively analyzed by Ronald Lyman (1985, 1991). Operative technological theories are created by applying the scientific method to human-artifice interactions, as has been considered at length by Herbert A. Simon in his *Sciences of the Artificial* (1969). Substantive theories tend to focus on technological making, operative theories on technological using. Psychology, economics, and administration are all operative technological theories in this sense.

Against Technology as Applied Science

Against Bunge's epistemology of technology as applied science there have developed at least two distinctive arguments. The first identifies one or more elements of technological knowledge as irreducibly distinct from science or even at odds with it. The second rejects the science/technology distinction implicit in Bunge's positivist-realist view based on some version of a "new" post-Kuhnian or phenomenological philosophy of science.

The first tack is well illustrated by diverse social science examinations of technology. As John Staudenmaier (1985) has ably shown in his analysis of the themes of articles in the first twenty years of *Technol-*

ogy and Culture (the quarterly of the Society for the History of Technology, internationally recognized as the premier journal in the field), historians and sociologists of technology, in order to establish their discipline as distinct from the history and sociology of science, have for over two decades been at pains to argue the distinctiveness of technology in relation to science. According to Staudenmaier, this distinctiveness has generally been attributed to some one or a combination of the following: the modification of scientific concepts, the use of problematic data, the uniqueness of engineering theory, and dependency on technical skill.

With regard to the technological modification of scientific concepts, although philosophers have sometimes made this point,[8] historians have done so in greater detail. Thomas M. Smith, for instance, in a study of the post–World War II "Project Whirlwind" at MIT, argues that "the idea that exogenous science provides a reservoir of knowledge essential to the continuing vitality of the R&D process may be a romantic notion of our time that is considerably overrated and that the Whirlwind experience severely qualifies."[9] The key concepts operative in the Whirlwind project to create a digital computer at MIT either were indigenous to the engineering field (block diagramming) or underwent substantial transformation in the course of being imported from other fields such as physics (principles associated with the development of magnetic core storage). The eight-year study of Project Hindsight (1966) on the extent to which the Department of Defense benefited from basic research—the study concluded that only a fraction of 1 percent of the events related to the development of twenty key weapons systems could be construed as basic science, while 91 percent were technological—provides further support for the idea that scientific concepts qua scientific concepts play only a small role in engineering.

Technological knowledge also utilizes a kind of problematic data quite distinct from the data of science. Historians and even engineers have been much more active than philosophers in arguing this issue. Indeed, aeronautical engineer Walter Vincenti's *What Engineers Know and How They Know It* (1990) is the most careful and extensive argument in this regard. Four original case studies in aeronautical design first published in *Technology and Culture* between 1979 and 1986 both pointed out how a technological problem can highlight an area of ignorance that science considers unimportant and showed that resolving engineering problems regularly requires the use of less than scientifically acceptable information. (Vincenti and other engineers even argue that seldom is a technology completely understood in the scientific sense, even after it has become part of normal practice.)

In his book Vincenti carries his analysis forward by identifying six key nonexclusive and nonexhaustive categories of distinctly engineering knowledge: (1) fundamental design concepts, (2) criteria and specifications, (3) theoretical tools, (4) quantitative data, (5) practical considerations, and (6) design instrumentalities. Some of these kinds of knowledge are what Vincenti calls descriptive, others prescriptive, still others tacit. It has, for instance, become an implicitly assumed given or fundamental design concept of aeronautical engineering that the fixed-wing airplane works by structuring a surface to "support a given weight by the application of power to the resistance of air."[10] As Vincenti summarizes the point: "Every device possesses an operational principle, and, once the device has become an object of normal, everyday design, a normal configuration" (p. 210). Criteria and specifications, by contrast, are usually explicitly determined, sometimes by nonengineers. Theoretical tools include a wide range of "intellectual concepts for thinking about design as well as mathematical methods and theories for making design calculations," some of which derive from science but many of which do not (p. 213).

Quantitative data include all sorts of "descriptive knowledge . . . of how things are"—meaning both nontechnological and technological things (p. 216). Here scientific information enters in, but much of the quantitative data in engineering handbooks is unique to the engineering field. Whereas quantitative data are the "precise and codifiable" results of "deliberate research," engineers also depend upon and deploy "an array of less sharply defined considerations derived from experience in practice" (p. 217). Finally, design instrumentalities include thinking by analogy and nonverbal or visual thinking. This last is, Vincenti argues, especially characteristic: "Outstanding designers are invariably outstanding visual thinkers" (p. 221). Notice that, in accord with his own characterization of engineering as proceeding without excessive concern for theoretical exactness, Vincenti's argument for "technology as an autonomous form of knowledge" (p. 4) is itself somewhat loose both in its use of the term "knowledge" and in its conclusions.

With regard to engineering theory, although there are certainly some formal parallels with scientific theory, modeling parameters and idealizing assumptions are almost always quite different. Discussing the ideologies of science and engineering, historian and engineer Edwin Layton is especially lucid on this point.

Engineering science often differs from basic science in important particulars. Engineering sciences often drop the fundamental ontology of natural philosophy, though on practical rather than metaphysical grounds. Thus, in solid mechanics, engineers deal with stresses in continuous media rather than a microcosm of atoms and forces. Engineering theory and experiment came to differ from those of physics because it was concerned with man-made devices rather than directly with nature. Thus, engineering theory often deals with idealizations of machines, beams, heat engines, or similar devices. And the results of engineering science are often statements about such devices rather than statements about nature. The experimental study of engineering involves the use of models, testing machines, towing tanks, wind tunnels, and the like. But such experimental studies involve scale effects. From Smeaton onward we find a constant concern with comparing the results gained with models with the performance of full-scale apparatus. By its very nature, therefore, engineering science is less abstracted and idealized; it is much closer to the "real" world of engineering. Thus, engineering science often differs from basic science in both style and substance. Generalizations about "science" based on one will not necessarily apply to the other. (1976, p. 695)

Two well-regarded case studies confirm Layton's point in relation to two quite different branches of engineering. Ronald Klein, director of the IEEE Center for the History of Electrical Engineering in New York, in a study of the invention of the induction motor, 1880–1900, shows that the attempt to make what might be called "top-down" applications of Maxwell's scientific theory of electricity actually inhibited the development of electrical motors. What was needed instead was "bottom-up" development, with some general guidance from Maxwell's equations, of "equivalent circuit" and "circle diagram" techniques.[11] Bruce Seely, in a study of transportation engineering research two decades later, likewise concludes that "infatuation with attitudes and experimental methods usually considered typical of science hindered the development of practical answers to engineering questions while failing to enhance theoretical understandings of the problems under investigation."[12]

Finally, with regard to technical skill, once again philosophers and social scientists have joined forces to emphasize an aspect of technology slighted by the idea of technology as applied science. This has been done in two different contexts. One focuses on premodern, the other on modern technology.

With regard to premodern technology—which he prefers to call craft or technics—even Bunge grants that skill exercises a defining influence. Technical skills are learned by experience through apprenticeship and imitation and are characteristic not only of laborers but even of master craftsmen. Historians of technology have pointed out, however, that even in the premodern context it is possible to subject intuitive skill and trial-and-error learning to nontheoretical rules or maxims for technical practice, and to some extent to articulate skill in descriptive mathematical form.[13] Even quite traditional craft technology can be much more sophisticated and complex than is sometimes recognized, and to some extent it may be emancipated from total dependence on transmission by lengthy apprenticeship.

A key characteristic of the development of modern technology, at least on the view of technology as applied science, however, is a two-fold de-skilling of making and using. First, there is the rationalization of skills known to the master craftsman, that is, their justification or explanation in terms of scientific knowledge rather than mythopoeic knowledge; second, there is the replacement of the skilled worker by the unskilled laborer.[14] The stronger argument of a number of social scientists and engineers is that skill, if not present to the same degree, is at least as essential to modern as to premodern technology. For engineer-historians such as Vincenti and Layton the importance of skill, even more than problematic data and the uniqueness of engineering theory, is the key to appreciating the extrascientific character of technology, whether ancient or modern. For them—and others such as philosophers James K. Feibleman and Joseph Agassi—modern technological knowledge is more than just applied theory, even engineering theory. Technology is not so much the *application* of knowledge as a *form* of knowledge, one persistently dependent on technical skill. Indeed, on the model of cognitive development provided by Piaget, such persistence of skill beyond the realm of craft practice should be expected, although in modern technology it will no doubt take new forms in a context transformed by the adaptation of scientific concepts, problematic data, and engineering theory.

A quite different criticism of the positivist idea of technology as applied science begins with a reconsideration of the character of science itself by asking whether science can be distinguished from other forms of cognition by virtue of some special claim to objectivity. The result is not so much to establish technology as independent of science as to blur distinctions in such a way as to undermine the ability to think of "application" as a one-way street. Insofar as technology is applied science, science is also applied technology. Although aeronautical engineering applies physics, physics applies particle accelerator technology.

For Thomas Kuhn and other proponents of a "new" philosophy of science, there are no strictly objective facts or observations that can uniquely determine some scientific law, which might then "ground" some technological rule. As Harold Brown has summarized the new perspective, it is characterized by a commitment to historical case study rather than to logical analysis. Furthermore, through its historical approach it argues that

> most scientific research consists . . . of a continuing attempt to interpret nature in terms of a *presupposed* theoretical framework. This framework plays a fundamental role in determining what problems must be solved and what are to count as solutions to these problems. . . . Rather than observations providing the independent data against which we test our theories, fundamental theories play a crucial role in determining what is observed, and the significance of observational data is changed when a scientific revolution takes place.[15]

Assuming that theories determine observation rather than the other way around, it readily becomes necessary to ask what factors do determine theories and theory choices. Historical traditions, aesthetic perceptions, economic constraints, national cultures—the same factors that affect technologies—all exercise strong influences over scientific theories. Scientific theory is also subject to the influence of machines and experimental apparatus[16] and the general "carpentered environment."[17] While not denying that science can influence technology, this position argues that the science-technology interaction is not simply one of scientific dominance but entails complex mutuality. This, for instance, is the richly detailed thesis of historian and metallurgist Cyril Stanley Smith over a lifetime of research and writing.[18]

Cybernetics

Against the background of discussions about the ways technology is or is not applied science, cybernetics can be considered in a manner that points toward larger epistemological and transepistemological issues. For Bunge, cybernetics is an example of what he calls an operative technological theory. It grows out of the application of the methods of science to the analysis of human-machine complexes. Others might well stress the special cybernetic transformation of "scientific" concepts (entropy, information), problematic data (noise in signal transmission), engineering theory (information theory, systems theory, automata theory), and technical skill (working with cybernetic devices such as computers depends on special training and habituation). Still others could emphasize the interaction of mathematics and various feedback control devices such as the steam engine fly-ball governor and the computer.

Alternatively defined as the science of "control and communication in the animal and the machine" in the subtitle of Wiener's *Cybernetics* (1948) and as the science of "all possible machines," whether electronic, mechanical, neural, or economic (Ashby 1956, p. 2), cybernetics surely claims to be a kind of technological knowledge. During the period of its initial formulation in the 1940s, cybernetics was closely associated with neurophysiology, on the hypothesis that negative feedback mechanisms are basic to the workings of the central nervous system. As a general theory of artifacts (from thermostats and self-tracking radar to prosthetic limbs and computers) and operations (from corrective neurosurgery to business management), cybernetics further proposes the possibility of a unified explanation of material, social, and mental phenomena.

Cybernetics, as Wiener's definition suggests, further implies a fundamental identity between animal and machine and thus proposes a fundamental expansion of the scope and pretensions of technological knowledge. Indeed, in its general account of reality it exhibits many of the features of technological philosophy or an engineering philosophy of technology. In traditional theory the difference between living and nonliving objects was that living things are self-moving, nonliving things are not. One aspect of the self-moving character of living things is that they are alleged to possess a source of motion within themselves or are otherwise able to draw energy out of the larger universe on their own initiative. Machines, despite their sometimes apparently self-moving character, cannot provide or acquire energy for themselves. Early modern technology is a power technology, focusing on ways of producing and transmitting energy. In cybernetics the emphasis shifts from sources of power to determinate operations; the availability of energy is taken for granted. Cybernetics is the science of "all forms of behavior insofar as they are regular, or determinate, or reproducible" (Ashby 1956, p. 1). Since both human beings and machines exhibit this regularity of behavior, cybernetics rejects traditional distinctions between human beings and machines, between the living and the nonliving.

In light of this reduction of animal and machine to patterns of determinate behavior, whether one views machines as extensions of animals (including humans), or animals as complex machines, seems to be a question of interpretation. Keith Gunderson (1971) has pointed out how issues raised here can be traced back at least to Julien Offray de La Mettrie's *L'Homme machine* (1747), which argued for a mechanistic interpretation of human behavior. Nineteenth-century debates between mechanists and vitalists in biology reflected similar issues in the human-machine question, as do current arguments about the implications of Kurt Gödel's incompleteness theorem, the limitations of artificial intelligence, and the validity of computer simulations of human cognitive processes.

The foundation of regular or ordered behavior is, in cybernetic theory, the technical concept of "information." In a somewhat circular definition, information may be described informally as a determination of the possibilities of behavior. In classical mechanics a machine is a mechanical linkage arranged so that any energy input into the system results in certain determinate motions with as little energy loss as possible due to resistance. A cybernetic device, by extension, is a communication linkage arranged so that any information input results in certain determinate information output with as little information loss as possible due to "noise." A machine can no longer be conceived as just a "closed kinematic chain" (Reuleaux) but could be redefined as a "closed information linkage."

Cybernetics, as the theory of the way information states interact with one another to produce certain behaviors, explains the nature of the technological in terms of information processing and proposes to provide a means to guide or direct this processing. The word "cybernetics" is derived from the Greek word for steersman and simply means knowledge of control. In this sense too cybernetics appears to be an archetype of modern technology.

In its orientation toward control, however, and in association with the development of artificial intelligence research, there arises another question related to the issue of technology as knowledge. The most advanced cybernetic devices are what are called smart artifacts (smart buildings, smart cars, smart aircraft) that regulate their own operation. Smart artifacts depend not just on cybernetic feedback loops but on expert systems and what is called knowledge engineering. But can knowledge be engineered? What happens to knowledge in the process, or what particular character does it take on? Such questions are also appropriate to discussions of the types of technology as knowledge.

Ancient and Modern Technology

Once again, then, one can state a difference between ancient and modern technology: the former relies for guidance primarily on sensorimotor skills, technical maxims, and descriptive laws, whereas the latter uses these resources plus technological rules and theories. One might maintain as well that this presence of technological rules and theories undermines the importance of skills and maxims. It would be useful, however, to explore the ways these technological rules and theories are made possible by modern science, and how they in turn make possible something like engineering design. There is a need, that is, for a more profound, not to say metaphysical, interpretation of the epistemology of technological knowledge.

Such interpretations have been suggested by Martin Heidegger's notion of modern technology as a special kind of truth or disclosure of the world as *Ge-stell*. The historicophilosophical studies of Hans Jonas on the development of early modern science and the inherently technological character of its theory provide further detail for such an interpretation. As Jonas says, comparing the Aristotelian-Thomist view of science with that of Bacon, for Aristotle and Thomas Aquinas

> the "speculative" (that is, theoretical) sciences . . . are about things unchangeable and eternal—the first causes and intelligible forms of Being—which, being unchangeable, *can* be contemplated only, not involved in action: theirs is *theoria* in the strict Aristotelian sense. The "practical sciences" . . . are "art," not "theory"—a knowledge concerning the planned changing of the changeable. Such knowledge springs from experience, not from theory or speculative reason.

With Bacon, however,

> Theory must be so revised that it yields "designations and directions for works," even has "the invention of arts" for its very end, and thus becomes itself an art of invention. Theory it is nonetheless, as it is discovery and rational account of first causes and universal laws (forms). It thus agrees with classical theory in that it has the nature of things and the totality of nature for its object; but it is such a science of causes and laws, or a science of such causes and laws, as then makes it possible

"to command nature in action." It makes this possible because from the outset it looks at nature *qua* acting, and achieves knowledge of nature's laws of action by itself engaging nature in action—that is, in experiment, and therefore on terms set by man himself. It yields directions for works because it first catches nature "at work."

A science of "nature at work" is a mechanics, or dynamics, of nature. For such a science Galileo and Descartes provided the speculative premises and the method of analysis and synthesis. Giving birth to a theory with inherently technological potential, they set on its actual course that fusion of theory and practice which Bacon was dreaming of.[19]

On this view technology might well be applied science, but in a quite different sense than Bunge thinks. Perhaps it is precisely the inherently technological character of science that makes technology in the modern sense possible.

CHAPTER NINE

• • • •

Types of Technology as Activity

Having examined the types of technology manifest in the modes of technology as object and as knowledge—the two most philosophically analyzed forms—it is appropriate to turn to a less commonly considered mode, technology as activity.

Technology as Activity

Technology includes more than material objects such as tools and machines and mental knowledge or cognition of the kind found in the engineering sciences. This is readily shown by the association of technology with such words as "industry" and "manufacture," "labor" and "work," "craftsmanship," "jobs," and "operations." Indeed, despite the quickness with which people think of physical objects or hardware when "technology" is mentioned, and the apparent etymological implications of the term itself, activity is arguably its primary manifestation. Technology as activity is that pivotal event in which knowledge and volition unite to bring artifacts into existence or to use them; it is likewise the occasion for artifacts themselves to influence the mind and will.

Technology as activity can thus be associated with diverse human behaviors, with distinctions among them often less clear than for either artifacts or cognitions. Technological activities inevitably and without easy demarcation also shade from the individual or personal into group or institutional forms, which call for a second if not wholly independent analysis. Philosophy of action does not solve the problems of political philosophy. Nevertheless, for present purposes analysis can reasonably be restricted, after the manner of previous discussions of technology as object and as knowledge.

Among the basic types of behavioral engagements of technology as activity one can readily include the following:

- crafting

- inventing

- designing

- manufacturing

- working

- operating

- maintaining

A cursory inspection of this overlapping diversity suggests that in active technological engagements with the world there are two broad themes: production and use. The former is an initiating "action" that establishes possibilities for the latter, recursive "process." Crafting, inventing, and designing are all actions in technology as activity; manufacturing, working, operating, and maintaining are processes in technology as activity.

The terms here are not wholly satisfactory or firmly fixed, but are merely loose linguistic connotations that hint at or prefigure certain distinctions subject to more detailed exploration and development.[1] One readily refers to the actions of crafting, inventing, and designing and the processes of manufacturing, maintaining, and operating. To talk about the "inventing process" strikes the ear as slightly off and evidently points to some special form of inventing; "manufacturing action" sounds even odder. The phrase "making process" best refers to a making action that uses complex technologies, while the "action of using" implies using to make.[2]

The Action of Making

The philosophy of action has largely ignored the unique character of making action and, as Andrew Harrison (1978) has argued, is desiccated as a result. In analyses of the rationality of human action, "the ideas of designing and constructing (except perhaps in somewhat specialized mathematical and related contexts and senses) figure rarely ... and the no less interesting notions of building, cobbling and bodging not at all" (p. 1). The philosophy of art considers aesthetic making, and creativity (a specific feature of one kind of making) has been subject to psychological, poetic, pedagogical, and related analyses. But the general approach has been to concentrate on newness, uniqueness, inventiveness, or the specific features of inventive cognition. Seldom has making even in this narrow artistic sense been considered simply qua activity. Since creative inventing is but one aspect of making in a broad sense, the present analysis focuses on and attempts to consider a broad spectrum of types of making as human actions.

Cultivating versus Constructing

Aristotle was the first to suggest a fundamental distinction between two types of making action, cultivating and constructing (see *Physics* 2.1.193a12–17; *Politics* 7.17.1337a2; and *Oeconomica* 1.1.1343a26–1343b2). Cultivation involves helping nature to produce more perfectly or abundantly things that she could produce of itself, and includes the *technai* or arts of medicine, teaching, and farming. Construction entails reforming or molding nature to produce things not found even in rare instances or under the best of circumstances, as with carpentry. As Andrew G. Van Melsen (1961) restates this distinction:

> In farming, although man performs all kinds of preparatory tasks, such as clearing, plowing, and sowing, nature itself has to do the rest. Once his preparatory task is done, man can only sit down and wait. It is the inner growing power of living nature which performs the work. (pp. 235–236)

By contrast,

> The craftsman gives natural materials forms which would not naturally arise in them. The technical object is something which is not cultivated but constructed, i.e., its component parts are arranged in an artificial pattern. The fashioning of these parts forces them into forms and functions which are not naturally present in them. (p. 236)

Thus, "in the work of construction there is a far more direct intervention in the natural order than there is in the work of cultivation" (p. 236).

Another version of this distinction might contrast technological actions that are in some way in harmony with nature with those that are not. The environmental and alternative technology movements can be interpreted as reviving this cultivation-construction distinction: intermediate or soft technologies assist or imitate nature by acting in harmony with it, whereas hard or high technologies depend on conditions and processes not found in nature. The former can also depend on "renewable resources," whereas the latter use up "nonrenewable resources." Note too how the difference at issue could be read as restating distinctions already present in terms of artifacts (tools versus machines), while pointing toward the possibility of another between types of technology as volition (the aim of harmony or peace versus control or domination).

Two other traditional distinctions to be noted but not confused with that between cultivating and constructing are those between the servile and liberal *technai* and between the useful and fine arts. The servile/liberal distinction depends on whether an action is primarily manual or mental; the distinction between useful arts and fine arts rests on an opposition between utilitarian and aesthetic functions. The utilitarian art of cooking and the fine art of painting (note the modern term "action painting") are both servile in the classical sense; those of theoretical philosophy and practical rhetoric are both liberal. Thus independent of the cultivating/constructing distinction there are also differentiations by means of media engaged or human functions served, yielding a complex, overlapping matrix of

- the servile arts of cultivating (agriculture but not pedagogy) and constructing (carpentry but not engineering)

- the liberal arts of cultivating (intellectual virtues) and constructing (propaganda)

- the useful arts of cultivating (agriculture and pedagogy) and constructing (carpentry and engineering)

- the fine arts of cultivating (flower growing) and constructing (industrial design)

Crafting

The second member of this matrix, servile constructing, together with Harrison's allusion to "cobbling and bodging" (or patching and jerry-rigging, to use equivalent American expressions), calls for further reflection on crafting and a kind of making action that is related to but distinct from craft—that is, bricolage. Etymologically "crafting" implies trickiness and cunning, but also a trickery or cunning of personal, even manual, action grounded in some direct and intuitive contact with the materials. Clearly it is a traditional form of the making action.

The term "bricolage"—close English words are "tinker" and "putter"—enters intellectual discourse through Claude Lévi-Strauss's attempt to define a practical correlate of mythopoeic thinking. Mythopoeic thinking is a kind of hodgepodge "science of the concrete"; in a similar manner bricolage is a heteronomous collection of specific skills.

> The "bricoleur" is adept at performing a large number of diverse tasks; but, unlike the engineer, he does not subordinate each of them to the availability of raw materials and tools conceived and procured for the purpose of the project. His universe of instruments is closed and the rules of his game are always to make do with "whatever is at hand," that is to say with a set of tools and materials which is always finite and is also heterogeneous because what it contains bears no relation to the current project, or indeed to any particular project, but is the contingent result of all the occasions there have been to renew or enrich the stock or to maintain it with the remains of previous constructions or destructions.[3]

Given some construction project, Lévi-Strauss goes on to say, the engineer "questions the universe" about how to achieve it. What resources are available? What principles can be utilized? The putterer, by contrast, simply picks around in a jumble of odds and ends left over from previous jobs. Here's something that might work. Let's try this. "The engineer is always trying to make his way out of and go beyond the constraints imposed by a particular state of civilization while the bricoleur by inclination or necessity always remains within them."[4] The engineer makes by means of concepts and analysis, the bricoleur by putziting and suggestive accidental conjunctions. Although not by necessary commitment, the bricoleur may be inherently more prone to wind up cultivating, the engineer constructing. One putters in the garden more readily than at building a house; the "putter-built" house will look a little like it grew up at the site from available materials, deficient in carpentered correctness.

In light of this contrast, craft or artisan making can be seen as intermediate between bricolage and engineering. Bricolage is perhaps only slightly more "efficient" than nature and "successfully" makes some form virtually by accident. The putterer patches the roof many times over before it stops leaking. An action painter such as Jackson Pollack (who, in rebellion against formalized or "engineered" art, adopts many of the practices of the bricoleur) has to throw out more paintings than he keeps. "Found art" is the product of much looking. Craft action begins to emancipate itself from such extreme subordination to its materials but examines or judges them from the point of view of certain sensuous forms rather than abstract concepts. Indeed, artisans seem to be attracted by sensuous complexity of form, as was visually exhibited in the 1964 "Architecture without Architects" show at the Museum of Modern Art.

> The untutored builders in space and time . . . demonstrate an admirable talent for fitting their buildings into the natural surroundings. Instead of trying to "conquer" nature, as we do, they welcome the vagaries of climate and the challenge of topography. Whereas we find flat, featureless country most to our liking (any flaws in the terrain are easily erased by the application of a bulldozer), [these vernacular architects] are attracted by rugged country. In fact, they do not hesitate to seek out the most complicated configurations in the landscape.[5]

Premodern making was and is apt to see all making as a kind of cultivation, whereas engineering action virtually abandons concern with specific sensuous form in favor of methods of construction that can meet the needs of clients or users and thus reconceive even traditional cultivation as a kind of construction (witness production agricultural and biomedical engineering, as well as educational technology).

Because of the way it operates within a framework provided by materials, it is appropriate that, before the rise of engineering and its abstract conception of making action, types of making should have been distinguished primarily according to material, cultural, and ritualized formations. Not only are making as bricolage and making as craft oriented toward cultivation of nature, but in themselves these activities become cultures. Ethos does not need to strive by means of ethics to impose itself on a technical action that only in rare instances transcends its specific roots. Indeed, perhaps it could be said that ancient making concentrated on cultivation in two dimensions, the natural and the human, whereas modern making becomes a construction of both the natural and the human.

Engineering Action

Engineering can be divided not only into various branches, according to what engineers actively engage *with*, but also into functions determined by *how* engineers act, that is, the roles they play in the production sequence as representatives of the various functional moments of the engineering method. In common parlance, engineering functions range from invention, research, and development through design, production, and construction to operation, sales, service, and management. Such distinctions are more significant than material divisions because they are repeated within any engineering field.

Invention is sometimes conceived as distinct from research and development and at other times as inclusive of them. Applied or mission-oriented (as opposed to pure) research uses scientific and mathematical knowledge plus appropriate experimentation to synthesize new materials or create new energy-generating or transforming processes. Development entails utilizing these materials, energies, and processes to design, fabricate, and assemble prototype products that solve particular problems or meet specific needs: "industrial research" is another name for this activity.[6]

Designing can be considered part development or an activity in its own right ordered toward construction and production. From some perspectives it is both. Designing is obviously necessary to the development of a new artifact or process, but once developed the designing for construction or production can also play a special role. In some instances the initial development can include specifications for production.

Production and construction are two kinds of making in a restricted sense. The former makes nonstationary artifacts (consumer goods), the latter stationary structures (houses, bridges, buildings). "Fabrication" and "assembly" can be synonyms for "making" in some circumstances.

Operation and management denote using processes, as do testing, service, maintenance, and sales—although testing can also be construed as a factor in development and design. The functions of planning, teaching, and consulting cut across these various distinctions.

The relationship among these functions can be schematized in a flow diagram (see fig. 3). As with the branches of engineering, however, there is no universally agreed-upon list of these functions; there is simply a spectrum of activities that can be divided and subdivided in numerous ways depending on the kind and degree of analytic detail desired. A flow diagram also oversimplifies. The point is that there are not just electrical engineers, but electrical research engineers (doing applied research on electrical energy or electrical energy-driven devices), electrical design engineers (designing either a specific electrical device for factory production or an electrical system for on-site construction), electrical service engineers (maintaining and servicing some electrical product or system), as well as other specialists.[7]

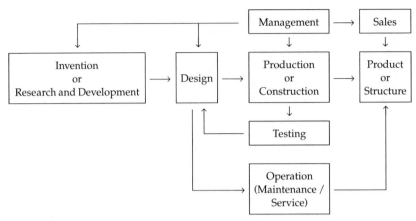

Figure 3. Relations between engineering functions.

Inventing

As an aspect of engineering action, inventing can be approached by means of selective contrasts: *as opposed to scientific discovery,* technological inventing refers to creating something new rather than finding something already there but hidden. Alexander Graham Bell invented the telephone; Isaac Newton discovered the law of gravity. The telephone did not exist before Bell's work; gravity existed but was not conceptualized in the form of scientific law before Newton. Adopting the realist epistemology characteristic of scientists and engineers, invention makes things come into existence from ideas, makes the world conform to thought; discovery, by deriving ideas from observation, causes thought to conform to the world.

One difficulty with this view is that scientific ideas (laws and theories) are underdetermined by observations and require some conceptual or imaginative creativity. *As opposed to conceiving or even imagining,* then, it is the concrete transformation of materials—making an imagined transformation physically real—that is the essence of inventing. George Cayley (1773–1857), founder of the science of aerodynamics, had an accurate conception of the airplane, but its invention (dating from the first successful flight at Kitty Hawk in 1903) had to await both the development of a suitable power plant and the Wright brothers' technical skills (of fabricating and operating).[8] Did Leonardo invent the parachute merely by imagining it, or Lenormand by fabricating and testing it? Inventing may begin with some conceptualization, but it does not finally take place until an artifact is operationally tested and proved able to perform its assigned task. It is this active, physical engagement that keeps invention from being just an element in technology as knowledge—although the kinds of ideas involved here certainly call for special epistemological analysis.

As conscious action originating in the mind and confirmed by worldly engagement, the concept of invention is *opposed to slow or incremental technical change* and is a distinctly modern notion.[9] Like scientific discovery, invention can take place over a short period in a single individual who introduces historical discontinuity or can occur through gradual development within a group. The slowed-down or spread-out invention through innumerable minor modifications that maintain historical continuity is sometimes termed "innovation." Other observers emphasize the historicosocial character of inventing,[10] crediting not individuals but technical communities, national groups, or historical periods with the invention of such artifacts as the astrolabe or compass. Indeed, from this perspective inventing originates not so much in the search for practical realization of ideas through material fabrication as in a haphazard alteration of the matter and form of artifacts over the course of time, with the eventual recognition of something useful. As such it is almost wholly devoid of the act of designing, a conspicuous and distinguishing feature of modern invention and innovation.

As opposed to designing, inventing appears as an action that proceeds by nonrational, unconscious, intuitive, or even accidental means. Designing implies intentionality, planning. On this account inventing is accidental designing—and as such highlights the element of insight and serendipity that plays a strong role in even highly systematized design work. Inventing also connotes a singularity of creation, whereas designing takes an invention and adapts it to circumstances of, say, mass production ("innovation" again). Although some inventors have been engineers, if existing materials and processes are adequate to the task an engineer is generally content to design around or with them, making only such refinements as circumstances immediately require. Inventing also connotes singularity of creation. Inventors are cowboys, engineers settlers.

From Inventing to Systematic Inventing

Modern engineering, as an attempt to settle and systematize the inventive process, has been called the "invention of invention": "The greatest invention of the nineteenth century was the invention of the method of invention. A new method entered into life. In order to understand our epoch, we can neglect all the details of change, such as railways, telegraphs, radios, spinning machines, synthetic dyes. We must concentrate on the method itself; that is the real novelty, which has broken up the foundations of the old civilization."[11]

A key figure in this development was Thomas Edison (1847–1931), who in the 1870s established what he called an "invention factory" that would make "inventions to order." For Edison, inventing was the product of organized purpose. Although by the early 1800s inventing was well recognized and culturally prized, it remained largely a matter of individual initiative and intuition, divorced from direct large-scale organization or financial backing. It was Edison who, especially with his massive, methodically directed trial-and-error search for a suitable filament for the incandescent light in conjunction with the systematic development of related elements necessary to its commercial exploitation (vacuum bulbs, parallel circuits, dynamos, voltage regulators, metering devices, etc.), first created the industrial research organization tied to capitalist economic structures.[12]

Inventing and engineered inventing can be contrasted by saying that an inventor creates the new whereas the engineer plans out the possible. An engineer remains within the familiar—does not venture into the unknown, only orders or reorders the known—so that, given a clearly specified problem, two equally competent engineers will come up with or "discover" solutions that differ only in the materials used.

Friedrich Dessauer, however, has argued that inventing or creating also involves the experience of discovery in a much stronger sense.[13] Indeed the word "to invent," from the Latin *invenire*, means "to come upon," "to find," or "to discover." Moreover, inventing is capable of exhibiting parallel histories and objective confirmation—as when two persons independently invent the same thing (as with Elisha Gray and Alexander Bell, who both invented the telephone, who even applied for patents on the same day). Gilbert Simondon (1958) provides a detailed mechanology or descriptive phenomenology of machines, documenting the tendency of inventing to generate certain stable forms, especially insofar as these forms are not obscured by the play of fashion and commercialization. This element of discovery and objectivity is so present in invention that Dessauer postulates a transcendent realm of preestablished solutions to technical problems to account for it. The natural or external world explains or accounts for the objectivity of science; but since inventing does not correspond to what already physically exists, there must be a transcendent existence to account for its discoveries.

Others, however, while admitting the importance of the moment of discovery in invention, offer less metaphysical explanations. David Pye, for instance, argues simply:

> Invention is the process of discovering a principle. Design is the process of applying that principle. The inventor discovers a class of system—a generalization—and the designer prescribes a particular embodiment of it to suit the particular result, objects, and source of energy he is concerned with.
>
> The facts which inventors discover are facts about the nature of the world just as much as the fact that gold amalgamates with mercury. Every useful invention is a discovery about the way things and energy can behave. The inventor does not make them behave as they do.[14]

The prior contrasts with inventing throw further light on the movement from simple or accidental to systematic invention. Whereas primitive inventing relies on accident, bricolage, fortuitous insight into possible relationships among elements in the given, invention research develops a calculus of such relationships that can be used to solve well-specified problems. That such a calculus may still rely at crucial moments on a cultivated serendipity (brainstorming sessions, etc.) and heuristics only reveals that irreducible essence of invention as creative insight that must so far remain as a circumscribed aspect of systematic invention.

For summary purposes the making action in the initial instance (it will be different with routine making) can be broken down into the sequence shown in figure 4. This is a logical, not historical, sequence. In fact, its various logical moments are existentially interrelated in considerably more complex fashion than can be schematically shown.

Inventing is, as it were, a bipolar concept, referring both to conceiving and to the discovering manifested in testing—hence its ambiguity. Imagination or concrete thinking is the moment linking these two aspects of inventing. This imaginative thinking or miniature constructing invariably runs up against certain barriers or problems that require new conceptions, a return to the conceptual moment, or questions that can be answered only by going larger scale—that is, by testing.

Thus to invent (in the full sense) invention is to conceive and to put into operation the inventive activity—or, to say the same thing in a different way, to consider the various conditions under which inventing readily takes place, to design (imaginatively plan) an institution that enhances those conditions, and to establish just such a working institution. Industrial research and development laboratories, applied (as op-

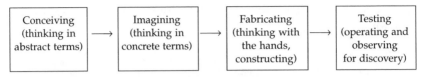

Figure 4. The moments of making action.

posed to basic) research institutions, are the result. But this introduces a question concerning the character of such designing or imaginative planning when operative in engineering.

Engineering Designing

The second moment in the sequence above can be identified as the location (if not yet the essence) of engineering design. As already indicated, virtually all general articles on engineering and all introductory engineering textbooks identify designing as the essence of engineering. The design project is typically the capstone of an undergraduate engineering education. The most well developed field of research on engineering qua engineering—as opposed to research within some branch of engineering—deals with engineering design and especially design methodology.[15] Admittedly, and as might be expected, there is considerable debate about the structure and the very possibility of an engineering design method. Nevertheless, when engineering method is contrasted with the scientific method, it is the method of design that is invoked.

But what, exactly, is engineering design? Designing (from the Latin *designare*, "to mark out") specifies some material object in sufficient detail to enable it to be fabricated. It is, as it were, reified intention. Indeed, the very word can be synonymous with intention, as in "His design was to make more money." The problem is that the standard engineering definitions of designing do little more than rephrase the standard definitions of engineering itself. Examples: "Engineering design is the process of applying the various techniques and scientific principles for the purposes of defining a device, a process or a system in sufficient detail to permit its physical realization."[16] Or engineering design is "an iterative decision-making activity to produce the plans by which resources are converted, preferably optimally, into systems or devices to meet human needs."[17] Or yet again, engineering design "is the intellectual attempt to meet certain demands in the best way possible."[18] Prescinding from implicit differences in such definitions, designing may thus be described as the attempt to solve in thought, using available knowledge, problems of fabrication that will save work (as materials or energy) in either the artifact to be produced, the process of production, or both.

Consider, for example, a foundation for some structure. Were a stonemason to construct this foundation working from experience and intuition alone, one of two things would be likely to happen: either it would be made too weak, so that the building eventually collapsed and had to be rebuilt; or, what is more likely, it would be made too strong, using more stone, concrete, and steel than necessary. In either case, more work than needed would have been done. When engineers design the same foundation, they attempt to calculate the weight of the building and other relevant forces. Then, using the principles of physics plus engineering geology (i.e., geological knowledge interpreted in terms of what kinds of structures various earth formations can support) and a socially specified safety factor, they describe a foundation to be constructed with neither more nor less than what is required. Although paradoxical, the right construction (like Aristotle's golden mean) is difficult to attain; it takes effort. But when this effort is expended at the right time, it saves effort in the long run. Engineering design is thus an effort (at first sight, of a mental sort) to save effort (of a physical sort).[19]

This mental effort is, however, something distinct from knowing or coming to know in a scientific or theoretical (or even technological) sense, because it does not terminate in an interior cognitive act. Its termination is construction—but construction of a special sort, construction in miniature or miniature making. Designing ends with "Aha! Let's make it this way. Let's go with this design." Scientists are more likely to reach the conclusion of their work with "Aha! Now I understand. I get it." Scientists often experience a tension between their knowledge and what they can express; they make discoveries and then have to push beyond what they feel is their proper sphere in order to write them up. But such tension is not a normal feature of design experience, because the construction of drawings or models (which also serve communication) is intimately bound up with the design process. The communication difficulty for design engineers is convincing others to invest money in their proposals.

Joseph Edward Shigley, for instance, in his *Theory of Machines*, asserts that "the use of the drawing board in kinematics instruction is very desirable and usually necessary" because "the most direct method of attacking a kinematic or dynamics problem is the graphical one."[20] If a design project is the capstone of an undergraduate engineering education, it ultimately rests on the foundation of engineering graphics, usually required of all first-year students. The importance of the physical act of drawing is also indicated by another author's contrast between sensorimotor skill, phantasmal capacity, and conceptual capacity. "The *phantasma,* or sensory representation at whatever level of complexity . . . is what we are concerned with."[21]

Engineering drawings, with the unique language and system for abstraction and representation, are not just means for communicating results arrived at by interior activity; they are part of the process and the means by which the results themselves are reached. At the same time, drawing may be only one way of performing the more general engineering action of modeling.

> One source of confusion in thinking about design is the tendency to identify design with one of its languages, drawing. . . . Design, like musical composition, is done essentially in the mind, and the making of drawings or writing of notes is a recording process. The designer, however, uses drawing for self-communication just as everyone uses words for thinking. This use of drawing as an extension of the mind, a sort of external (and reliable) memory can be a very important part of the design process. Drawing should be taught not primarily to give the student facility in the use of tools—pencil, triangle, tee square, and most important, the eraser—but to give . . . practice in pictorial extension of the mind. It is not to be expected that all students are equally endowed with the ability to think pictorially any more than to think mathematically. Somehow educators tend to look upon mathematical ability as a more desirable quality than the ability to think in terms of spacial relations. Before dismissing the latter as something of lesser merit, it may be well to reflect that one of the greatest engineers of all time, Leonardo da Vinci, was essentially a draftsman, not a mathematician.
>
> Pictorial language is especially well adapted to expressing particular physical form and physical space relationships. Functional relationships are often better expressed by a symbolic language. Such languages have particular facility in expressing generalizations without specifying detail. The chemical engineer uses the flow sheet, the electrical engineer the circuit diagram, and all kinds of engineers use the block diagram as important tools in the conceptual process.
>
> The designer often uses the symbolic languages of mathematics but usually in connection with the analysis of a design rather than directly in the conceptual process.[22]

Although it starts out by stating a position ostensibly at odds with the idea of design as essentially involved with drawing or miniature construction, by arguing that drawing is merely one form of modeling or picturing, this passage actually supports and generalizes the present argument.[23]

The point at issue is confirmed by yet another analysis of "Technology as Knowledge" (Layton 1974). After noting that for artists thinking means something different than for philosophers, and that "technologists display a plastic, geometrical, and to some extent nonverbal mode of thought that was more in common with that of artists than that of philosophers" (p. 36), Edwin T. Layton Jr. describes designing in the following terms:

> The first stages of design involve a conception in a person's mind which, by degrees, is translated into a detailed plan or design. But it is only in the last stages, in drafting the blueprints, that design can be reduced to technique. And it is still later that design is manifested in tools and things made. Design involves a structure or pattern, a particular combination of details or component parts, and it is precisely the gestalt or pattern that is of the essence for the designer.
>
> We may view technology as a spectrum, with ideas at one end and techniques and things at the other, with design as a middle term. Technological ideas must be translated into designs and tools to produce things. (pp. 37–38)

Layton's mistake here is to call this designing activity primarily a kind of knowledge and to fail to notice that modeling in one form or another goes on not just at the stage of making blueprints, but at virtually all stages of engineering action. Indeed, this activity can be described as the creation of a series of designs, first quite general (freehand sketches, perhaps simply a block diagram analysis of the problem), but progressively detailed and specific (working drawings), until it terminates in the actual construction (a process overseen by engineers in which the carpenters and other "technicians" act to some extent like "living pens and pencils" scaling up a drawing or design one last time).

With regard to the last suggestion, although the actual execution of a plan does not involve designing, except insofar as the plan may continue to be modified to meet originally unanticipated situations, such continuing design through final execution is in fact the norm. On large construction projects a draftsman will be continuously at work revising drawings in the light of exigencies and changed circumstances, thus seeking to anticipate their further consequences. Drawing is a kind of testing or interrelating of various factors[24] by miniature building. It is not thinking in the sense of conceptualizing or relating concepts; it is thinking as picturing or imagining, and relating specific materials and energies. The designer solves problems of relating parts the way an artist does, by seeing them in practice. It is activity, only on a reduced scale, made as free of physical labor as possible, but nevertheless not entirely free. It is still an effort (of a miniature physical sort) to save effort (of a gross physical sort).[25] (This particular miniaturization of construction is, however, intimately related to special kinds of knowledge, especially the engineering sciences.)

One version of this argument about the primacy of visual or graphic action in engineering has been extended to book length, with copious illustration, by Eugene S. Ferguson's *Engineering and the Mind's Eye* (1992). Ferguson, like Layton, continues to call this a kind of "thinking," and to stress its importance as a means of communication. But his simultaneous emphasis on the dependence of effective engineering design on skillful engagement with the technical world and his inventory of the "tools of visualization" belies his own terminology. In the words of Shigley, the aim of design is *"the creation of an end result by taking definite action, or the creation of something having physical reality."*[26] Although Shigley describes this as the purpose of the design, it is something also present right in designing itself in the form of drawing and modeling. Although both Shigley and Ferguson strongly criticize the scientization of engineering, both nevertheless seems so influenced by science that they fail sufficiently to see what they themselves describe, that the drawing and modeling central to engineering designing are inherently miniature makings. Engineering simply substitutes for craft trial-and-error making at the level of the finished artifact, trial-and-error making at the level of drawing—often assisted by mathematical calculation and engineering science.

Ferguson's deft sketch of the historical development of engineering drawing only serves once more to emphasize this point. Engineering drawing emerges not from theory but from practice. From Leon Battista Alberti's 15th century invention of the transparent drawing grid through Albrecht Dürer's 16th century first orthographic projections to Gaspard Monge's 18th century formulation of descriptive geometry engineering drawing is linked to activities of the hand, although it is manual action increasingly divorced from sensory engagement with the physical world and constrained by clearly formulated rules and technical devices.[27]

Efficiency, Effectiveness, Economy

Engineering design is a systematic effort to save effort. But what, more precisely, is this saving of effort? Can it perhaps be further explicated as efficiency? As a guiding principle of engineering design, engineers themselves repeatedly refer to the ideal of efficiency. Jean-François Lyotard even refers to all technology as "a game pertaining not to the true, the just, or the beautiful, etc., but to efficiency."[28] But what is efficiency? As Stanley Carpenter (1983) and others have well pointed out, efficiency is a context-dependent notion.[29] Context dependency, however, need not deprive a concept of all formal characteristics. Legal justice, for instance, is equally context dependent but nevertheless denotes a special way of looking at behavior—that is, in terms of its conformity with a set of rules articulated and enforced by the state. Engineering likewise can be said to have its own special approach to the activity of designing—in terms of efficiency.

The term "efficiency" has its roots in the Latin *efficere* (to produce, effect, or make). The derived adjective, *efficiens*, modifying *causa*, indicates one of the Aristotelian four "causes of motion" (*Physics* 2.3.194b16–195b30). In contrast to formal, material, or final causation, the efficient cause is the "principle of change" that unites the other three. In English, "efficiency" traditionally meant the operative agency or power of something or someone to get something done, to produce results. In Christian theology, God as creator ex nihilo is described as the supremely efficient or "most effective" cause. In military parlance of the late 1800s a soldier who could do his job was "efficient" or, it would be said today, "effective." In all such uses there is no sense of efficiency in the technical or engineering sense as a comparison of outputs with inputs.

The first use of the word "efficiency" in what has come to be called the "technical sense" occurs, appropriately enough, in mechanical engineering during the second half of the nineteenth century.[30] In the technical sense, efficiency is defined as a ratio of outputs to inputs. It is difficult for us to recognize how unusual this perspective is.

The idea of looking at or judging any object or process in terms of a relationship between outputs and inputs is not, for instance, operative in our primordial engagements with such quotidian practices as speaking, dressing, or eating, nor in any premodern system of morality. Common speech and traditional rhetoric are prolix, wasteful, inefficient in their effusiveness; people customarily go on and on saying the same thing over and over again in different ways. Only modern rhetoric is pared down to essentials, aspiring to say no more than is necessary. We normally eat what tastes good or what is appropriate for the occasion, not what gives us the right amount of nourishment for the least expenditure of money, time, or effort. For Plato one should strive to imitate an ideal; for Aristotle the goal is a more immanent virtue or perfection of operation. Natural law theory likewise stresses acting in harmony with some larger order.

Thinking in terms of an input-output relationship is first explicitly presented in Pseudo-Aristotle's "Mechanical Problems"[31] and subsequent Archimedean analyses of simple machines (i.e., tools). Indeed, the very concept of an *instrumentum* as an object to be judged in terms of its "use" independent of user and context—a notion related perhaps even to the idea of the sacraments as instruments of salvation—implicitly entails input-output considerations. But the idea begins to take on clearly definable form only with early modern mechanics,[32] the advent of double-entry bookkeeping, and the formulation of theories of political economy. Certainly the notion plays no role in Aristotle's *Oeconomica* or other premodern economic texts. Utilitarianism as moral theory can be read as related. The amplification and application of efficiency as a technical concept in economics was also undertaken by an engineer turned social scientist (Vilfredo Pareto), and its popularization as a social ideal was promoted by engineers turned social-political activists.[33]

The philosopher who has made the most effort to elucidate the character of efficiency as a technical ideal in engineering is Henryk Skolimowski.[34] According to Skolimowski, progress in science is demonstrated by better theories (increases in knowledge), progress in technology by better artifacts or the processes for making artifacts (increases in effectiveness or efficiency). Skolimowski then tries to show the specific forms that effectiveness takes in different branches of engineering. In surveying, for example, it is accuracy of measurement, in civil engineering durability of structures.

But Skolimowski's analysis is weak in two respects. It does not properly distinguish effectiveness and efficiency. The two terms are not interchangeable, as they are often treated, nor is efficiency "a measure of effectiveness." A less efficient but more powerful bomb could easily be more effective than a more efficient but less powerful one. Skolimowski further fails to note the more subtle and sophisticated forms of efficiency that can be embodied in engineering design methods. For this one can turn to the work of Herbert Simon (1969), who describes the science of design as including the methods of optimization and "satisficity" (also called "bounded rationality"). A logic of optimization, for instance, distinguishes between what are called command variables (means or inputs), fixed parameters (laws or rules), and constraints (ends or outputs). "The optimization problem is to find an admissible set of values of the command variables [e.g., kinds and quantities of food], compatible with the constraints [nutritional requirements], that maximize the utility function" for some given environment or situation (1981, p. 135).

This need not always produce an "absolute efficiency" or some one best way.[35] Given the complexities of most real-world problems, it is sufficient if there exists a design method that provides a way for choosing between x number of alternatives or for doing what Simon calls "satisficing," that is, achieving a satisfactory if not perfect solution to a problem—one that is more efficient than others within the bounds of those that can be compared. This process can entail cost-benefit analyses even of the process of design itself.[36]

As Simon says elsewhere, "Within the behavioral model of bounded rationality, one doesn't have to make choices that are infinitely deep in time, that encompass the whole range of human values, and in which each problem is interconnected with all the other problems in the world." Instead, rationality can "focus on dealing with one or a few problems at a time" (1981, p. 20).

With regard to engineering, "the skillful designer . . . develops a design which is close to being an optimum design—that is, it results in the best product from some stated point of view."[37] To make the same point in different words, "In optimum-design procedures, the significant desirable effect is explicitly maximized, or the significant undesirable effect minimized." And within the "regional constraints" of any "stated point of view" or explicitly defined effect there now exist a number of formal decision-procedure techniques such as maximum/minimum differentiation, dual-variable analysis, and numerical search.[38] All design textbooks include chapters on such optimization techniques. Although to say that engineers search for "the one best way" may be a rhetorical overstatement, it remains the case that when they examine "a variety of possible [design] solutions" in order to identify "one that [is] good or satisfactory"[39] what they are doing is attempting to determine the best way within certain clearly defined constraints.

The creation of this "science of the artificial" has transformed engineering design from an "intellectually soft, intuitive, informal, and cookbooky" (Simon 1981, p. 130) activity to one characterized by "optimizing algorithms, search procedures, and special-purpose programs for designing motors, balancing assembly lines, selecting investment portfolios, locating warehouses, designing highways, and so forth" (p. 80). The counterargument of Donald Schön that engineer designing retains the character of "a reflective conversation with the materials" of a situation (1983, p. 175) overlooks the distinctive character of what this conversation is about and its special structure. In like manner, Billy Vaughn Koen (1985) stresses how the role of heuristics in engineering design fails to acknowledge the systematic character of engineering heuristics. Koen is no doubt correct that there is not always some universal solution to a problem, but more because that problem does not have a truly unique form than because the engineering method has room for multiple solutions. Were Koen's argument pressed, it would return engineering to craft if not bricolage.

Although it is true that what can be counted as inputs and outputs may be virtually unlimited and is often socially determined, within some specified input-output parameters, engineering design searches for ways to minimize the input-output difference[40]—by means of miniature construction. In German the very word for engineering design, *Konstruktiontätigkeit* (literally, "construction activity"), confirms this point. This miniature construction most often proceeds by means of visual representation, but also by modeling and by mathematical analysis of the resulting drawing or model. It is the visual or schematic representation[41] of this input-output conceptualization that definitively characterizes engineering design and allows it to proceed under the ideal of efficiency—that is, making choices between alternatives based on comparisons of input-output relationships—thus distinguishing it from other types of designing.

Designing in Engineering and in Art

In contemporary theory of engineering design, the structure of this action is sometimes thought to be a method, differing from, yet analogous to, the scientific method of knowing. Design is put forth as a method of practical action. As such it has been argued to underlie all practical activity, not only in engineering but in business, education, law, politics, art, and the like—if not all human action. The method is one; the only differences are in goals pursued and, perhaps, materials employed. One designs business ventures to make money, educational curricula to impart information and knowledge, laws to be enforceable and alter public behavior, political campaigns to win votes. But it is primarily within the engineering field that the methodology of design has been most seriously investigated.[42]

Without detailing the exact character of this alleged method or its relation to the method of science, one can nevertheless distinguish engineering design from artistic design according to ideals or ends in view. The engineering design ideal of efficiency stands in contrast to the artistic design ideal of beauty. Beauty is not so much a question of materials and energy as of form.[43] About this the whole subject of aesthetics has more to say, whereas it is ethics or politics that would incorporate a philosophical evaluation of efficiency.

Yet the difference between these two types of design does not remain at the level of ideals; it penetrates to the design activity itself. Efficiency refers to a process—is a criterion for choosing between processes or products conceived as functioning units—whereas beauty is in the primary instance a property of stable objects. Does a potter aim at efficiency in creating a beautiful pot? No, the aim is a good work, one of proportion and harmony; efficiency in production, while not to be wholly ignored, is a distinctly secondary consideration. For the engineer, however, beauty is of secondary importance. While not to be ignored, in industrial design beauty is judged in terms of its contribution to function or efficiency, perhaps even to marketability or sales.[44] The ends of artistic design must be formal, whereas the final causes of engineering actions are justified in terms of human needs, wants, or desires.

A further observation: Engineering design limits itself to material reality (metaphysically, matter and energy are both matter as contrasted with form). This limitation is to be grasped or approached, however, by means of a mathematical calculus of forces closely associated with classical physics (Galileo and Newton) and its specific mathematical abstraction. The picturing or imagining that goes on in engineering design is done, as it were, through a grid derived from this physics—the grid itself being articulated, in the first instance, as the engineering science of mechanics. This viewing of matter and energy through the grid of classical physics gives engineering design a rational character not found in art. Engineering images, unlike other images, are subject to mathematical analysis and judgment; this is their unique character and one that sometimes leads people to confuse them with thinking in a deeper sense.

Art also is concerned with imagining, but its images cannot be quantitatively analyzed—they are not subject to any well-developed calculus. Thus art, in contrast to engineering, appears as both more intuitive and more dependent on the senses. Although artists too are concerned to design artifacts, they necessarily do so in drawings and models that remain much closer in their reality to the final product.[45] Compare, for instance, a Rembrandt sketch for a painting with an engineering drawing of a building. Even the Rembrandt sketch is art; the engineering drawing is simply thrown away. (But note also that architectural drawings do exhibit a kind of aesthetic character similar to that of the building drawn.)[46]

The Process of Using

Despite discussions of practice in Marx, Dewey, Polanyi, and others, the philosophical analysis of using is slighted. Although the philosophy of action and ethics have something to say about using, the concept of use is conspicuous by its absence as a theme in all major texts. By and large what is said does not contribute directly to the clarification of using as a type of technology as activity.[47] It is thus necessary to begin with quite preliminary observations.

Types of Use

The English word "use" has no recognized roots deeper than the Latin *usus* (meaning "use," "exercise," "practice"), the past participle of *uti* ("to use," "to have a relationship with"). The adjective form is *utilis*, that is, "useful," "serviceable," "beneficial." Its etymological shallowness and limitation to the Roman family of languages may even suggest certain technological overtones.

In English the verb "to use" commonly denotes "to bring or to put into service" and "to employ for some purpose"—hence the "useful" arts and crafts, in the sense of making things that are to be employed. As an auxiliary verb it denotes habitual practice or familiarity ("He used to work" or "He is used to going to work"). As a noun "the use" of a thing indicates the regular way it is put into service or employed. The adjective "used" implies both previously put into service and exploitation for personal gain that might ignore proper use. Related words include "usual" (meaning "common" or "regular"), "to usurp" ("to take possession of," often illegitimately), and "to abuse" ("to misuse" or "use up"). Although it is possible to speak of "using" a person as well as an object, one means that the person either has been improperly treated or is simply a representative of some role. The word is more oriented toward engagements with things, even artifacts. One does not speak of "using" trees or rocks quite as comfortably as of "using" tools or instruments. Furthermore, because of its connotations of regularity or commonness "use" seems associated more appropriately with repetitive, not to say mechanical, processes than with creative or original ones, that is, putting into practice as opposed to bringing into existence.

Using is more inclusive than making. Virtually all making involves some using of artifacts (tools and machines), but not all using results in new artifacts. One type of using that does not directly produce artifacts, for instance, is living within. It may well be that there is a different type of use for each type of technological object—containers by being filled, structures by being lived in, tools by being handled, machines by being operated or "driven," art objects by being viewed, systems by being managed.

From a user's point of view, these various usings are subordinate to certain ends. These ends can include producing or maintaining as well as certain doings that are ends in themselves—hence the relevance of ethics to the analysis of using. At the same time, one of these ends in itself could on occasion include the pleasure of the activity of making as such. From the point of view of the object being used, using processes can be distinguished into those that make it, maintain it, or wear it out.

From the perspective of the object, again, the "use" of a technology can have at least three different but overlapping meanings. Consider the example of a gun. First, the "use of a gun" can refer to its technical function, which can itself be described in less or more abstract terms: "Guns are used to kill" or, more abstractly, "The use of the gun is to propel small objects at speeds greater than that of sound and with high accuracy to distances greater than arm's length." Second, the "use of the gun" can refer to the purpose or end to which the technical function is put. The use of the gun is to kill animals or enemies or people one robs or dislikes. It can be used to attack or defend any number of interests. Third, the "use of the gun" can indicate the act of using the gun to perform its technical function, that is, pulling the trigger, or to realize some purpose—for example, to ward off a robbery or to rob a bank.[48]

(As an aside, note that to speak of a technology as neutral, that is, able to be used for good or bad purposes, employs the term "use" solely in the third sense, prescinding from any acknowledgment of the first sense and the ways that technical use can and must in itself have consequences that are the same no matter what its purposes. The polluting car pollutes whether it is used to take sick people to the hospital or to rob a bank. In the same way, to stress the nonneutrality of technology focuses strongly on the first sense of "use.")

In the analysis of using as a technological activity, however, one must be careful not to let the term become so expansive as to turn all human behavior into technology. Although one *can* speak of walking on a sidewalk as using the sidewalk, of living in a house as using the house, of looking at a painting as using the painting, of reading a book of poetry as using the book, of playing the violin as using the violin, and of driving a car as using the car, in each case the connotations are quite different. Those human activities that have a self-contained quality about them, such as looking at a painting, reading a book, or playing the violin, seem most incorrectly described simply as use; indeed, to do so is common only when the user has missed the point of the objects concerned, that is, has failed to engage them in the proper manner. If a person is described as "using a book" one would be likely to think that he was doing something other than reading it—sitting on it, maybe. It is noteworthy that many usings, perhaps the less technological ones, have their own proper names, as with looking at works of art, reading books, or playing musical instruments.

The case of using the sidewalk raises different issues. Consider walking barefoot in the forest, walking barefoot on a forest path, walking barefoot on a sidewalk, walking shod on the sidewalk, walking barefoot or shod back and forth (partly on and partly off the sidewalk) while constructing some artifical walkway. It would be strange if using artifacts were to be interpreted so that walking became a technological activity simply by stepping from the forest onto the forest path or the concrete sidewalk. At the same time, walking on a wood floor and walking a hard-surfaced walkway do require subtly different motions and have a slightly different impact on the feet and legs. Perhaps one should distinguish between weak and strong senses of using, the former denoting more passive interactions between humans and artifacts, the latter active manipulations.

From the point of view of the object being used, again, the weaker sense of passive using seldom if ever does more than use up or wear out the engaged artifice. (It need not do even this much; looking at a painting qua looking in no way tends to wear it out, although its being made available to be looked at may do so.) But active manipulations, using processes in the strong sense, do tend, at a minimum, to wear out the objects they are most immediately engaged with, and indeed can often be distinguished as those that maintain or service another artifact and others that produce artifacts. All such meanings are implicit in the general description of technology as the making and using of artifacts.

Maintaining

A pivotal form of using, one that cuts across kinds of objects involved, is maintaining. Indeed, maintaining is in some sense intermediary between making and using and thus deserves special consideration.

Maintaining fails to put an object to use in the straightforward sense of engaging its technical or end uses, and might even be said to extend or prolong making, since it protects or retains the object made. To maintain a library or its books does not require reading the books, and to maintain a car does not require driving the car. Instead, maintaining preserves library making by keeping dust off the books and the humidity at the right level to keep the pages from becoming brittle; it prolongs automobile manufacture by changing the oil and keeping the water in the battery at the right level. In many instances maintaining also requires replacing parts—tires on cars, windows in buildings. In this it even becomes a kind of remaking.

At the same time this remaking is subordinate not to making but to using, and is a kind of using in that its aim is to keep something usable. Dusting the books and changing the oil in the car are repetitive processes like making the bed or washing the dishes. Indeed, one is almost tempted to say that beds are not just made to be slept in but also made "to be made," that the second making is itself using, and that using dishes entails not only eating off them but also washing and drying them. In this sense maintaining can be seen as a natural prolongation of using, an adjunct or secondary using, certainly a preparation for reusing.

Although maintaining cuts across divisions between objects and types of using, there nevertheless are also distinctions to be observed about types of maintaining. Maintaining tools and quotidian, traditionally crafted artifacts is different in character from maintaining machines, industrial complexes, and computer networks. The former tend to be maintained by the users themselves; those who sleep in beds or eat off dishes are also those who make the beds and wash the dishes. Using skills or techniques can be readily prolonged into maintaining techniques, and can exhibit the hands-on character of crafts. The need for and quality of maintenance is self-governing.

Maintaining machines, industrial complexes, and computer networks, by contrast, is done by a special group of persons who are users only by accident or supplementally. Those who maintain computer networks, for example, do not as maintenance personnel or systems operators use a network for the purpose it was established for. The maintaining of such high-tech artifacts becomes divorced from direct using; it is pursued by a special group or class of users who become part of a bureaucratic system established for monitoring and governing the maintaining function. Maintenance schedules have to be set up, checklists of maintenance tests and procedures developed, maintenance technicians distinguished from supervisory personnel, and so on. High-tech maintenance even points toward a special kind of high-tech using in which the same or similarly complex using bureaucracies are engaged.

In the world of high-tech artifice there is also a shift in the balance between maintaining and other forms of making and using. Consider, for illustration, the constructing and maintaining of a series of buildings. Imagine that it takes one person eight hours a day for one hundred days to construct a building of a certain type, which once constructed requires two hours a day of upkeep and maintenance. By means of the following schedule one can readily see that the life of this person will rapidly be transformed from one of constructing to one of maintaining:

Building 1   To construct: 8 hrs/day for one person for 100 days = 800 hrs
             Then 2 hrs/day upkeep maintenance
Building 2   To construct: 800 hrs, but now the builder only has 6 hrs/day to spend, so construction takes 133.3 days
             Then 2 hrs/day upkeep maintenance
Building 3   To construct: 800 hrs, but now the builder only has 4 hrs/day to spend on construction (since 4 hrs/day are spent on maintaining the two previous buildings), so construction takes 200 days
             Then 2 hrs/day upkeep maintenance
Building 4   To construct: 800 hrs, or 400 days, followed by full 8 hrs/day on maintenance. No more construction is possible.

Maintenance inevitably tends to overwhelm construction. Solutions to this problem are fourfold: More people can be employed, construction can become more efficient, maintenance can become more efficient (or less necessary), or buildings can simply be allowed to deteriorate. The high-tech society employs all four tactics.

Techniques of Using

Usings are to be distinguished not only by the kinds of artifacts they are engaged with and by their impact on some related objects, but also by their own internal character as skills or techniques.[49] Usings can be more or less skillful, more or less technically proficient, even more or less technical. The word "technique" here also raises another conceptual issue—the need to articulate the intuitive basis of a contrast between technique and technology.

One limited distinction is that of the nineteenth century, in which "technology" means a systematic knowledge of the industrial arts, with "technique" being the means of practical application. Although this distinction continues to influence French (*technologie* vs. *technique*) and German (*Technologie* vs. *Technik*) usage, it has broken down in English, for good reasons. It appreciates neither the inherently practical character of "technology" (as knowledge) nor the generality of "technique" (as skill, which can be in playing the piano or even reading a book).

Another proposed distinction argues that technological practice involves only interactions with artifacts, whereas technique can involve interaction with artifacts, natural objects, or human beings. There are techniques of swimming, wrestling, politics, computer programming, and automobile construction and maintenance; but there are only technologies of computer programming and automobile construction and maintenance.[50] In other words, there are techniques of both making and doing, but there are only technologies of making and using (when use involves artifacts), and there could be no technology of making in the most primitive sense, making with the hands.

This way of distinguishing technology and technique has immediate commonsense appeal, yet by stressing a material differentia it glosses over a number of difficulties. First, it fails to explain how we sometimes wish to speak not about the technology of, say, computer programming but of the technique. It is not just the presence or absence of artifacts in a human activity that determines whether it becomes a technology; the question is, instead, one of prominence or relationship to the human. In drawing, writing, or piano playing, the development and training of the human body is much more central than any particular artifact, even though artifacts are undoubtedly employed. In assembly-line production, by contrast, the tools or machines are themselves more central. Thus tools or hand instruments tend to engender techniques, whereas machines engender technologies—although even with machines, when one wishes to focus on the human manipulative processes, one speaks of techniques. (In some cases, of course, the prominence of the tools is unclear and, as with glassblowing, processes can be spoken of as both techniques and technologies.)

Second, "technique," and especially its adjective "technical," connotes singular making, whereas technology connotes multiplicity of production. An object may, for instance, be said to be technically feasible but not technologically feasible—meaning it can be made but not mass-produced. Finally, there is a sense in which technology, as opposed to technique, involves the greater use of rules, consciously articulated procedures, and guidelines. As just suggested, technique is more involved with the training of the human body and mind (which is why one can speak of the "techniques of logic" but not so easily of the "technology of logic"), whereas technology is concerned with exterior things and their rational manipulation. Techniques involve a large unrational or at least unrationalized (better still, unconscious) component. Techniques rely more on intuition than on discursive thought. At the same time technologies are more tightly associated with the conscious articulation of rules and principles (which is why, in another sense, it is possible to speak of logic as a technology). Sometimes these rules are forced to remain at the level of heuristic principles. But at the core of technology can be found a desire to transform the heuristics of technique into algorithms of practice. When this is achieved, however, techniques become bound up with technology as knowledge as much as with processes.[51]

Economics of Using

Each of these types of using can be subject to economic analysis and constraint. From an economic perspective the following technological usings or processes are commonly distinguished: laborsaving (capital-intensive) processes, capital saving (labor-intensive) processes, neutral processes;[52] potential versus realized technology; invention versus innovation; and material versus social technology.[53]

With regard to the invention/innovation distinction: The classic analysis is that developed by Joseph Schumpeter, who distinguished sharply between inventions (both patented and nonpatented) and innovation or their incorporation into commercial activities.[54] At the same time, innovation can denote "any thought, behavior, or thing that is new because it is qualitatively different from existing forms" or the activity of engendering such thought, behavior, or thing.[55] This readily includes invention or the creation of new material objects. But this discussion is limited to social science usage; for the engineer, innovation is more likely to denote small-scale or minor invention.[56]

As terms of contrast, invention = creation of a new artifact; innovation = the economic development and exploitation of some artifact, new or existing, by means of a reorganization of goods, methods of production, sources of supply, industrial structures, marketing, and such. (Does this include political development and exploitation as in warfare or the United States space program?) Innovation as a conscious process is thus a kind of using requiring systematic technical assistance. It might even be characterized as "technological use." Hence potential technology = a technological invention awaiting economic (or political?) exploitation by innovation or technological use.

One further distinction along this line is invention versus technological change. Invention commonly defines novelty; technological improvements in existing hardware based on what is already known do not qualify as inventions. For example, the original four-stroke-cycle internal combustion engine (the Otto silent engine, 1876) was an invention; combining two or more four-stroke cylinders into one engine was merely technological change. (Patent law, as an aspect of political economy, usually protects the former but not the latter, on the grounds that the former is more expensive and requires longer to realize a return on its investment.)

According to Stephen Toulmin's evolutionary analysis, innovation is part of a three-stage process: "(1) the phase of *mutation*, (2) that of *selection*, and (3) that of *diffusion* and eventual dominance."[57] The first is a conceptual or mental activity, the second involves practical testing, and the third is dependent on economic exploitation. "The phase of *mutation* corresponds to the first half of the research and development operation, during which new techniques and processes are devised and prepared for testing and costing; the phase of *selection* is the one at which, within some specific area of application, the techniques or processes in question are shown to be feasible, both in technical and in economic terms; while the phase of *diffusion* and *dominance* . . . is that in which these skills spread into the general body of industrial and engineering techniques."[58] Whereas utilization in the broad sense may well have this sequential structure (with possible interrelationships),[59] questions of innovation are more often directly concerned with stage 3, that of diffusion. The exact character of this diffusion process itself can vary, however, from an unconscious long-term adaptation to a consciously stimulated acquisition. The complex changes in, say, traditional agriculture over long periods are quite different from the well-advertised exploitation of new products in a consumer-oriented society.

Innovation also involves shifts in use and thus is related to what in sociological literature is termed technological transfer. The transfer at issue can be from laboratory to production line, from one country to another, or even from one technology to another. To some extent the literature on technology transfer illustrates the structures of innovation presented by Toulmin; to some extent it also provides bases for criticism.

Contrary to Toulmin, mutation can be either accidental or planned: that is, it can originate either in the artifacts themselves—as a result of wear, accidental variations in materials and fabrication techniques—or in the influence of larger cultural changes or interactions, or else it can take place consciously as the result of a systematic process both in the mind of an inventor and at the level of miniature fabrication (design). In the first case, it is possible that physical diffusion could even occur before recognition of utility in the object; in any case it is a recognition (or discovery) of utility that will be primary. Yet if utilization is grounded in this recognition of utility rather than in novelty of conceptualization, then utility will not be nearly as subject to conscious development. When mutation takes place as a result of creative conceptualization, however, utilization (testing and innovation) is likely to have to be planned. A mental framework at the beginning has implications for the mental structuring of diffusion. The conscious structuring of mutation, testing, and diffusion is what in some circumstances is termed management.

Management as a Technological Using

The extent to which management is a technological process depends, first, on relationships between economics and technology and, second, on a theory of bureaucracy.

By way of historical background: classical economics identified land, labor, and capital (with technological objects, machines, etc. = fixed capital) as three factors in the production of wealth. The end of the nineteenth century saw identification of a fourth (by Alfred Marshall): business enterprise and organization. From this fourth element has evolved the modern concept of management, or the organizing and directing of a business enterprise.

Some management theorists present this fourth element as the essence of modern technology. Peter F. Drucker (1970), for instance, argues that "technology is not about tools, it deals with how Man works" (p. 45) and conceives management as the decisive determinant of this "how." In Drucker's view, the twentieth century has seen "technological activity . . . become what it never was before: an organized and systematic discipline" (p. 55). Moreover, the "one fundamental insight underlying all management . . . is that the business enterprise is a *system* of the highest order," which the manager is called on to organize and direct (p. 193).

According to one influential classification, there exist at least eleven overlapping approaches to this organizing and directing.[60] Different schools of management emphasize empirical case studies, interpersonal behavior, group behavior, cooperative social systems, sociotechnical systems, decision theory, systems, mathematical or management science, contingency or situational management, managerial roles, or management operations. Management operations, which will necessarily occupy a central place in any theory, include planning, organizing and staffing, leading, and controlling.

Now, any of these types of management can be thought of as technological activities, because what they manage is the making and using of artifacts, through human organizations. Moreover, technology as organization is part of the study of technology precisely insofar as the organization is oriented toward making and using, especially as a structured process. Indeed, the set of management operations exhibit distinct similarities to those of engineering designing. But decision theory and mathematical approaches can be considered technologies in an even stronger sense, according to the contrast between technique and technology set forth earlier, since these approaches seek to formulate explicit procedures for correctly using tools and machines. In this they attempt to do for using what design does for making; they constitute an engineering of use. (Note also how innovation is the managerial equivalent of invention; thus invention is to making as innovation is to use.)

As for the issue of bureaucracy, this can appear (ironically) to be not so much a kind of making as a doing—an activity pursued for its own sake. But given the dependence of bureaucracy on the modern technological infrastructure (telephones, copy machines, computers, etc.), its rise in conjunction with attempts to control highly technological social orders, and its tendency to become a kind of technocracy, this too can be identified as a facet of technology as a using process.

One further suggestion: technology assessment (TA) can also be viewed as a type of management in the broad sense. TA attempts to articulate (somewhat eclectic) procedures on which to base decisions about making and using.[61]

Work: From Alienated Labor to "Action into Nature"

Management is a form of technological using, but what is managed is work. Is work itself a kind of using process or a making action? In different senses it is both—as well as being the most prominent form of technology as activity. Sociohistorical analyses contrast premodern societies, in which members identify themselves by means of kinship relations or village residence, with modern societies, where members define themselves in terms of jobs or work; they further describe historical transformations in the meaning of work.[62] In engineering the technical concept of work as force multiplied by distance provides the basis of mechanics. Economic analyses describe the transformation of craftwork into industrial labor through the division of labor (Adam Smith) and its related decontextualizing or disembedding (Karl Polanyi).[63]

The disembedding of work takes place on two planes. First, the introduction of nonhuman sources of energy, particularly steam power, disengages tools from their dependence on the human body. What the traditional small-scale supplementing of human and animal power (mostly on the farm and in transportation) and the derivation of power from natural elements (water on land, wind at sea) had done in only limited ways, the steam engine advanced with a vengeance.

Second, and with an opening provided by this disengagement, a new *functional* division of labor takes production out of the home and places it apart from other life activities in a factory setting. What may be called *substantive* division of labor is a traditional feature of the workplace, with some artisans specializing in the making of shoes, others of pottery, and so on. There might even be further specialization between those making pots for the dining table and those making pots for food storage. But within such substantive division of labor artisans were, with the help of assistants, still regularly involved with one product from start to finish. With the coming of an intensified functional division of labor, in which pins are made by one person cutting wires to length, another sharpening one end, and still another soldering heads on the other end, labor becomes transformed into the production line.[64]

In premodern philosophy humanly and socially embedded work existed on the margins of reflection. In modern philosophy disembedded work is analyzed under two contrasting descriptions: positively as the production of wealth and negatively as alienated human activity. Both descriptions implicitly distinguish labor (repetitive making devoid of the action of designing) from work (making that incorporates designing and producing), although there is a strong tendency to identify the two (insofar as the essence of work has become labor).[65]

Despite its modern significance, however, there is surprisingly little by way of explicitly philosophical analyses of work, and what exists emphasizes work as labor.[66] The distinctive wealth and power that result from the modern functional division of labor can be examined from the perspective either of labor or of use. The labor theory of value, although it originates with John Locke, plays a pivotal role in Marxist thought, where it becomes the foundation for a negative criticism of alienation. The alternative theory of value as utility is the basis for a positive description of labor that nevertheless cuts wealth and power free from the laboring process. Goods or commodities have value only insofar as they are useful, which relates them more to using processes, if not to entrepreneurship and advertising.

Alienated Labor

Historically speaking, the original critical reflections on modern technological activity grew out of transformations in the nature of work during the Industrial Revolution. Economic oppression of the worker and the psychological consequences of a mechanized division of labor became primary themes in discussions of work during the eighteenth and nineteenth centuries. But as poverty ceased to be an overriding issue, attention has centered more on the "problem of alienation" and the uses of leisure.

Alienation constitutes a multifaceted issue grounded in reflection on the complexities and ambiguities of making and using, against the background of a traditional reflection on the complexities and ambiguities of thinking and moral doing. In Platonic philosophy, for instance, alienation can refer to a worldly self-diremption that leads to perfection and unification with the transcendent. Augustine speaks of *alie-*

natio mentis a corpore to signify that state in which the human soul is elevated above herself to become one with God, a positive good. The Hebrew prophets, by contrast, speak of immoral activities as separating or alienating people from God, something to be rejected by changes in personal behavior.

With regard to technological work, alienation can be compared to what takes place in thinking. Just as thinking does not automatically terminate in understanding but often throws up ideas that are puzzles to their formulators, so making, instead of leading directly to appropriation and humanization of the world, involves a moment of alienation or estrangement. Such alienation is not a perfection of making, nor is it a separation that can be overcome simply by changes in behavior. It is a necessary moment in making action.

The first philosopher to deal explicitly with the dialectic of alienation is Georg W. F. Hegel, who approaches consciousness as a kind of self-creative practical activity, thus deepening the modern epistemological principle that to know involves being able to make, and showing how in many instances knowledge develops in ways analogous to that by which craftworkers discover themselves and take satisfaction in their work. The making of consciousness for Hegel involves an initial self-alienation, the separation and objectification of an unconscious part of one's self. Once objectified, however, this element is available to be brought into consciousness; alienation is overcome by recognizing its origins in the creative self. Alienation is thus a process for the immanent enrichment of a creative subject through the differentiation and appropriation of its own content.

Marx rejects alienation as a means to a higher and more comprehensive unity within the self or with the transcendent as an idea that distorts the human essence. For Marx thinking, even when understood as a kind of making, is not the essence of humanity. The human essence is making of both world and self; human nature is realized in work and its transformation of the world. But this possibility is denied by the capitalist economic system. Under capitalism, work is coercive rather than spontaneous and creative; workers have little control over work processes; the products of their labor are expropriated by others to be used against them; and workers themselves become commodities in the labor market. "All these consequences result from the fact that the worker is related to the *product of his labor* as to an *alien* object."[67]

Subsequent to Marx, sociologists and social philosophers have expanded the concept of alienation. The 1960s especially witnessed a rediscovery of the concept, as well as a tendency to link alienation with romantic criticisms of technology as separating humanity from nature and the affective life, with the sociological categories of anomie (Durkheim) and "disenchantment" (Weber), and with Freud's psychological theory of repression.[68] What is most clear to Marxist critics of alienation is that it cannot be assuaged either by a return to craft production or by ameliorative worker benefits, from increased wages to subsidized health care. For worker-critic Harry Braverman, for instance, "The worker can regain mastery over collective and socialized production only by assuming the scientific, design, and operational prerogatives of modern engineering; short of this, there is no mastery of the labor process."[69]

Action into Nature

What is insufficiently recognized by Braverman, however, is the extent to which alienation applies not just to work but to the engineering design activity as well, and thereby prepares the way for what Hannah Arendt calls technological "action into nature." All traditional design, since it was limited by materials and energies given in nature, could not introduce into the terrestrial lifeworld any product or process fundamentally at odds with or alien to it. With the coming of steam, electric, and nuclear power, and the development of the ever-expanding chemical industry, such is no longer the case.

In her extended examination in *The Human Condition* (1958)—which is structured around a delineation of relations between labor, work, and action—Arendt observes a persistent attempt to replace political action and its inherent contingencies with making and its definitive results. Although the modern interpretation of the human as essentially *Homo faber* has brought this attempt to an apotheosis, "The substitution of making for acting and the concomitant degradation of politics into a means to obtain an allegedly 'higher' end—in antiquity the protection of the good men from the rule of the bad . . . , in the Middle Ages the salvation of souls, in the modern age the productivity and progress of society—is as old as the tradition of political philosophy" (p. 229).

The unprecedented success of this degradation of politics—that is, the clear subordination of politics to an end other than that of political life itself, in the modern instance to the pursuit of scientific technology and especially technologically mediated work—has transformed the political realm. This transformation has been examined as leading to technocracy (see Hans Lenk, ed., 1973) or technological politics (Langdon Winner 1977), and as establishing a kind of "pseudogovern-ment."[70] In each case there is an attempt to replace the give-and-take contingencies and consequences of politics with the order if not certainties of technological management.

Human action with others exhibits a persistent tendency to introduce into the web of human relationships chains of events that outstrip original intentions. Efforts to seize power lead to losses of power; efforts to limit power lead to its expansion. Especially when politics falls under the sway of technology is there an effort to control this persistent devolution of events by transforming the variegated web of human relations into the sociologically and bureaucratically managed state.

But at the same time as this devolution ostensibly frees politics from human action and its contingencies, it also transforms technological processes into a kind of action. For Arendt, "The attempt to eliminate action because of its uncertainty and to save human affairs from their frailty by dealing with them as though they were or could become the planned products of human making has first of all resulted in channeling the human capacity for action . . . into an attitude toward nature" manifested in an exploration of the laws of nature and their use to fabricate new objects and processes. In this way, according to Arendt, human beings "have begun to act into nature" (p. 230–231). Such action into nature takes in big science experiments such as nuclear explosions, large-scale technological projects such as dams and transportation systems that alter whole ecologies, and the mass production of synthetic chemicals and products.

Such action into nature is dependent on technological work in the modern sense and indeed can be argued to reveal its fundamental character. As Peter Drucker has succinctly stated, the purpose of technology as work

> is to overcome man's own natural, i.e., animal, limitations. Technology enables man, a land-bound biped, without gills, fins, or wings, to be at home in the water or in the air. It enables an animal with very poor body insulation, that is, a subtropical animal, to live in climate zones. It enables one of the weakest and slowest of the primates to add to his own strength that of elephant or ox, and to push his life span from his "natural" twenty years or so to the threescore years and ten; it even enables him to forget that natural death is death from predators, disease, starvation, or accident, and to call death from natural causes that which has never been observed in wild animals; death from organic decay in old age. (1970, p. 45)

In contrast to Drucker's enthusiasm, Barry Cooper (1991) pursues Arendt's insight into the dialectic transformation of work from alienation to action into nature. As Cooper explains, the pivotal feature of political action is that, unlike labor (the repetitive process of meeting daily natural needs like chopping wood and drawing water, an activity that tends to overwhelm the distinctly human world of politics) and work (the fabricating of that stable but limited artifice, built with and within nature, in which the human world comes to birth and dwells), action constitutes a new beginning and initiates phenomena with consequences that regularly outstrip human knowledge and intention. In a paradoxical cross between these two, modern "technology shares the process character of labor but also the formative character of work. Unlike work, technology does not stop when the work is done; unlike labor, technological processes are not guided by nature" (p. 134).

By transforming what was once a making action (work inclusive of design) into a making process (mass production work using machines), political action (into the web of human relations) is transformed into technological action (into nature). As a result of its new powers,

> Modern technology can do in the realm of nature what Vico thought could be done only in the realm of history. Technological human beings have shown themselves capable of starting natural processes that would never have existed without human initiative. Technological action has the inevitable consequence of carrying human unpredictability in that realm of being that used to be conceptualized in terms of inexorable laws such as the law of gravity. The final and puzzling consequence of acting into nature is that we have succeeded in "making" nature. . . . To state the obvious, that this was not intended simply affirms once again the unpredictability of human action. (p. 146)

The web of relationships in nature is as complex and indeterminate as those in the human world. The indeterminacy introduced into human action by the transformation from tribal to societal levels of organization is mirrored by what happens in technology with the shift first from craft to industrial production and then even more decisively with the development of nuclear weapons, computers, and biotechnology. These new technologies readily introduce into the web of ecological relationships chains of events that outstrip original intentions not unlike those previously exhibited in the political realm. The rise of technology assessment institutions and methodologies and contemporary attempts to develop an earth system science and the planetary technology for a global management of Spaceship Earth are efforts to extend management from work to world.

Again, Ancient versus Modern Technology

More clearly perhaps than any other technological activity, work is distinguished into ancient and modern forms. But as was noted with regard to technological objects and implied in the analysis of technology as knowledge, fundamental divisions in technology as activity can also be indicated by the differential presence of other aspects of this mode of manifestation. Making by cultivation is more prominent in premodern work than is making by construction. As artifice becomes more prominent in the lifeworld, maintenance takes precedence over making of all types. And as already suggested, the presence of bureaucracy as a distinctive institution for the coordination of technological making and using is a distinguishing feature of modern technology as activity, transforming the action of inventing, the process of maintaining, and work in its characteristically modern form.

Another indicator of the great divide in the history of technology is the dominance of artistic design in the ancient world and engineering design in the modern. Before the development of modern mechanics and its calculus of forces, artisan and architect tended to focus on formal, not to say aesthetic, properties in their structures. With the development of the science of mechanics attention shifts toward concern for material, energy, and spatial efficiencies in products and structures as well as in processes of fabrication and construction.

Indeed, this new focus on efficiency was a major contributor to the economic expansion characteristic of the Industrial Revolution, which might have taken place to some extent independent of the development of new sources of power. The energy calculus alone makes possible precise assessments and increases in efficiency, especially once sources can be priced. In more ways than one the scientific revolution of the seventeenth century contained within its conceptual formulations the technological revolution of the eighteenth century. One final social indicator of the importance of the shift from artistic to engineering design is that before, say, 1750 technological advances strengthened the artisan class; after that they undermined and eventually destroyed it.[71]

CHAPTER TEN

• • • •

Types of Technology as Volition

Engineering includes distinctive perspectives on and analyses of technology as object, as knowledge, and as activity. It has, however, nothing to say about technology as volition. The closest approximations are discussions of cybernetic control (Wiener) and decision theory (Simon), both of which nevertheless reduce deciding and choosing to acts of rational analysis. The turn to technology as volition thus constitutes a turn away from engineering and a return to philosophy.

But whereas technology as object, as knowledge, and as activity can readily be engaged by philosophical traditions of reflection, it is difficult to get a philosophical purchase on technology as volition. Partly this is because willing, although clearly a theme, is itself so poorly articulated by philosophy; will is the elusive Proteus of the philosophy of mind. What follows is thus different in character from previous analyses and necessarily reappropriates the tradition of historicophilosophical reflection in an attempt to use various philosophies of technology and of the will to engage technology as volition.

Philosophies of Technology as Volition

The protean character of volition is implicit in many philosophies of technology. Technologies have been associated with diverse types of will, drive, motive, aspiration, intention, and choice. For example, technology has been described as

- the will to survive or to satisfy some basic biological need (Spengler, Ferré)

- the will to control or power (Mumford 1967)

- the will to freedom (Grant, Walker, Zschimmer)

- the pursuit of or will to efficiency (Skolimowski 1966)

- the will to realize the *Gestalt* of the worker (Jünger) or almost any self-concept (Ortega)

each of which could arguably be expected to produce different types of technology.

For Oswald Spengler, reflection of the relation between *Der Mensch und die Technik* (1931) reveals "technics as the tactics of living" (p. 2): *"Technics is not to be understood in terms of the implement.* What matters is not how one fashions things, *but what one does with them;* not the weapon, but the battle. . . . This battle *is* life—life, indeed, in the Nietzschean sense, a grim, pitiless, no-quarter battle of the Will-to-Power" (pp. 10 and 16).

Frederick Ferré in his general introduction to the *Philosophy of Technology* (1988) characterizes technology in volitional terms similar to Spengler's. Defining technology first as "practical implementations of intelligence" (p. 30), Ferré describes practical intelligence as "mental self-discipline in the service of the urge of life" (p. 36).

> Motivated by the urge to live and to thrive, practical intelligence sorts . . . possibilities into orders of relevance for realization and attempts to guide action into the fruitful channels of regular method. Once a fortunate method is found to serve the urge to live and thrive, it may be remembered and retained. . . . The long history of practical intelligence embodied in culture and perpetrated by tradition allowed for gradual improvement of methods, since the urge of life to thrive—"to live, to live well, to live better"—continues to motivate within the framework of tradition. (pp. 36–37)

Indeed, for Ferré what unites tradition-based and theory-based practical intelligence—that is, craft technology and scientific technology—is this unity of motivation, of technology as will "to live and to thrive."

What Spengler and Ferré see as the life force, Lewis Mumford sees as a restricting urge to control. From the biotechnic or polytechnic world of a plurality of volitional drives, modern technology singles out the pursuit of physical power to create an all-encompassing monotechnics. For Mumford the problem with "monotechnics" is that it exemplifies a will to power at odds with a will broadly oriented toward life; a single purpose dominates and excludes all others. According to Timothy Walker and Eberhard Zschimmer, however, the control of monotechnics is legitimate because it grows out of and realizes a will to freedom.

For Skolimowski the issue is not power, but efficiency. Technology is almost always distinguished from science by ends or intentions: science is said to aim at knowing the world, technology at controlling or manipulating it. Skolimowski asserts that "technology is a form of human knowledge," then distinguishes it from scientific knowledge by saying that "science concerns itself with what *is;* technology with what *is to be"* because "our technological pursuits consist in providing means for constructing objects according to our desires and dreams" (1968, p. 554, italics added).

For Ernst Jünger, technology is "the mobilization of the world through the *Gestalt* of the worker" that in the first instance reveals a nihilistic will to destroy and in the second a will to reconstitute along the lines of power and brute rationality. For Ortega, technological objects, knowledge, and processes are grounded in some willed self-realization, but not necessarily that of the worker. Creation at the level of material invention is preceded by a self-creative affirmation. To adapt Jean-Paul Sartre's well-known formula: existence (as a willing subject) precedes essence (the willing of specific subjectivity). Whatever is willed calls forth its appropriate technology.

Related discussions of the volitional aspects of technology can be found in economists from Joseph Schumpeter, Max Weber, and Friedrich von Gottl-Ottlilienfeld to Daniel Bell and Nathan Rosenberg, who regularly link technology with the entrepreneurial spirit.[1] Otto Ullrich's *Technik und Herrschaft* (1977) provides further analysis, from the perspective of an engineer, of the structural affinities between technology and capital, while Ernst Bloch (1959) identifies a similar drive in the utopian will. Thomas F. Tierney's *The Value of Convenience* (1993) argues that the "desire for ease" is the genealogical source of technology. By contrast, historian Lynn White Jr. finds technology grounded in Christian charity and temperance or what might also be described as an altruistic, disciplined will. Will is a regularly mentioned if undeveloped part of analyses of technology.[2]

One important extended analysis of technology in volitional terms is provided by the French philosopher Jean Brun. As Daniel Cérézuelle (1979) has summarized his thought, for Brun technology grows out of Western ontological aspiration to merge subject and object. In *Les Conquêtes de l'homme* (1961) Brun traces from Aristotle to the contemporary technician the progressive emergence of a desire to master "the spatiotemporal limits of existence, making them no longer marks of his deprivation but of his power" and becoming thereby through free will a "self-creative being" (p. 59). In *Le Main et l'esprit* (1963) Brun shows how this existential commitment to technology transformed the philosophical interpretation of the meaning of the hand and its technological projections. In *Le Retour de Dionysos* (1969) he further analyzes how scientific ascesis and technological positivism nourish what he terms "technological Dionysianism." "Now Dionysian humanity escapes itself through machines, seeks to explore all that is outside itself by means of the new resources of an exo-organism, even of an artificial exo-sex" (p. 72). With *Les Masques du désir* (1981) and *Le Rêve et la machine* (1992) Brun further examines the transformations of desire and dream as they engage the modern technological project.

As George Grant summarizes the implications of such theories,

> All descriptions or definitions of technique which place it outside ourselves hide us from what it is. This applies to the simplest accounts which describe technological advance as new machines and inventions as well as to the more sophisticated which include within their understanding the whole hierarchy of interdependent organisations and their methods. Technique comes forth from and is sustained in our vision of ourselves, as creative freedom, making ourselves, and conquering the chances of an indifferent world. (1969, p. 137)

This modern notion that human beings make themselves, or will what they are to become, has serious practical consequences throughout the lifeworld. Not to mention other examples, it seems to be behind the desire to "conquer space" and, in the biomedical field, to overcome "genetic roulette." Genetic counseling, followed up by abortion when necessary, is designed to keep the spontaneity of nature from interfering with an individual's self-creative volition.

Volition as a Conceptual Problem in Relation to Technology

The diversity of views about the volitional character of technology points up a threefold problem. First, volition is the most individualized and subjective of the four modes of manifestation of technology. There may well be a sense in which each person's motivation, being unique, becomes connected to artifacts, knowledge, and making and using in different ways. But surely this not only is the least philosophically interesting observation, it is also the least practically meaningful. Such individuality seldom has public consequences—except perhaps as terrorism—until linked with similar volitions to produce what might be called a social or cultural act of willing.

Perhaps no one has argued this issue in regard to technology more carefully than Steven Goldman (1984). In an original proposal to bring "to philosophy that in technology which is at once essential to technology and of central concern to philosophy" (p. 115), Goldman argues that it is a mistake to limit our conception of technology to artifacts, forms of the production of artifacts, or technical knowledge. Instead, he says, "'Technology' should be understood as referring . . . to a particular social process, to a form of action that is decision-ruled, one in which specific 'captive' knowledge bases in engineering and science (primarily) and craft skills (secondarily) are put at the disposal of people who in general are not themselves competent in those knowledge bases and who wield them on behalf of ends reflecting a parochial interpretation of prevailing personal, institutional, and social values" (p. 121).

Although admitting that technology often exhibits a kind of logic of its own, Goldman maintains that the actual "determinants of action derive from non-unique [that is, not simply individual] values judgments rather than from either the products of technology or its knowledge base" (p. 123). That is, what human beings do with technology can be explained only by appeal to "economic, institutional, political, social, and [only] sometimes personal value judgments" (p. 124). Given the valuational and volitional dimension of technology, "to understand technology fully, we need to understand how will enters into its various dimensions, not least into the level of managerial decision-making where willfulness is particularly manifest in the non-unique, interpretive, and partially arbitrary character of these decisions" (p. 136). "Acknowledging the elementarity of will . . . is one condition of philosophical analyses of technology" (p. 137). In furthering this interpretative project Goldman has examined "the social captivity of engineering" (1991), the role of engineering in Western culture (1990), and the prospects for a democratically deliberative technology policy (1992).

But second, in volition there is always the problem of correspondence between subjective intentions and "objective intentions," that is, the objective tendencies of any particular technological means. An act of willing, except one's own, cannot be directly known (some would even argue against its being known directly by oneself); it can only be inferred from action (including, of course, speech). But is the external action or means chosen an adequate expression of a particular inner intention, so that one can legitimately infer from the character of the one the character of the other?

A comparable question arises with regard to knowing and ideas. Do one's ideas adequately correspond to what one knows in reality? In the case of knowledge this issue is at least heuristically negotiated, first, by clarifying concepts so that individuals may judge for themselves, from their own experiences of an object, which concept might be the best mental correlate. Following such individual judgment there is, second, an attempt to rely on informed group consensus as some warrant for the adequacy of an idea.

One sensitive analysis of the will that recognizes the volitional correlate of this cognitive issue notes that "when attempting to put a lofty goal . . . into action, we must consider various means and proceed step by step accordingly. To consider means in this manner is to seek harmony and accord with the object. If in the long run we fail to discover appropriate means, we have no recourse other than to alter the goal. [But] when the goal is close to the given actuality, as in the habitual conduct of everyday life, the desire immediately turns into performance."[3] By comparison, in attempting to judge the adequacy of a proposed volition in relation to technology, one seems obliged (on the same heuristic grounds) to subject technological objects and activities to experimental examination with regard to their real-world tendencies or implications. Indeed, much of the popular discussion about "technology and values" is vacuous precisely because it does not attempt to do this. Instead, it assumes that technology as object, as knowledge, and as activity is value- or intention-neutral—that one can take any value or volition, attach it to an existing artifact or activity, and create a new technology.

But is the object, knowledge, or activity really commensurate with the volition? Sometimes it is, sometimes it is not. The problem is obviously recognized on one level, when people do not try to use guns as toothpicks. Yet people do say things such as "Technology does not have a will of its own" or "The problem is not technology but what people want to do with it"—believing, apparently, that technologies can be magically transformed by differential volitions.

Consider, for instance, one extreme illustration of this kind of discussion, that regarding the harnessing of nuclear explosives for peaceful purposes—to dig canals and such. The whole idea is both unrealistic and misleading. Nuclear explosives have so many characteristics that make them inherently oriented toward military use, otherwise recognized in the requirements for extreme security, that talk of peaceful civilian appropriation requires virtually ignoring their real-world properties. Such talk further undermines empirical attempts at technology assessment—historical, economic, sociological, political, ecological—that aim to identify those intrinsic implications of technological objects, knowledge, and activities that influence whatever intention is attached to them.

Third, there is the problem of self-understanding and levels of the will. According to Alexander Pfänder's *Phenomenology of Willing and Motivation*,[4] willing in the general sense is only awareness of striving— a psychological phenomenon that impinges on the ego but does not involve its center or core. In its lowest form, striving can be experienced as a biological urge or instinct, although it might also be felt as a peripheral wish, hope, longing, desire, or fear. Striving is simply characterized by an awareness of something absent that attracts, and it can comprise numerous, even conflicting impulses.

What converts this striving into an act of willing is its being taken into the center of the ego. Willing in the strong sense is constituted only after one comes to believe that the goal of striving can be realized through one's own actions, and when the ego spontaneously or reflectively sides or identifies with such striving. In other words willing, like knowing, emerges out of memory, the memory not of representations from which we as egos seem to stand aside but of representations in which we and especially our bodies are intimately involved. Not memories of watching a log burn in the stove or simply observing a cup of water at rest on the table, but of being warmed by the fire or transporting the cup to one's mouth and slaking thirst are what can give rise to an act of the will. Faced with a diversity of such memories— cutting wood, drinking water—a decision after the consideration of various desires, like the formation of a cognitive judgment after deliberation, establishes an internal unity. Now I think this, or it is time to do that.

As Pfänder says, "Thus willing, but not striving, includes the immediate consciousness of self." In other words, "The act of willing is . . . a *practical act of proposing filled with a certain intent of the will* which issues from the ego-center and, penetrating to the ego itself, induces in it a certain future behavior. It is an act of self-determination in the sense that the ego is both the subject and the object of the act."[5] But since it is of "the very essence of the performance of an act of willing" that the ego "appears as the agent,"[6] willing is dependent on the self-concept possessed by the ego. Only if one sees oneself in a certain way can one identify with some particular striving.

Technology as volition, on this account, could manifest itself, first, as a general striving and, second, as an ego-affirmed project. In this context, the questions for the philosophy of technology concern (1)

What strivings are able to be realized by what particular technologies? and (2) What self-concepts thus emerge from or enable and perhaps incline one to identify with these technologies? The former is again related to the issue of the empirical implications of certain technologies, the latter to historico- and psychophilosophical analysis.

Philosophies of Volition in Relation to Technology

The chief difficulty in formulating a theory of technology as volition is that the concept of will is one about which the history of philosophy provides precious little consensus. The term "will" does not occur as such in Greek thought; it enters Western intellectual history by way of the Christian philosophical tradition. In modern times it has been subject to both excessive emphasis (Friedrich Nietzsche) and extreme criticism (Gilbert Ryle). Nevertheless, as has been argued by representatives of both the existential-phenomenological and analytic traditions of philosophy, the issue of volition remains central to any philosophy of action—and thereby to any comprehensive understanding of technology.

According to Hannah Arendt, for instance, in an analysis slightly at odds with that presented above, "the Will, if it exists at all . . . is as obviously our mental organ for the future as memory is our mental organ for the past. [But] the moment we turn our mind to the future, we are no longer concerned with 'objects' but with *projects*. . . . And just as the past always presents itself to the mind in the guise of certainty, the future's main characteristic is its basic uncertainty." [7]

Although she does not mention technology as such, Arendt points out how the Aristotelian analysis of making as bringing potentiality into actuality "implicitly denies the future as an authentic tense." For Aristotle, the future is only "a consequence of the past, and the difference between natural and man-made things is merely between those whose potentialities necessarily grow into actualities and those that may or may not be actualized." [8] Indeed, Arendt's chronicle of the discovery of the will as the creation of "a future that in principle is indeterminable and therefore a possible harbinger of novelty" [9] parallels the modern discovery of technology, and she concludes with a discussion of Heidegger's interpretation of technology as "the will to will."

In a more detailed phenomenological description of willing that deepens Pfänder's psychology and Arendt's historical study, Paul Ricoeur[10] identifies three levels: "I will" can mean "I desire," "I move my body," or "I consent." Technology as volition can thus be analyzed in at least these three volitional senses: as technological desire, as technical motivation or movement, and as consent to technology.

Such an analysis could begin to elucidate the dialectic of "technological eros" (Jakob Hommes 1955). As is documented by the sociological literature, technological desires engender and are reinforced by technological motivations, which through the creation of objects, knowledge, and activity, reflect back onto and are supported by consent to the technological presence. It is this feedback process and its large-scale institutional formations that give rise to what has been referred to as technological momentum, technological autonomy, or a technological imperative.

One analysis in the phenomenological tradition that contains deeper suggestions about technology as volition is that of Martin Heidegger. In his early work Heidegger makes few explicit references to either technology or volition as such. Yet it is against the background of his early phenomenological description of human existence that Heidegger's later explicit arguments are developed. Heidegger begins with a phenomenological description of the existential features of *Dasein* ("to be here"), a kind of Cartesian consciousness nonetheless characterized by worldly involvement. In *Being and Time* (1927) Heidegger presents human existence or *Dasein* as inherently a being-in-the-world and then proceeds to investigate different aspects of this being-in-the-world in order to disclose the distinctive character of human existence.[11]

The world of *Dasein* is characterized primarily by a practical concern for or worry about (*Besorgen*) manipulating things and putting them to use. Objects encountered in the process Heidegger calls "equipment" or "gear." "In our dealings we come across equipment for writing, sewing, working, transportation, measurement." Breaking down the writing equipment totality into its constituents discloses "inkstand, pen, ink blotting pad, table, lamp, furniture, windows, doors, room" (1927 [trans., 1962], p. 68). The distinguishing character of all equipment is that it is fundamentally context dependent. It is not individual utensils that are first grasped and then assembled into equipment totalities. Equipment is present initially as part of a larger framework or system; then within this framework individual items can be recognized as pieces of equipment.

The world within which human existence as *Dasein* operates is a system of relationships that Heidegger calls readiness-to-hand. Within this system, materials are defined by their usability, tools by their serviceability. Common sense tells us that things are first simply present-at-hand as things in their own right; but according to Heidegger this is not the case. The description of a hammer as "for hammering" is more primordial than any conceptual description of it as being of some particular size, shape, weight, and color. To conceptualize a utensil as a neutral object that is present-at-hand is to abstract from its given state. Generalized, this insight means that science is an abstraction from technology, knowledge for its own sake an abstraction from practical knowledge. The difficulties encountered by a philosopher such as Descartes—who begins with pure consciousness and tries to derive the practical world from it—confirm this insight. Pure consciousness derives from practical consciousness, not vice versa.

In the human realm, *Dasein*'s being-in-the-world turns into a being-with-others. Practical involvement takes on the new form of solicitude for (*Fürsorge*) members of the social community.

Finally, in its relation of being-in, *Dasein* is characterized by two fundamental structures, or "existentialia" as Heidegger calls them: mood and understanding. Heidegger's special use of the term "understanding"—not as something theoretical but as something practical, related to mood—underscores again the centrality of technological activity in his analysis of human existence. Heidegger is at pains to present making and using skills as true forms of knowledge, with conscious theoretical knowledge being derived from this preconscious nontheoretical base. The reason is that he is trying to analyze knowledge as constitutive of the world in the sense required by the distinctly modern conception of the human as maker. This view necessitates his interpretation of what in traditional terms has been called practical knowledge—the kind of knowledge that is intimately tied up with volition (and therefore temporality)—as the fundamental form of knowledge. Only this kind of knowledge forms both the world and the human beings in it. More deeply than any previous philosopher, Heidegger reveals the essential features of the human being conceived as *Homo faber*.

At the end of the first part of *Being and Time*, Heidegger concludes his existential analytic by arguing that care (*Sorge*) is the essence of *Dasein*. Being-in-the-world necessarily entails a multifaceted practical involvement with the entities of the world. Although "care" cannot be translated simply as volitional activity, Heidegger explicitly states that it is care that makes willing possible. Although Heidegger does not use the terms "volition" and "will" frequently, *Being and Time* presents technology as object, knowledge, and activity as fundamentally related to volition.

In Heidegger's celebrated turn away from the phenomenological method of *Being and Time*, his thinking does not retreat on this point. It becomes even more explicit. In an essay titled "European Nihilism" in his *Nietzsche* book (lectures from the 1940s, but not published until 1961), Heidegger argues that modern technology could arise only in a world that has become nihilistic or forgetful of Being. The nihilism of Nietzsche's "will to power," Heidegger claims, is the culmination of Western subjectivism and leads to the pure "will to will" of the technological age.

Heidegger's most important later discussion of the subject is "The Question concerning Technology." In this essay, which grew out of lectures given in part to engineers in 1949 and 1950, Heidegger rejects the common ideas of technology as pure means and human activity, arguing instead, in a view still very much in harmony with *Being and Time*, that technology is a kind of truth, a revealing or disclosing of what is. Where he advances beyond earlier analyses is in his strong distinction between ancient and modern technology and in stating the essence of modern technology. Ancient technology reveals by means of the "bringing-forth" of art and poetry. Modern technology, by contrast, is a "challenging" revealing, a "setting-upon." In less cryptic language, whereas premodern technology cooperated with nature to bring forth artifacts, modern technology imposes on nature, forcing it to yield up materials and energies that are not otherwise to be found.

The foundation of this challenging setting up of nature Heidegger calls *Ge-stell*. *Ge-stell* can be interpreted as an impersonal cognitive frame of mind. But according to Heidegger, *Ge-stell* is more fundamentally an impersonal volition. Not only does *Ge-stell* "set upon" and "challenge" the world—a description that already hints at volitional elements, it also sets upon and challenges humanity. Ultimately, it is not just human desires turned into motivations that give rise to modern technology. There is consent to a movement, a historical movement, that is transhuman.

Ge-stell is thus presented as a historical destiny or fate that calls on humanity to act in a particular way. It is not, however, a fate that compels in some crude sense. Elsewhere, having contrasted calculative and meditative thinking, Heidegger suggests that associated with such ways of thinking are two kinds of consent.

For all of us, the arrangements, devices, and machinery of technology are to a greater or lesser extent indispensable. . . .

We depend on technical devices; they even challenge us to ever greater advances. But suddenly and unaware we find ourselves so firmly shackled to these technical devices that we fall into bondage to them.

Still we can act otherwise. We can use technical devices, and yet with proper use also keep ourselves so free of them, that we may let go of them any time. . . .

But will not saying both yes and no this way to technical devices make our relation to technology ambivalent and insecure? On the contrary! Our relation to technology will become wonderfully simple and relaxed. . . . I would call this comportment toward technology which expresses "yes" and at the same time "no," by an old word, *releasement toward things* [*Gelassenheit*]. (1959 [trans. 1966], pp. 53–54)

In different words, releasement or detachment from technology transforms a consent to become an active or dominating will into consent to receptivity, or a receptive will. The will to will is transformed into a will not to will.

Toward Ethics

According to Aristotle (*Nicomachean Ethics* 1.5), there are at least four opinions about the good, or happiness, or what in relation to the present analysis might be called self-concepts. These are that happiness consists in physical pleasure (the hedonistic life), in honor (the political life), and in knowledge or wisdom (the philosophic life), plus a spurious opinion that happiness consists in making money (leading to the life of business). Each of these quite traditional volitional options can engender different technologies. The hedonistic life technologies are oriented toward food, clothing, sexuality, and perhaps medicine. The political life technologies are oriented toward both military affairs and communication. The philosophic life technologies involve, again, education, communication, and perhaps even scientific experimentation. (Money, as something approaching a pure means, can be found in different forms in all the others.) Aristotle leaves out of consideration at least two other traditional alternative ideals—those that would identify the good with beauty or with religious practice—and hence their distinctive ways of life and corresponding technologies.

The more or less traditional ethical consideration of different ways of life thus begins to provide another basis for analyzing and accounting for many of the possible types of technology as volition. To consider technology as volition thus points toward the need for an ethical analysis of technology. But it also suggests the need to go beyond traditional ethical analyses in at least two respects. First, traditional moral philosophy fails to provide an adequate account of those technological volitions described by phrases such as "will to control" or "will to power" or even "the pursuit of efficiency." Second, although traditional ethical analyses consider the relationship between different understandings of the good and certain human institutions (states, educational curricula, family structures, economic systems, etc.), they do not, except in quite limited ways, begin to address the correspondence between different understandings of the good and of technology.

Speculatively and impressionistically, precisely because of such weaknesses, it is perhaps permissible to suggest that the pursuit of efficiency or the will to control might even be termed a historically unique volition that can be associated with technology in a new way. Traditional or premodern technologies seldom if ever place much stress on efficiency of operation, not only because there is seldom a method for calculating efficiency in any systematic manner, but also because the technical activity is taken as having some inherent value. As Mircea Eliade and other historians of religion note, crafts are commonly "valorized" by becoming adjuncts of various religious rituals. A focus on evaluating the work in terms of effectiveness or efficiency requires rejecting such valorization on the basis of an idea of the human being as something like a machine with inputs and outputs to be husbanded and controlled. This new way can be correlated with the transition from craft making to science-based making.

Technology and Weakness of the Will

One special problem raised by reflection on types of technology as volition comprises not only the radical alternative of nonwilling as opposed to willing, but also the issue of willing as undermined or weakened by technology. This can be related to what is traditionally known as the problem of weakness of the will or incontinence.

"Incontinence" is a word that calls for clarification. Although commonly used to designate the medical pathology of being unable to retain urine or feces, in Scholastic moral philosophy it had a more general reference to the absence of *contentia* or self-control—of which the medical pathology is only a specific instance. Here the term is used to indicate a hiatus between knowledge and action, in an effort to avoid terms such as *akrasia* or "weakness of the will," which prejudge the interpretation of the phenomenon.

It is helpful to begin by situating the problem of incontinence within the broader context of ethical discussion regarding modern technology. If power or the ability to act increases, then so must intelligent control—otherwise power will eventually lead to disaster. Everyone probably would admit the soundness of this argument and agree that such a formulation is general enough to apply to a wide range of technologies—from automobiles and chain saws to nuclear weapons and feats of biomedical engineering.

But what are the preconditions for the full exercise of such intelligent control? The intelligent control of technology depends on (1) knowing what we *should do* with technology, the end or goal toward which technological activity ought to be directed; (2) knowing the consequences of technological actions before the actual performance of such actions; and (3) acting on the basis of or in accord with both types of knowledge—in other words, translating intelligence into active volition. Most discussions concerning the responsible use of technology focus on (1) and (2), or both. Insofar as (3) is recognized, it is subsumed under questions of societal organization (cultural lag) or observed as a psychological pathology (alienation, etc.).

At the same time, one encounters daily any number of examples of the problems related to (3), the issue of incontinence. There is the nurse or physician who is well aware that smoking causes cancer and any number of other health problems yet continues to smoke. There is the automotive safety engineer who knows full well the importance of seat belts but fails to buckle up. Genetic counselors tell horror stories of persons who, even when apprised of the near certainty of passing on disabling or fatal genetic defects, nevertheless choose to bear children. In each case conditions (1) and (2) are clearly met. The individuals in question know what they should do and how to do it. They should pursue health; indeed, they actively do so in many aspects of their lives. Moreover, they know the consequences of particular actions that are diametrically opposed to the good they desire. Yet they do not perform the actions dictated by such knowledge of ends and means. They know the good but do not do it. Shouldn't an analysis of the dimensions of such behavior form a substantial part of the ethics of technology?

An appraisal of the dimensions of the problem at issue must distinguish two major versions of incontinence. In the weak version there is what might be described as the resistance of matter to intelligence. Discussions of this go back to the works of Plato and Aristotle. In the strong version there is an opposition of intelligence to intelligence. This is associated with theological discussions of freedom of the will and the possible ability of a creature to at once know and reject or turn away from God.

The locus classicus for a discussion of incontinence in the weak sense is Aristotle, *Nicomachean Ethics*, book 7. According to a common historical distinction (based on Aristotle's own remarks), Socrates identifies knowledge with virtue and thus rejects the problem of incontinence, whereas Aristotle argues that this is patently contradicted by experience. The central difficulty in analyzing *akrasia* or weakness of the will is thus to explain in what sense a person could know the good and still not do it.

According to Aristotle, there are four senses in which a person can know but not do the good (see *Nicomachean Ethics* 7.3): (1) A person can know in the sense of being able to remember, but not at present remembering. (In New York City it is illegal to turn right on a red light—although it is legal in the state as a whole—but one can just space out and forget.) (2) A person can have universal knowledge that includes some particular without being fully aware that it does so. (One can know that sugar is unhealthful without realizing that ketchup is laced with sugar.) (3) A person can have knowledge that is obscured by sleep, drunkenness, or some other physiological state. Finally (4), a person can have two kinds of knowledge, and the lower can overcome the higher by virtue of an accidental feature of one's individual state (or social condition).

Cases (1) through (3) cannot, however, really account for the experience of incontinence. In none of these instances will a person experience a struggle to resist temptation. After the fact one may well look back and recognize a failure to act rationally or in accord with the good. Yet at the time, the person who forgets, or fails to recognize how a particular falls under a universal, or is drunk does not experience a struggle for the good that ultimately fails. It is only case (4) that offers some explanation for this experience.

Aristotle's highly condensed presentation of this fourth case may be elaborated with the following example. Suppose a person knows both that working in an asbestos factory is bad for her health and that working in such a plant will pay her a high wage. She needs work and is offered the job. But she hesitates. If she takes it, she will get her first paycheck in two weeks. The bad effects of asbestos exposure will be much more distant, to say the least. It will be years before asbestos can take its toll on her health. In fact, the first two weeks of work all by themselves may take no toll at all. Only if they become combined with hypothetical additional weeks and years of exposure will they be bad for her. Besides, what is bad is not the work; as an income-producing job, that is good. Given the remoteness of the bad and the immediacy of the good, is it any wonder that people in such circumstances will, after some hesitation, take the job—knowing in some sense that they should not?

Notice what this implies. Science is an attempt to replace ignorance with knowledge by bringing the more knowable but less known closer to our range of experience. This is why books are better teachers of science than the world; books can explain first principles. Ethics, likewise, tries to make the less influential but higher good more immediately influential in our lives. Thus, following his analysis of the moral experience of incontinence, Aristotle asks, How is the "ignorance" of an incontinent person to be dispelled?

The incontinent person does not act against knowledge in the truest sense, says Aristotle (agreeing now with a view earlier attributed to Socrates). If she had been as vividly aware of the long-range effects of asbestos exposure as she was of the short-range benefits of a paycheck (and as able to act on such awareness), she would not have taken the job. The overcoming of incontinence becomes a function of education and moral training (and perhaps the restructuring of society). The "artifice" of the polis is a better teacher of ethics than is nature. Incontinence loses its force as a conundrum and becomes merely an indicator of the need to transcend nature with culture.

This is the understanding that animates much of the practice in our society relative to raising consciousness about the dangers of certain technologies. To discourage smoking, its long-range effects are made as vivid and as immediate as possible by means of computer-assisted epidemiological studies, warning labels, newspaper articles and books, films, public service television ads, and glass-encased sections of dead smokers' lungs fashioned into ashtrays. Similarly, as war is made more horrible by advances in weapons technology, argue those who are encouraged by this approach, media technologies drive the horrors of war home to everyone. The depressing thing is that this approach neither lowers the number of people who smoke below one-third of the population nor seems to have much effect on the number of wars in the world.

Supplemental to this information-oriented approach—an approach especially characteristic of our "information society"—are two other strategies. One is to structure the environment so that it "artificially" reflects the long-range consequences of smoking. This brings remote knowledge down to the level of everyday experience. Large tax burdens are placed on smokers. Laws are passed against smoking in public places. Parents withhold allowances and even resort to corporal punishment. Human beings—perhaps most human beings—are not guided by information alone.

A second strategy is the "technological fix," which would separate "bad" artifacts and actions from their long-range consequences so that only the short-range consequences really matter. The American Tobacco Institute is determined to invent a cigarette that does not cause cancer. Short of that, we have a national effort to find a medical cure for cancer.

One problem with the information-oriented response to weakness of the will is that an information-rich society can sometimes aggravate the weakness at issue through an inclination to postpone action in favor of the pursuit of more information, or otherwise undermine the ability to perform difficult and heroic actions. Certainly the former tactic has been used most effectively by the tobacco lobby to thwart and delay antismoking legislation; a bias in favor of more information can also be used to protect established elites against rapid social change. As for the latter possibility, George Will has pointed out that if there had been television news cameras at Gettysburg, the Civil War would have had a different ending.[12]

The weakness of Aristotle's analysis of the relation between knowing the good and doing it is that despite his professed intention to preserve the experience of incontinence, this experience is subverted by his defense of the power and primacy of true knowledge. The gap between knowledge and action is closed by distinctions between different types of knowing, and by an affirmation of the power of at least some kind of knowing to fully determine human behavior. The gap opens up only in the presence of a weak or inadequate form of knowledge (or in a social situation that deprives that knowledge of its efficacy). What is commonly referred to as Aristotle's analysis of the weakness of the will is really an analysis of the weakness of certain kinds of knowledge.

The locus classicus for a discussion of incontinence in a much stronger sense is Augustine's *De libero arbitrio voluntasis*, especially book 3. For Augustine, the issue of the power of knowledge arises in relation to the question of the origin of evil. If God created the world out of nothing and did not give human beings truly independent agency, then God would have to be the remote cause of all evil. But if he *did* give human beings truly independent agency, how can such agency be seated in the intellectual act of knowing, since knowledge must always bear on what already is? Insofar as the intellect "chooses" evil, the choice (as Aristotle rightly observed) comes about because the intellect somehow fails in its act of knowing. Yet if this limitation is built into the intellect by its creator, once again God must be ultimately responsible. Faced with this difficulty, Augustine turns to the will as the source of evil and the cause of a gap between knowledge and action, which can only be called a stronger form of incontinence.

It is crucial to note that Augustine in no way "solves" the problem of incontinence in this stronger sense. He cannot explain how it is that the will could freely choose to do what is known to be a lesser good. He simply tries to acknowledge what he sees as a fact of experience. To explain how it happens would once again subordinate the will to the intellect. If it is a truly independent element, the will must be to some degree unintelligible to the intellect. All Augustine does is recognize how the occasionally radical independence of the will explains the paradoxical relationship between a good God and a world stained with evil.

"Why do we have to inquire into the origin of this movement by which the will is turned from immutable to transitory goods?" complains Augustine at one point (3.1.11). The will and its ability to do evil are facts of experience (3.1.12). Moreover, "What cause of the will could there be [when the will does evil], except the will itself? It is either the will itself—and it is not possible to go back to the root of the will—; or else it is not the will, and there is no sin. Either the will is the first cause of sin, or else there is no first cause" (3.17.168).[13] That is, either the human intellect or the human will is the cause of evil. But it cannot be the intellect; therefore it must be the will. It is as simple as this disjunctive syllogism. But Augustine admits that he cannot plumb the depths of the free will. Nor does he propose any method for dealing with the sinful will other than preaching religious conversion and developing political sanctions against criminal behavior.

The contemporary manifestation of the election of evil in the form of terrorism, employing modern technologies, appears equally intractable. Despite the fact the electronic media and computers are swiftly being adopted for religious uses, it remains highly doubtful that the electronic church realizes any net gain in the effectiveness of preaching, especially given the pervasive secularization to which the media also contribute. Although information technologies may make it theoretically more possible to identify terrorists and apply socially protective countermeasures, in practice the civil liberties traditions of the West limit such possibilities at the same time that existing information systems provide any potential terrorist with enhanced access to a plenitude of technological means and destructive powers.

Having distinguished two versions of the problem of incontinence and noted different ways information technologies contribute to their melioration, it remains to point out an incontinence-related volitional contradiction at the heart of the modern technological project. Ever since Augustine, the philosophical scandal of free will has occupied a place in the spectrum of philosophical conundrums. But the modern period, by identifying the will rather than the intellect as the highest aspect of humanity (Descartes), and by making freedom rather than justice the primary aim of politics (Rousseau), has given the conundrum a unique twist. It is exactly this identification of freedom as the human essence that can be argued to ground the technological project (the aim of which is to realize that freedom), while the project itself (the Enlightenment pursuit of a union of science and politics in knowledge-based power) presumes the impossibility of incontinent freedom.

"We hold these truths to be self-evident," proclaims the American Declaration of Independence, "that all men . . . are endowed by their creator with [inalienable rights to] life, liberty, and the pursuit of happiness [and] that to secure these rights governments are instituted among men." Not life and intelligence, but life and liberty or free will become the key characteristics of human beings. Popular discussions of the differences between computers (artificial intelligence) and human beings often confirm this position: machines may be more intelligent than human beings, but they do not have a will. It is within the worldview reflected in such principles that modern technology has taken its firmest hold.

The problems and paradoxes raised by this identification of the human essence with freedom of the will began to be explored by such nineteenth-century thinkers as Arthur Schopenhauer and Fyodor Dostoyevsky. In *Notes from Underground* Dostoyevsky creates a protagonist who, though well aware of the rational recommendations of a utilitarian calculus, consciously chooses to act against it—out of a desire to preserve or affirm free will in an increasingly rationalized and technological setting. For the underground man, incontinence is no longer a vice, but becomes the essential virtue. It is this idea that, in one form or another, one can find present in existentialist discussions of free will from Nietzsche to Sartre. In light of Nazism, even Freud is forced to postulate a subconscious death instinct. Human action is ultimately not determined by reason. There is something more fundamental, more basic, more real—namely the will. This is witnessed by the fact of incontinence; knowing what is good on a rational level, human beings nevertheless often do something else. The challenge of such a phenomenon is heightened by the manifestation of technology as volition.

CONCLUSION
· · · ·
Continuing to Think about Technology

By way of conclusion it is appropriate to review and reflect on the argument and to consider possible extensions. Having come this far, how far are we—and is it possible to go farther?

The Argument Revisited

The guiding concern of this book has been to identify the stance and distinctions proper to thinking about technology philosophically, in a way that does not exclude engineering discourse. Indeed, the argument is that philosophy of technology must engage that discourse while remaining open as well to the widest possible spectrum of humanities discussions. The intention has been to seek a purchase not just on the complex reality that is technology, but also on the sometimes cacophonous exchange that constitutes the modern debate about a multitude of issues related to the nature and meaning of technology. It might even be said by way of conclusion that philosophy *of* technology, no matter whether "of" is interpreted as subjective or objective genitive, is too narrow a term. Studies of philosophy *and* technology, with both entering into the fray, are what is needed.

To this end the introduction sketched a historical and organizational background most immediately relevant to contemporary philosophy and technology studies. Highlighted were certain tensions between theory and practice and a suggested need to reaffirm discussions that are not immediately ethical in import.

The work then began in earnest with part 1 by distinguishing two traditions in the philosophy of technology—engineering and humanities philosophy of technology—and proceeded to argue that the latter is more inherently philosophical than the former. It followed with a summary overview of the issues in humanities philosophy of technology in terms of the traditional branches of philosophy. Part 1 concluded by extending the basic questioning into an investigation of the possibility of a premodern philosophy of technology.

A crucial other side of the argument of part 1, however, was that humanities philosophy of technology needed to demonstrate its ability to take seriously and to comprehend engineering-technological experience, to some extent even on its own terms. The need for this shift in emphasis was restated at the beginning of part 2, in which the aim is not just to defend humanities philosophy of technology but to broaden and deepen it—that is, to take technology seriously, philosophically seriously. The chapters of part 2 thus intentionally wallow in the details of engineering texts. Although neither necessary nor sufficient, such a close and extended engagement with engineering discourse is surely one way to respond to potential engineering complaints that the humanities do not take engineering seriously. Unlike previous humanities philosophy of technology, part 2 attempts to reconnoiter technology as experienced and practiced. Although by no means exhaustive, many of the analyses here should be able to bear inspection by engineers themselves.

The thesis of part 2 is that we need to develop a well-articulated understanding of technology in its diversity of manifestations. Initially, this was done with a provisional distinction between four modes of manifestation. In contrast to previous humanities philosophies of technology, this framework encourages an extensive encounter with technology and engineering discourse, and discloses numerous differences where humanities philosophers have tended to find only sameness. Technology as object can be distinguished according to types of objects (utilities, tools, machines), technology as knowledge according to types of knowledge (maxims, rules, theories), technology as activity according to types of activity (making, designing, maintaining, using), and technology as volition according to types of volition (active will, receptive will).

One can argue that the hypothesis of this analysis has been supported or provisionally confirmed by the results of its utilization. More completely than other frameworks, it begins to map out and appreciate the philosophical richness of technology. Indeed, although at times different types of technology as object, as knowledge, as activity, and as volition may appear to be only more or less fully present and thus perhaps inappropriately classified within any one modality, this can now be viewed as the result of an ability of the framework to reveal a plethora of significant elements. At some level, all four modalities are immediately present and provide, as it were, different perspectives on or entrées into technology. Moreover, between the fourfold manifested presence of a "complete" technology one can postulate such theoretical and sometimes practical possibilities of overlapping technologies

- as object and as activity (in the absence of technology as knowledge and as volition)—that is, as play with toys;

- as knowledge and as activity (in the absence of technology as object and as volition)—that is, the engineering design of imaginary cars as a hobby;

- as activity and as volition (in the absence of technology as object and as knowledge in the strong sense)—that is, pure technical skill;

- as object, as activity, and as volition (in the absence of conscious technology as knowledge)—that is, a mass production assembly line.

Lest this modal framework appear too much like a rationalist construction, it can and ought also to be interpreted simply as a set of quasi-empirical categories for speaking about technology. About technology one can speak in many ways. Such speakings fall out into four genera or kinds not wholly unlike those itemized by Aristotle for that much more general level of phenomena known as words. But as Aristotle's failure to provide an invariable list of his categories indicates, no such reflective abstraction should be reified. It must remain a flexible framework for both thought and action.

In this manner the elaborated framework provides a way to begin to engage the philosophical stance of questioning with a multitude of engineering concepts and discussions. The proof or defense of this hypothesis is thus not one that can be stated directly but is present only insofar as it contributes to the elucidation of what is going on in numerous other discussions—as it begins to bridge, from the side of philosophy, the gap between engineering and humanities philosophy of technology. Only as philosophy of technology is transformed into philosophy and technology studies does it fully realize this ideal.

This book is meant as a kind of celebration of and contribution to pluralistic philosophy and technology studies. In the effort to put forth a comprehensive, engineering-sensitive analysis, this celebration further aspires to engage in a truly international dialogue. On both counts, the argument aspires to make common cause with such groups as the Society for Philosophy and Technology (SPT). SPT presidents—

Larry Hickman (1993-1995)   Alex Michalos (1983-1985)   Carl Mitcham (1981-1983)

José Sanmartin (1995-1997)   Kristin Shrader-Frechette
(1985-1987)   Marx Wartofsky (1987-1989)

Langdon Winner (1991-1993)   Joseph Pitt (1989-1991)

Presidents of the Society for Philosophy and Technology. Drawings by Emilie Jaffe.

coming from the United States, Canada, and Spain—have included defenders (Joseph Pitt) and critics (Langdon Winner) of technological rationality, proponents (Alex Michalos) and questioners (Kristin Shrader-Frechette) of quantative and analytic technology assessments, Marxists (Marx Wartofsky) and pragmatists (Larry Hickman). Members from both engineering and the humanities regularly meet both in the United States and abroad, somewhat on the periphery of the academic limelight, where in fact more serious work can be done.

In response to the possible charge that such peripheral celebration only opens up problems without solving them, one can observe that its work is nevertheless grounded in an implicit vision of what philosophy and technology studies should be—a vision growing out of an expansive belief about philosophy itself. Philosophy has been most vital when its representatives were actually engaged with the kinds of things and experiences it was talking about, not simply operating as the specialized discipline of a professional class. Socrates and Plato participated in the political life of the polis, to the profit of both political life and political philosophy. Aristotle was a "scientist" (primarily what today might be termed a biologist) as much as a philosopher. Augustine was not just a philosopher (or theologian) but a pastoral bishop. Thomas Aquinas's theological scholarship is enriched by his Scholastic teaching and front-line development of Christian doctrine. Descartes and Leibniz were working scientists (primarily physicists) as well as philosophers. Rousseau and Marx promoted the revolutionary political movements of their time. Only since the classic phase of the Industrial Revolution have many philosophers ceased to be engaged with the world they live in—except as journalists or culture critics.

Without in any way implying that this book is comparable to the works of such exemplary figures, one can still try to imitate their model and keep the philosophy of technology from becoming an arid, sterile discipline after the manner of so much recent philosophy of science and linguistic analysis. Independent of what engineers might charge, it is good for philosophy, if it wishes to reflect on technology, to engage engineering practice and take it seriously, and that is the idea behind the perhaps clumsy and bookish analysis of technical texts in much of part 2.

Critics might well suggest, however, that to engage engineering and technology after the manner of Plato's participation in politics or Aristotle's investigations of nature, one needs to be an engineer rather than simply an analyst of engineering texts. There is some truth to this. But even without actually practicing engineering—an exercise that, because of the character of modern technology, it would be difficult to combine with substantial study of philosophy—an informed and sympathetic reading of its texts is not without significance.

Science, Technology, and Society Studies

Another dimension of real-world engagement that can benefit, and benefit from, a deepened humanities philosophy of technology is the field of Science, Technology, and Society (STS) studies. STS studies received their practical impetus from the 1960s rising awareness of environmental pollution and the consumer movement, but they were given supplementary stimulus from concerns about nuclear weapons, the social impact of rapid technological change (e.g., automation), the space race with the Soviet Union, the energy crisis, advances in biomedical engineering, problems of technology transfer, and similar issues. At a more theoretical level a breakdown took place in the pursuit of a logical reconstruction or foundationalist philosophy of science.

The year 1962 witnessed publications from both these areas that were to have widespread consequences: Rachel Carson's *Silent Spring* and Thomas Kuhn's *The Structure of Scientific Revolutions*. A ferment of discussion began to open up about the place of science and technology in society. In 1925 Alfred North Whitehead could confidently propose that "more and more it is becoming evident that what the West can most readily give to the East is its science and its scientific outlook," that it can do so almost effortlessly "wherever there is a rational society" and "without the wanton destruction of [some indigenous cultural] inheritance."[1] By 1975 this proposition was, to say the least, highly doubtful, and it was increasingly obvious that the interrelations between science, technology, and society were complex in the extreme.

The pedagogical reaction was to establish what have come to be known as STS programs at a number of major universities, especially in the United States. More recently the idea of STS studies has been expanded and promoted as a necessary component of the liberal arts in a highly technological society, even a new kind of literacy, and extended to institutions of higher learning without formal STS programs as well as to secondary and primary curricula.[2]

Within the STS community, however, one can readily identify two approaches. Simplifying somewhat, the first is that of engineers, who argue that the basic problems associated with technology are based on a lack of understanding about technology itself. Those who criticize technology for (say) causing environmental pollution fail to appreciate what science and technology have done and can do. What is needed is more knowledge about science and technology and its social relationships so that people can act more effectively to realize their ends. The problems of technology will be solved not by less technology, but by more and better technology. For such persons STS courses are a means to promote greater awareness of and appreciation for science and technology.

Another approach is that of the humanities. Here the argument is sometimes that certain problems are caused by the inherent nature of technology itself, that not more but less technology or alternative technologies are required to deal with the problems of environmental pollution and societal change. Furthermore, the powers of technology, while unable to prescribe the values they should serve, nevertheless transform social structures in ways that tend factually to predetermine their uses. Literature, drama, poetry, history, even the social sciences have all developed means for exploring such aspects of the STS relationship and thereby disclosing the limits of the technological way of being in the world.

As an academic discipline, philosophy has, until quite recently, been among the less responsive and creative in this respect. Philosophers are often uneasy viewing themselves as part of a larger movement like STS studies. They want to defend their own disciplinary or subdisciplinary interests, which is perhaps why the "applied turn" in philosophy has led to little more than another kind of philosophy and the promotion of bioethics, computer ethics, and such. But situating the philosophy of technology as part of the larger STS field may be a way of engaging with the world that befits philosophy by providing philosophy a broader framework within which to develop its special insights. Insofar as philosophy is at the heart of the humanities, bringing technology into philosophy constitutes placing technology in a broader context. But if technology is understood as part of a science-technology-society relationship, and philosophy takes up an active role in culture, then thinking through *technology* might become *thinking through* technology.

Along with the more informed turn to ethics adumbrated at the end of part 2 and the bringing of technology into the humanities, are there other general implications of this approach? One must admit that the ultimate payoff from many of the distinctions in this volume is not clear and cannot be known in advance. Theory seldom reveals its own implications. The inherent attractiveness and influence of theory, however, is attested by the general impression that philosophy and technology courses have on occasion encouraged engineers to change their majors to the humanities and, although more rarely, even brought students of the humanities into engineering. In each instance the engineering-humanities and humanities-engineering scholars are different from their colleagues.

On the argument that human beings are committed to technology by their very nature as *Homo faber*—an argument reenunciated by Samuel Florman in *The Civilized Engineer* (1987)—this seems somewhat inexplicable. As Florman explains it, "I think that engineering is what human beings, deep down, want to do. Not the only thing, but one of the most basic and satisfying things. Engineering is an activity that is fulfilling—*existentially*. . . . To be human is to be technological. When we are being technological we are being human" (pp. 19–20). But even were the human properly defined by technological abilities, Florman's argument fails to acknowledge that there are at least two fundamentally different types of technology. One implication of the analysis of part 2 is thus to raise basic questions about this kind of pro-technology argument, but in a manner that does not require straightforward confrontation and opposition. We do not have to argue that human beings are not essentially technological in order to question whether they are necessarily involved with modern technology.

One final point: In light of the manifold aspects of technology revealed by the typology, the development of engineering ethics, bioethics, computer ethics, and the like may never be sufficient to "control" technology in its modern form. Certainly, engineering ethics as whistle-blowing seldom comes to grips with what engineers themselves take to be the essence of engineering—that is, design. But such ideas, again, point beyond this book—toward ethics—for which it must remain a prolegomenon. This work remains no more than a beginning to thinking through technology.

EPILOGUE

• • • •

Three Ways of Being-with Technology

In serious discussions of relations between technology and humanity there readily arises a general question about the primary member in this relationship. It is difficult to deny that we exercise some choice over the kinds of technics we live with—that we control technology. But it is equally difficult to deny that technics exert profound influences on the ways we live—that they structure our existence. "We shape our buildings," Winston Churchill once remarked (apropos of proposals for a new Parliament building); "thereafter they shape us."[1] But which comes first, logically if not temporally—the builder or the buildings? Which is primary—humanity or technology?

This is of course a chicken-and-egg question, one not subject to any straightforward, definitive answer. But it is not therefore insignificant, nor is it enough to propose as some kind of synthesis that there is simply a mutual relationship between the two, that humanity and technology are always found together. Mutual relationship is not some one thing; mutual relationships take many different forms. There are, for instance, mutualities of parent and child, of husband and wife, or of citizens. Humanity and technology can be found together in more than one way. Rather than argue the primacy of one or the other factor or the cliché of mutuality in the humanity-technology relationship, we can better pursue understanding through a structural examination of three forms the relationship itself can take, three ways of being-with technology.

Being-with: From Persons to Technics

To speak of three ways of being-with technology is necessarily to borrow and adapt a category from Martin Heidegger's *Being and Time* (1927) in a manner that deserves acknowledgment. In his seminal work, Heidegger proposes to develop a new understanding of being human by taking the primordial human condition, being-in-the-world, and subjecting this given to what he terms an existential analysis. The analysis proceeds by elucidating three equiprimordial aspects of this condition of being human: *the world* within which the human finds itself, the *being-in* relationship, and the *being* who is in the relationship—all as a means of approaching what, for Heidegger, is the fundamental question, the meaning of Being.

The fundamental question need not, on this occasion, be addressed. What we can briefly consider instead is the central place of technics in Heidegger's analysis and the disclosure of being-with as one of its central features. For Heidegger the "worldhood of the world," as he calls it, comes into view through technical engagements, which reveal a network of equipment and artifacts ready-to-hand for manipulation, and other human beings likewise so engaged. These others are neither just technically ready-to-hand (like tools) nor even scientifically present-at-hand (like natural objects); on the contrary, they are *like* the very human being who notices them in that "*they are there too, and there with it*" (1927 [trans. 1962, p. 154]).

The being-with relationship thus disclosed through technical engagements is therefore primarily social; it refers to the social character of the world that comes to light through technical practice. Such a world is composed not solely of tools and artifacts, but of tools used with others and artifacts belonging to others. Technical engagements are not just technical but have an immediately and intimately social dimension. Indeed, this is all so immediate that it requires a labored stepping back even to recognize and state—a distancing and articulation which are in large part precisely what philosophy is about.

The present attempt to step back and examine various ways of being-with technology rather than being-with others (through technology) takes off from but does not proceed in the same manner as Heidegger's social analysis of the They and the problem of authenticity in the technological world. For Heidegger, being-with refers to an immediate personal presence in technics. Social being-with can manifest itself, however, not only on the level of immediate or existential presence but also in ideas. Indeed, the social world is as much a world of ideas as of persons, if not more so. Persons hold ideas and interact with others and with things through them. These ideas can even enclose the realm of technics—that is, become a language or *logos* of technics, a "technology."

The idea of being-with technology presupposes this "logical" encompassing of technics by a society and its philosophical or proto-philosophical articulation. For many people, however, the ideas that guide their lives may not be held with conscious awareness or full articulation. They often take the form of myth. Philosophical argument and discussion introduce into such a world of ideas a break or rupture with the immediately given. This rupture need not require rejecting or abandoning that given, but it will entail bringing the given into fuller consciousness or awareness, from which it must be accepted (or rejected) in a new way or on new grounds.

Against this background, then , it is possible to develop historicophilosophical descriptions, necessarily somewhat truncated, of three alternative ways of being-with technology. The first is what may be called ancient skepticism; the second, Renaissance and Enlightenment optimism; and the third, romantic ambiguity or uneasiness. Even in the somewhat simplified form of ideal types in which they will be presented, considering the issues that divide these three ways of being-with technology may help illuminate the difficulties we face in trying to live with modern technology and its manifest problems.

Ancient Skepticism

The original articulation of a relationship between humanity and technics, an articulation that in its earliest forms is coeval with the appearance of recorded history, can be stated boldly as "technology is bad but necessary" or, perhaps more carefully, as "technology (that is, the study of technics) is necessary but dangerous." The idea is hinted at by a plethora of archaic myths—such as the story of the Tower of Babel or the myths of Prometheus, Hephaestus, and Daedalus and Icarus. Certainly the transition from hunting and gathering to the domestication of animals and plants introduced a profound and disturbing transition into culture. Technics, according to these myths, although to some extent required by humanity and thus on occasion a cause for legitimate celebration,[2] easily turns against the human by severing it from some larger reality—a severing that can be manifest in a failure of faith or shift of the will, a refusal to rely on or trust God or the gods, whether manifested in nature or in Providence.[3]

Ethical arguments in support of this distrust or uneasiness about technical activities can be detected in the earliest strata of Western philosophy. According to the overlooked works of the Greek military hero and historian Xenophon, for instance, his teacher Socrates (469–399 B.C.E.) considered farming, the least technical of the arts, to be the most philosophical of occupations. Although the earth "provides the goods things most abundantly, farming does not yield them up to soft-

ness but . . . produces a kind of humanity. . . . Moreover, the Earth, being a goddess, teaches justice to those who are able to learn" (*Oeconomicus* 5.4, 12). This idea of agriculture as the most virtuous of the arts,
one in which human technical action tends to be kept within proper
limits, is repeated by representatives of the philosophical tradition as
diverse as Plato,[4] Aristotle,[5] Thomas Aquinas,[6] and Thomas Jefferson.[7]

Elsewhere Xenophon notes Socrates' distinction between questions
about *whether* to perform an action and about *how* to perform it, along
with one between scientific or technological questions concerning the
laws of nature and ethical or political questions about what is right
and wrong, good and bad, pious and impious, just and unjust. In elaborating on the whether/how distinction, Socrates stresses that human
beings must determine for themselves *how* to perform their actions—
that they can take lessons in "construction [*tektonikos*], forging metal,
agriculture, ruling human beings, and . . . calculation, economics, and
military strategy" (*Memorabilia* 1.1.7) and therefore should not depend
on the gods for help in "counting, measuring, or weighting" (*Memorabilia* 1.1.9); the ultimate consequences of their technical actions are
nonetheless hidden. His initial example is even taken from agriculture:
the man who knows how to plant a field does not know whether he
will reap the harvest. Thus whether we should employ our technical
powers is a subject about which we must rely on guidance from the
gods (cf. also *Memorabilia* 4.7.10, and *Anabasis* 3.1).

At the same time, with regard to the science/ethics distinction, Socrates argues that because of the supreme importance of ethical and
political issues, human beings should not allow themselves to become
preoccupied with scientific and technological pursuits. In the intellectual autobiography attributed to him in the *Phaedo*, for instance, Socrates relates how he turned away from natural science because of the
cosmological and moral confusion it tends to engender (cf. also *Memorabilia* 4.7.6–7). In the *Memorabilia* it is similarly said of Socrates that
"he did not like others discuss the nature of all things, nor did he
speculate on the 'cosmos' of the sophists or the necessities of the heavens, but he declared that those who worried about such matters were
foolish. And first he would ask whether such persons became involved
with these problems because they believed that their knowledge of
human things was complete or whether they thought they were obli-
gated to neglect human things to speculate on divine things" (*Memorabilia* 1.1.11–12).

Persons who turn away from human things to things having to do
with the heavens appear to think "that when they know the laws by

which everything comes into being, they will, when they choose, create winds, water, seasons, and anything else like these that they may need" (*Memorabilia* 1.1.15) (cf. Empedocles, frag. 111; see also *Academica* 1.4.15). As "the first to call philosophy down from the heavens and place it in the city and . . . compel it to inquire about life and morality and things good and bad" (Cicero *Tusculan Disputations* 5.4.10–11),[8] however, Socrates' own conversation is described as always about human things: What is pious? What is impious? What is good? What is shameful? What is just? What is unjust? What is moderation? For, as Xenophon says on another occasion, Socrates "was not eager to make his companions orators and businessmen and inventors, but thought that they should first possess moderation [*sophrosune*]. For he believed that without moderation those abilities only enabled a person to become more unjust and to work more evil" (*Memorabilia* 4.3). The whether/how distinction grants technical or how-to questions a realistic prominence in human affairs but recognizes their ambiguity and uncertainty; the science/ethics one subordinates any systematic pursuit of technical knowledge to ethical and political concerns.

Such uneasiness before the immoderate possibilities inherent in technological powers is further elaborated by Plato. Near the beginning of the *Republic*, after Socrates outlines a primitive state and Glaucon objects that this is no more than a "city of pigs," Socrates replies:

> The true state is in my opinion the one we have described—a healthy state, as it were. But if you want, we can examine a feverish state as well. . . . For there are some, it seems, who will not be satisfied with these things or this way of life; but beds, tables, and other furnishings will have to be added, and of course seasonings, perfume, incense, girls, and sweets—all kinds of each. And the requirements we mentioned before can no longer be limited to the necessities of houses, clothes, and shoes; but [various *technai*] must also be set in motion. . . . The healthy state will no longer be large enough either, but it must be swollen in size by a multitude of activities which go beyond the meeting of necessities. (372d–373b)

As this passage implies, and as can be confirmed by earlier references to Homer and the poets, classical Greek culture was shot through with a distrust of the wealth and affluence that the *technai* or arts could produce if not kept within strict limits. For according to the ancients such wealth accustoms people to easy things. But *kalepa ta kala*, difficult is the beautiful or the perfect; the perfection of anything, including human nature, is the opposite of what is soft or easy. Under conditions of affluence human beings tend to become accustomed to ease, and thus to choose the less over the more perfect, the lower over the higher, both for themselves and for others.

With no art is this more prevalent than with medicine. Once drugs are available as palliatives, for instance, most individuals will choose them for the alleviation of pain over the more strenuous paths of physical hygiene or psychological enlightenment. The current *techne* of medicine, Socrates maintains to Glaucon later in *Republic* 3, is an education in disease that "draws out death" (406b); instead of promoting health, it allows the unhealthy to have "a long and wretched life" and "to produce offspring like themselves" (407d). That Socrates' description applies even more strongly to modern medical technology than to that current in Athens scarcely need be mentioned.

Another aspect of this tension between politics and technology is illustrated by Plato's observations on the dangers of technical change. In the words of Adeimantus, with whom Socrates in this instance evidently agrees, once change has established itself as normal in the arts, "it overflows its bounds into human character and activity and from there issues forth to attack commercial affairs, and then proceeds against the laws and political orders" (424d–e). It is desirable that obedience to the law should rest primarily on habit rather than force. Technological change, which undermines the authority of custom and habit, thus tends to introduce violence into the state. Surely this is a possibility that the experience of the twentieth century, one of the most violent in history, should encourage us to take seriously.

This wariness about technological activity on moral and political grounds can be supplemented by an epistemological critique of the limitations of technological knowledge and a metaphysical analysis of the inferior status of technical objects. During a discussion of the education of the philosopher-king in *Republic* 7, Socrates considers what kind of teaching most effectively brings a student "into the light" of the highest or most important things. One conclusion is that it is not those *technai* that "are oriented toward human opinions and desires or concerned with creation and fabrication and attending to things that grow and are put together" (533b). Because it cannot convert or emancipate the mind from the cares and concerns of the world, technology should not be a primary focus of human life. The orientation of technics, because it is concerned to remedy the defects in nature, is always toward the lower or the weaker (342c–d). A doctor sees more sick people than healthy ones. Eros or love, by contrast, is oriented toward the higher or the stronger; it seeks out the good and strives for transcendence. "And the person who is versed in such matters is said to have spiritual wisdom, as opposed to the wisdom of one with *technai* or low-grade handicraft skills," Diotima tells Socrates in the *Symposium* (203a).

Aristotle agrees, but for quite different, more properly metaphysical reasons. According to Aristotle and his followers, reality or being resides in particulars. It is not some abstract species Homo Sapiens (with capital H and capital S) that *is* in the primary sense, but Socrates and Xanthippe. However, the reality of all natural entities is dependent on an intimate union of form and matter, and the telos or end determined thereby. The problem with artifacts is that they fail to achieve this kind of unity at a very deep level and thus can have a variety of uses or extrinsic ends imposed upon them. As Aristotle observes, if a bed sprouts what grows is not a bed but a tree (*Physics* 193b10). Insofar as it truly imitates nature, art engenders an inimitable individuality in its products, precisely because its attempt to effect as close a union of form and matter as possible requires a respect for or deference to the materials it works with. In a systematized art or technology matter necessarily tends to be overlooked or relegated to the status of an undifferentiated substrate to be manipulated at will. Indeed, in relation to this Aristotle suggests a distinction between the arts of cultivation—for example, medicine, education, and agriculture, which help nature to produce more abundantly things that she could produce of herself—from those of construction or domination—arts that bring into existence things nature would not produce (compare Aristotle *Physics* 2.1.193a12–17; *Politics* 7.7.1337a2; and *Oeconomica* 1.1.1343a26–1343b2).

The metaphysical issue here can be illustrated by observing the contrast between a handcrafted ceramic plate and Tupperware dishes. The clay plate has a solid weight, rich texture, and explicit reference to its surroundings not unlike that of a natural stone, whereas Tupperware exhibits a lightness of body and undistinguished surface that only abstractly engages the environment of its creation and use. As an advertising argument might say, since synthetic products are "better than the real thing," the word "synthetic," which implies a "pallid imitation," ought to be discarded. But whether this is true depends heavily on a prior understanding of what is real in the first place. For Aristotle there is a kind of reality that can be found only in particulars and is thus beyond the scope of mass production, function-oriented polymer technology.

For Plato and the Platonic tradition, too, artifice is less real than nature. Indeed, in *Republic* 10 there is a discussion of the making of beds (to which Aristotle's remarks from the *Physics* may allude) by god or nature, by the carpenter, or *tekton*, and by painter or artist. Socrates' argument is that the natural bed, the one made by the god, is the primary reality; the many beds made in imitation by artisans are a secondary reality; and the pictures of beds painted by artists are a tertiary reality. *Techne* is thus creative in a second or "third generation" sense (597e)—and thus readily subject to moral and metaphysical guidance.

In moral terms artifice is to be guided or judged in terms of its goodness or usefulness. In metaphysical terms the criterion of judgment is proper proportion or beauty. One possible disagreement between Platonists and Aristotelians with regard to one or another aspect of making is whether the good or the beautiful, ethics or aesthetics, is the proper criterion for its guidance. Such disagreement should nevertheless not be allowed to obscure a more fundamental agreement, the recognition of the need to subject *poiesis* and *technai* to certain well-defined limitations. Insofar as technical objects or activities fail to be subject to the inner guidance of nature (*phusis*), nature must be brought to bear upon them consciously, from the outside as it were, by human beings. Again, the tendency of contemporary technical creations to bring about environmental problems or ecological disorders to some extent confirms the premodern point of view.

The ancient critique of technology thus rests on a tightly woven, fourfold argument: (1) the will to technology or the technological intention often involves a turning away from faith or trust in nature or Providence; (2) technical affluence and the concomitant processes of change tend to undermine individual striving for excellence and societal stability; (3) technological knowledge likewise draws human beings into intercourse with the world and obscures transcendence; (4) technical objects are less real than objects of nature. Only some necessity of survival, not some ideal of the good, can justify setting aside such arguments. The life of the great Hellenistic scientist Archimedes provides us (as it did antiquity) with a kind of icon or lived-out image of these arguments. Although, according to Plutarch, Archimedes was capable of inventing all sorts of devices, he was too high-minded to do so except when pressed by military necessity—yet even then he refused to leave behind any treatise on the subject because of a salutary fear that his weapons would be too easily misused by humankind (Plutarch, "Life of Marcellus," near the middle).

Allied with the Judeo-Christian-Islamic criticism of the vanity of human knowledge and of worldly wealth and power,[9] this premodern distrust of technology dominated Western culture until the end of the Middle Ages, and elements of it can be found vigorously repeated by numerous figures since—from Samuel Johnson's neoclassicist criticism of Milton's promotion of education in natural science[10] to Norbert Wiener, who in 1947, like Archimedes twenty-three hundred years before, vowed not to publish anything more that could do damage in the hands of militarists.[11] In one less well-known allusion to another aspect of the classical moral argument, John Wesley (1703–1791), in both private journals and public sermons, ruefully acknowledges the paradox that Christian conversion gives birth to a kind of self-discipline that easily engenders the accumulation of wealth, which then readily undermines true Christian virtue. "Indeed, according to the natural tendency of riches, we cannot expect it to be otherwise," writes Wesley.[12]

In contemporary versions of other aspects of the premodern critique, Lewis Mumford has criticized the will to power manifested in modern technology, and Heidegger, following the lead of the poet Rainer Maria Rilke, has invoked the metaphysical argument by pointing out the disappearance of the thinghood of things, the loss of a sense of the earth in mass-produced consumer objects. From Heidegger's point of view, nuclear annihilation of all things would be "the mere final emission of what has long since taken place, has already happened." [13]

From the point of view of the ancients, then, being-with technology is an uneasy being-alongside-of and working-to-keep-at-arms-length. This premodern attitude looks on technics as dangerous or guilty until proven innocent or necessary—and in any case, the burden of proof lies with those who favor technology, not those who would restraint it.

Enlightenment Optimism

A radically different way of being-with technology—one that shifts the burden of proof from those who favor to those who oppose the introduction of inventions—argues the inherent goodness of technology and the consequent accidental character of all misuse. Aspects of this idea or attitude are not without premodern adumbration. But in comprehensive and persuasive form arguments to this effect are first fully articulated in the writings of Francis Bacon (1561–1626) at the time of the Renaissance and subsequently become characteristic of the Enlightenment philosophy of the eighteenth century.

Like Xenophon's Socrates, Bacon grants that the initiation of human actions should be guided by divine counsel. But unlike Socrates, Bacon maintains that God has given humanity a clear mandate to pursue technology as a means for the compassionate melioration of the suffering of the human condition, of being-in-the-world. Technical knowhow is cut loose from all doubt about the consequences of technical action. In the choice between ways of life devoted to scientific-technological or ethical-political questions, Bacon further argues that Christian revelation directs men toward the former over the latter. "For it was not that pure and uncorrupted natural knowledge whereby Adam gave names to the creatures according to their propriety, which gave occasion to the fall. It was the ambitious and proud desire of moral knowledge to judge of good and evil, to the end that man may revolt from God and give laws to himself, which was the form and manner of the temptation" (*The Great Instauration*, preface).

Contrary to what is implied by the myth of Prometheus or the legend of Faust, it was not scientific and technological knowledge that led to the Fall, but vain philosophical speculation concerning moral questions. Formed in the image and likeness of God, human beings are called on to be creators; to abjure that vocation and pursue instead an unproductive discourse on ethical quandaries brings about the just punishment of a poverty-stricken existence. "He that will not apply new remedies must expect new evils" ("Of Innovations"). Yet "the kingdom of man, founded on the sciences," says Bacon, is "not much other than ... the kingdom of heaven" (*Novum Organum* 1.68).

The argument between Socrates and Bacon is not, it is important to note, simply one between anti- and pro-technology partisans. Socrates allows technics a legitimate but strictly utilitarian function, then points out the difficulty of obtaining a knowledge of consequences on which to base any certainty of trust or commitment. Technical action is circumscribed by uncertainty or risk. Bacon, however, although he makes some appeals to a consequentialist justification, ultimately grounds his commitment in something approaching deontological principles. The proof is that he never even considers evaluating technical projects on their individual merit, but simply argues for an all-out affirmation of technology in general. It is right to pursue technological action, never mind what might look like dangerous consequences. Intuitions of uncertainty are jettisoned in the name of revelation.

The uniqueness of the Baconian (or Renaissance) interpretation of the theological tradition is also to be noted. For millennia the doctrines of God as creator of "the heavens and the earth" (Gen. 1:1) and of human beings as made "in the image of God" (Gen. 1:27) exercised profound influence over Jewish and later Christian anthropology, without ever being explicitly interpreted as a warrant for or a call to technical activity. Traditional or premodern interpretations focus on the soul, the intellect, or the capacity for love as the key to the *imago Dei*.[14] The earliest attribution to this doctrine of technological implications occurs in the early Renaissance. The contemporary theological notion of the human as using technology to prolong creation or co-create with God depends on just the reinterpretation of Genesis adumbrated by Bacon.

The Enlightenment version of Bacon's religious argument is to replace the theological obligation with a natural one. In the first place, human beings simply could not survive without technics. As d'Alembert puts it in the "Preliminary Discourse" to the *Encyclopedia* (1751), there is a prejudice against the mechanical arts that is a result of their accidental association with the lower classes. In truth,

> the advantage that the liberal arts have over the mechanical arts, because of their demands upon the intellect and because of the difficulty of excelling in them, is sufficiently counterbalanced by the quite superior usefulness which the latter for the most part have for us. It is their very usefulness which reduced them perforce to purely mechanical operations in order to make them accessible to a larger number of men. But while justly respecting great geniuses for their enlightenment, society ought not to degrade the hands by which it is served.[15]

In the even more direct words of Immanuel Kant, "Nature has willed that man should, by himself, produce everything that goes beyond the mechanical ordering of his animal existence, and that he should partake of no other happiness or perfection than that which he himself, independently of instinct, has created by his own reason."[16] Nature and reason, if not God, command humanity to pursue technology; the human being is redefined not as *Homo sapiens* but as *Homo faber*. Technology is the essential human activity. In more ways than Kant explicitly proclaims, "Enlightenment is man's release from his self-incurred tutelage."[17]

Following a redirecting (Bacon) or reinterpreting (d'Alembert and Kant) of the will, Bacon and his followers explicitly reject the ethical-political argument against technological activities in the name of moderation. With no apparent irony, Bacon maintains that the inventions of printing, gunpowder, and the compass have done more to benefit humanity than all the philosophical debates and political reforms throughout history. It may, he admits, be pernicious for an individual or a nation to pursue power. Individuals or small groups may well abuse such power. "But if an man endeavor to establish and extend the power and dominion of the human race itself over the universe," writes Bacon, "his ambition (if ambition it can be called) is without doubt both a more wholesome and a more noble thing than the other two." And, of course, "the empire of man over things depends wholly on the arts and sciences" (*Novum Organum* 1.129).

Bacon does not expound at length on the wholesomeness of technics. All he does is reject the traditional idea of their corrupting influence on morals by arguing for a distinction between change in politics and in the arts. "In matters of state a change even for the better is distrusted [Bacon observes], because it unsettles what is established; these things resting on authority, consent, fame and opinion, not on demonstration. But arts and sciences should be like mines, where the noise of new works and further advances is heard on every side" (*Novum Organum* 1.90). Unlike Aristotle and Aquinas, both of whom noticed the same distinction but found it grounds for caution in technology,[18] Bacon thinks the observation itself is enough to set technology on its own path of development.

Bacon's Enlightenment followers, however, go considerably further and argue for the positive or beneficial influence of the arts on morals. In the *Encyclopedia*, for instance, having identified "luxury" as simply "the use human beings make of wealth and industry to assure themselves of a pleasant existence," with its origin in "that dissatisfaction with our condition . . . which is and must be present in all men," Saint-Lambert undertakes to reply directly to the ancient "diatribes by the moralists who have censured it with more gloominess than light."[19] Critics of material welfare have maintained that it undermines morals, and apologists have responded that this is the case only when it is carried to excess. Both are wrong. Wealth is, as we would say today, neutral. A survey of history reveals that luxury "did not determine morals, but . . . it took its character rather from them."[20] Indeed, it is quite possible to have a moral luxury, one that promotes virtuous development.

But if a first line of defense is to argue for moderation, and a second to urge neutrality, a third is to maintain a positive influence. David Hume (1711–1776), for instance, in his essay "Of Commerce," argues that a state should encourage its citizens to be manufacturers rather than farmers or soldiers. By pursuit of "the arts of *luxury*, they add to the happiness of the state."[21] Then, in "Of Refinement in the Arts," he explains that the ages of luxury are both "the happiest and the most virtuous" because of their propensity to encourage industry, knowledge, and humanity." "In times when industry and the arts flourish," writes Hume, "men are kept in perpetual occupation, and enjoy, as their reward, the occupation itself, as well as those pleasures which are the fruit of their labour."[22]

Furthermore, the spirit of activity in the arts will galvanize that in the sciences and vice versa; knowledge and industry increase together. In Hume's own inimitable words: "We cannot reasonably expect that a piece of woolen cloth will be wrought to perfection in a nation which is ignorant of astronomy."[23] And the more the arts and sciences advance, "the more sociable men become." Technical engagements promote civil peace because they siphon off energy that might otherwise go into sectarian competition. Technological commerce and scientific aspirations tend to break down national and class barriers, thus ushering in tolerance and sociability. In the words of Hume's contemporary, Montesquieu, "Commerce is a cure for the most destructive prejudices; for it is a general rule, that wherever we find tender manners, there commerce flourishes; and that wherever there is commerce, there we meet with tender manners" (*Spirit of the Laws*, 1.20.1).

The ethical significance of technological activity is not limited to its socializing influence, however. Technology is an intellectual as well as a moral virtue, because it is a means to the acquisition of true knowledge. That technological activity contributes to scientific advance rests on a theory of knowledge that again is first clearly articulated by Bacon, who begins his *Novum Organum*, or "new instrument," with the argument that true knowledge is acquired only by a close intercourse with things themselves: "Neither the naked hand nor the understanding left to itself can effect much. It is by instruments and helps that the work is done, which are as much wanted for the understanding as for the hand" (*Novum Organum* 1.2). Knowledge is to be acquired by active experimentation and ultimately evaluated according to its ability to engender works. The means to true knowledge is what Bacon candidly refers to as the "torturing of nature"; left free and at large, nature, like human beings, is loath to reveal its secrets.[24] The result of this new way will be the union of knowledge and power (*Novum Organum* 1.3). Bacon is, quite simply, an epistemological pragmatist. What is true is what works. "Our only hope," he says, "therefore lies in a true induction" (*Novum Organum* 1.14).

The very basis of the great French *Encyclopedia, or Rational Dictionary of Sciences, Arts, and Crafts* is precisely this epistemological vision of unity between theory and practice. Bacon is explicitly identified as its inspiration and is praised for having conceived philosophy "as being only that part of our knowledge which should contribute to making us better or happier, thus . . . confining it within the limits of useful things [and inviting] scholars to study and perfect the arts, which he regards as the most exalted and most essential part of human science."[25] Indeed, in explicating the priorities of the *Encyclopedia*, the "Preliminary Discourse" goes on to say that "too much has been written on the sciences; not enough has been written well on the mechanical arts."[26] The article "Art" in the *Encyclopedia* further criticizes the prejudice against the mechanical arts, not only because it has "tended to fill cities with . . . idle speculation,"[27] but even more because of its failure to produce genuine knowledge. "It is difficult if not impossible . . . to have a thorough knowledge of the speculative aspects of an art without being versed in its practice," although it is equally difficult "to go far in the practice of an art without speculation."[28] It is this new unity of theory and practice—a unity based more in practice than in theory[29]—that is at the basis of, for instance, Bernard de Fontenelle's eulogies on the practice of experimental science as an intellectual virtue as well as a moral one and the Enlightenment reconception of Socrates as having called philosophy down from the heavens to experiment with the world.[30]

Bacon's true induction likewise rests on a metaphysical rejection of natural teleology. The pursuit of a knowledge of final causes "rather corrupts than advances the sciences," declares Bacon, "except such as have to do with human action" (*Novum Organum* 2.2). Belief in final causes or purposes inherent in nature is a result of superstition or false religion. It must be rejected in order to make possible "a very diligent dissection and anatomy of the world" (*Novum Organum* 1.124). Nature and artifice are not ontologically distinct. "All Nature is but Art, unknown to thee," claims Alexander Pope.[31] The Aristotelian distinction between arts of cultivation and of construction is jettisoned in favor of universal construction.

With regard to Pope, although it is not uncommon to find comparisons of the God/nature and artist/artwork relationships in Greek and Christian, ancient and modern authors, there are subtle differences. For Plato (*Sophist* 265b ff. and *Timaeus* 27c ff.) and Saint Augustine (*De civitate Dei* 11.21), for example, there is a fundamental distinction to be drawn between divine and human *poiesis*, both of which must be differentiated from *techne*. Also, even though made by a god, the world is not to be looked upon as an artifact or something that functions in an artificial manner. Thomas Hobbes, Bacon's secretary, however, proposes to view nature not just as produced by a divine art but as itself "the art whereby God hath made and governs the world" (*Leviathan*, introduction). Indeed, so much is this the case that for Hobbes human art itself may be said to produce natural objects. Or, to say the same thing in different words, the whole distinction between nature and artifice disappears.

This last point also links up with the first; metaphysics supports volition. If nature and artifice are not ontologically distinct, then the traditional distinction between technics of cultivation and technics of domination disappears. There is no technics that helps nature to realize its own internal reality, and human beings are free to pursue power. If nature is just another form of mechanical artifice, it is likewise reasonable to think of the human being as a machine. "Man is a machine and . . . in the whole universe there is but a single substance variously modified," concludes La Mettrie.[32] "For what is the heart," wrote Hobbes a century earlier, "but a spring; and the nerves, but so many strings; and the joints, but so many wheeles" (*Leviathan*, introduction). But the activities appropriate to machines are technological ones; *Homo faber* is yet another form of *l'homme-machine*, and vice versa.

Like that of the ancients, then, the distinctly modern way of being-with technology may be articulated in terms of four interrelated arguments: (1) the will to technology is ordained for humanity by God or by nature; (2) technological activity is morally beneficial because, while stimulating human action, it ministers to physical needs and increases sociability; (3) knowledge acquired by a technical closure with the world is more true than abstract theory; and (4) nature is no more real than artifice—indeed, it operates by the same principles. It is scarcely necessary to illustrate how aspects of this ideology remain part of intellectual discourse in Marxism, in pragmatism, and in popular attitudes regarding technological progress, technology assessment and public policy, education, and medicine.

Romantic Uneasiness

The premodern argument that technology is bad but necessary characterizes a way of being-with technology that effectively limited rapid technical expansion in the West for approximately two thousand years. The Renaissance and Enlightenment argument in support of the theory that technology is inherently good discloses a way of being-with technology that has been the foundation for a Promethean unleashing of technical power unprecedented in history. The proximate causes of this radical transformation were, of course, legion: geographic, economic, political, military, scientific. But what brought all such factors together in England in the mid-eighteenth century to engender a new way of life, what enabled them to coalesce into a veritable new way of being-in-the-world, was a certain optimism regarding the expansion of material development that is not to be found so fully articulated at any other point in premodern culture.[33]

In contrast to premodern skepticism about technology, however, the typically modern optimism has not retained its primacy in theory even though it has continued to dominate in practice. The reasons for this are complex. But faced with the real-life consequences of the Industrial Revolution, from societal and cultural disruptions to environmental pollution, post-Enlightenment theory has become more critical of technology. Romanticism, as the name for the typically modern response to the Enlightenment, thus implicitly contains a new way of being-with technology, one that can be identified with neither ancient skepticism nor modern optimism.

Romanticism is, of course, a multidimensional phenomenon. In one sense it can refer to a permanent tendency in human nature that manifests itself differently at different times. In another it refers to a particular manifestation in nineteenth-century literature and thought. Virtually all attempts to analyze this particular historical manifestation interpret romanticism as a reaction to and criticism of modern science. Against Newtonian mechanics, the romantics propose an organic cosmology; in opposition to scientific rationality, romantics assert the legitimacy and importance of imagination and feeling. What is seldom appreciated is the extent to which romanticism can also be interpreted as a questioning—in fact, the first self-conscious questioning—of modern technology.[34] So interpreted, however, romanticism reflects an uneasiness about technology that is nevertheless fundamentally ambivalent; although as a whole the romantic critique may be distinct from ancient skepticism and modern optimism, in its parts it nevertheless exhibits differential affinities with both.

Consider, to begin with, the volitional aspect of technology. On the ancient view, technology was seen as a turning away from God or the gods. On the modern view, it is ordained by God or, with the Enlightenment rejection of God, by nature. With the romantics the will to technology either remains grounded in nature or is cut free from all extrahuman determination. In the former instance, however, nature is reconceived not just as mechanistic movement but as an organic striving toward creative development and expression. From the perspective of "mechanical philosophy," human technology is a prolongation of mechanical order; from that of *Naturphilosophie* it becomes a participation in the self-expression of life. When liberated from even such organic creativity, technology is grounded solely in the human will to power, but with recognition of its often negative consequences; the human condition takes on the visage of gothic pathos.[35] The most one can argue, it seems, is that the technological intention—that is, the will to power—should not be pursued to the exclusion of other volitional options, or that it should be guided by aesthetic ideals.

William Wordsworth (1770–1850), for instance, the most philosophical of the English romantic poets, in the next-to-last book of his long narrative poem *The Excursion* (1814), describes how he has "lived to mark / A new and unforeseen creation rise" (8.89–90).

> Casting reserve away, exult to see
> An intellectual mastery exercised
> O'er the blind elements; a purpose given,
> A perseverence fed; almost a soul
> Imparted—to brute matter. I rejoice,
> Measuring the force of those gigantic powers
> That, by the thinking mind, have been compelled
> To serve the will of feeble-bodied Man.
>
> (8.200–207)

Here the rejoicing in and affirmation of technological conquest and control is clearly in harmony with Enlightenment sentiments.

> Yet in the midst of this exultation
> I grieve, when on the dark side
> Of this great change I look; and there behold
> Such outrage done to nature.
>
> (8.151–153)

And afterward he writes,

> How insecure, how baseless in itself,
> Is the Philosophy whose sway depends
> On mere material instruments;—how weak
> Those arts, and high inventions, if unpropped
> By virtue.
>
> (8.223–227)

Here Enlightenment optimism is clearly replaced by something approaching premodern skepticism.

Clarifying his position in the last book of the poem, Wordsworth admits that although he has complained, in regard to the factory labor of children, that a child is

subjected to the arts
Of modern ingenuity, and made
The senseless member of a vast machine
(9.157–159)

he is not insensitive to the fact that the rural life is also often an "un-
happy lot" enslaved to "ignorance," "want," and "miserable hunger"
(9.163–165). Nevertheless, he says, his thoughts cannot help but be

turned to evils that are new and chosen,
A bondage lurking under shape of good,—
Arts, in themselves beneficent and kind,
But all too fondly followed and too far.
(9.187–190)

In such lines Wordsworth no longer maintains with any equanimity
the Enlightenment principle that the arts are "in themselves beneficent
and kind." With his suggestion that the self-creative thrust has in technology been followed "too fondly" and "too far," and that bondage
has been created under the disguise of good, he introduces a profound
questioning. But unlike the ancients, who called for specific limitations
on technics, with the romantics there is no clear outcome other than a
critical uneasiness—or a heightened aesthetic sensibility.

Later, in a sonnet titled "Steamboats, Viaducts, and Railways"
(1835), having observed contradictions between the practical and aesthetic qualities of such artifacts, Wordsworth concludes that

In spite of all that beauty may disown
In your harsh features, Nature doth embrace
Her lawful offspring in Man's art; and Time,
Pleased with your triumphs o'er his brother Space,
Accepts from your bold hands the proffered crown
Of hope, and smiles on you with cheer sublime.

Once again technology, in Enlightenment fashion, is viewed as an extension of nature and even described in Baconian terms as the triumph
of time over space.[36] The "lawful offspring" is nevertheless ugly, full of
"harsh features" that beauty disowns. Yet from the "bold hands" of
technology temporal change is given the "crown of hope . . . with cheer
sublime" that things will work out for the good. In Wordsworth's own
commentary on *The Excursion*, the problem "is an ill-regulated and excessive application of powers so admirable in themselves."[37] But it is

precisely this ill-regulated and excessive technology that also gives birth to a new kind of admiration, the admiration of the sublime.

With regard to the moral character of technology, ambivalence is even more apparent. Consider, for instance, the arguments of Jean-Jacques Rousseau (1712–1778), a man who is in important respects the father of the romantic movement, and whose critique takes shape even before the inauguration of the Industrial Revolution itself, strictly in response to ideas expressed by the philosophes. In his 1750 *Discourse on the Sciences and Arts* Rousseau boldly argues that "as the commodities of life multiply, as the arts are perfected and luxury extended, true courage falls away, the militant virtues fade away."[38] But what might sound at first like a simple return to the moral principles of the ancients is made in the name of quite different ideals. Virtue, for Rousseau, is not the same thing it is for Plato or Aristotle—as is clearly shown by his praise of Francis Bacon as "perhaps the greatest of philosophers."[39] In agreement with Bacon, Rousseau criticizes "moral philosophy" as an outgrowth of "human pride" as well as the hiatus between knowledge and power, thought and action, that he finds to be a mark of civilization; instead, he praises those who are able to act decisively in the world, to alter it in their favor, even when these are men the Greeks would have considered barbarians. Virtue, for instance, lies with the Scythians who conquered Persia, not with the Persians; with the Goths who conquered Rome, not the Romans; with the Franks who conquered the Gauls, the Saxons who conquered England. In civilized countries, he says, "There are a thousand prizes for fine discourses, and none for good action."[40] Action, even destructive action, particularly on a grand (or sublime) scale, is preferable to inaction.[41]

With Bacon, Rousseau argues the need for actions, not words, and approves the initial achievements of the Renaissance in freeing humanity from a barren medieval Scholasticism.[42] But unlike Bacon, Rousseau sees that even scientific rationality, through the alienation of affection, can often weaken the determination and commitment needed for decisive action. Thus, in a paradox that will become a hallmark of romanticism, Rousseau turns against technology—but in the name of ideals that are at the heart of technology. He criticizes a particular historical embodiment of technology, but only to advance a project that has become momentarily or partially impotent.

It was in England, however, where the Industrial Revolution found its earliest full-scale manifestation, that this paradoxical critique achieved an initial broad literary expression. Such expression took a realistic turn, rejecting classical patterns in favor of the specific depiction of real situations, often in unconventional forms. Works such as William Blake's poem "London" (1794) and Charles Dickens's novel *Hard Times* (1854), in their presentation of the dehumanizing consequences of factory labor, illustrate equally well the force of this approach. Wordsworth, again, may be quoted to extend the issue of the alienation of affections to the social level. In a letter from 1801 he writes,

> It appears to me that the most calamitous effect which has followed the measures which have lately been pursued in this country, is a rapid decay of the domestic affections among the lower orders of society. . . . For many years past, the tendency of society, amongst all the nations of Europe, has been to produce it; but recently, by the spreading of manufactures through every part of the country . . . the bonds of domestic feeling . . . have been weakened, and in innumerable instances entirely destroyed. . . . If this is true, . . . no greater curse can befall a land.[43]

Romantic realism is allied with visionary symbolism, however, and through this with epistemological issues. Consider, for instance, another aspect of Blake's genius, his prophetic poems. Over a century before, John Milton had in *Paradise Lost* (1667) already identified Satan with the technical activities of mining, smelting, forging, and molding the metals of hell into the city of Pandemonium.[44] Following this lead, in *Milton* (1804) Blake identifies Satan with the abused powers of technology—and Newtonian science. Satan, "Prince of the Starry Hosts and of the Wheels of Heaven," also has the job of turning "the [textile] Mills day & night" (1.4.9–10). But in the prefatory lyric that opens this apocalyptic epic, Blake rejects the necessity of "these dark Satanic Mills" and cries out

> I will not cease from Mental Fight,
> Nor shall my Sword sleep in my hand,
> Till we have built Jerusalem
> In England's green & pleasant Land.

This lyric, "And Did Those Feet in Ancient Time," is set to music and becomes the anthem of British socialism. A visionary, imaginative—not to say utopian—socialism is the romantic answer to the romantic critique of the moral limitations of technology. Mary Shelley's *Frankenstein* (1818), in another instance, likewise presents a love-hate relationship with technology in which what is hated is properly redeemed not by premodern delimitation but by the affective correlate of an expansive imagination—that is, love.

Industrialization, then, undermines affection—feeling and emotion—at both the individual and social levels. And this practical fact readily becomes allied with a more theoretical criticism of the Enlightenment emphasis on reason as the sole or principal cognitive faculty. The Enlightenment argued for the primacy of reason as the only means to advance human freedom from material limitations. The romantic replies that not only does such an emphasis on reason not free humanity from material bonds (witness the evils of the Industrial Revolution), but in itself it is (in the words of William Blake) a "mind-forged manacle." The focus on reason is itself a limitation that must be overcome; and through the consequent liberation of imagination the historical condition of technical activity can in turn be altered. In the "classic" epistemological defense and definition of Samuel Taylor Coleridge:

> The imagination . . . I consider either as primary, or secondary. The primary imagination I hold to be the living power and prime agent of all human perception, and as a repetition in the finite mind of the eternal act of creation in the infinite I AM. The secondary I consider as an echo of the former, co-existing with the conscious will, yet still as identical with the primary in the kind of its agency, and differing only in degree, and in the mode of its operation. It dissolves, diffuses, dissipates, in order to re-create; or where this process is rendered impossible, yet still, at all events, it struggles to idealize and to unify.[45]

Indeed, it is this power that Blake also appeals to as the source of his social revolution when he proclaims, "I know of no other Christianity and of no other Gospel than the liberty both of body & mind to exercise the Divine Arts of Imagination, the real & eternal World of which this Vegetable Universe is but a faint shadow, & in which we shall live in our Eternal or Imaginative Bodies when these Vegetable Mortal Bodies are no more."[46]

Finally, with regard to artifacts, the romantic view is again both like and unlike that of the Enlightenment. It is similar in the belief that nature and artifice operate by the same principles. Contra the Enlightenment, however, the romantic view takes nature as the key to artifice rather than artifice as the key to nature. The machine is a diminished form of life, not life a complex machine. Furthermore, nature is no longer perceived primarily in terms of stable forms; the reality of nature is one of process and change. Wordsworth and other English ro-

mantics are taken with the "mutability" of nature. Lord Byron, for instance, at the conclusion of *Childe Harold's Pilgrimage* (1818), when he
aspires "to mingle with the Universe, and feel / What I can ne'er express" (4.177), describes nature as the

> glorious mirror, where the Almighty's form
> Glasses itself in tempests; in all time,
> Calm or convulsed—in breeze, or gale, or storm—
> Icing the Pole, or in the torrid clime
> Dark-heaving—boundless, endless, and sublime—
> The image of Eternity.
>
> (4.183)

Nature, thus reconceptualized, reflects its new character onto the
world of artifice.

For the Enlightenment, at their highest levels of reality nature and
artifice both exhibit various aspects of mechanical order, the inter-
locking of parts in a mathematical interrelation of the well-drafted
lines of a Euclidean geometry. The metaphysical character of such reality is manifest to the senses through a "classical" vision of the beautiful—although there develops an Enlightenment excitement with the
great or grandiose (and the consequent projecting of art beyond nature) that contradicts the models of harmonious stability within nature
characteristic of classical antiquity and thus intimates romantic sensibilities. For romanticism, by contrast, the metaphysical reality of both
nature and artifice is best denoted not by stable or well-ordered form
but by process or change, especially as apprehended by the new aesthetic category of the sublime or the overwhelming and what Byron
refers to as "pleasing fear" (4.184).

As an aesthetic category, the idea of the sublime can be traced back
to Longinus (third century C.E.), who departed from classical canons
of criticism by praising literature that could provoke "ecstasy." But the
concept received little real emphasis until Edmund Burke's *Philosophical Enquiry into the Origin of Our Ideas of the Sublime and Beautiful* (1757).
For Burke, beauty is associated with social order and is represented
with harmony and proportion in word and figure; the sublime, by contrast, is concerned with the individual striving and is proclaimed by
magnitude and broken line. "Whatever is fitted in any sort to excite
the ideas of pain, and danger, whatever is in any sort terrible, or is
conversant about terrible objects, or operates in a manner analogous
to terror, is a source of the *sublime*" is Burke's famous definition.[47] Certainly modern technological objects and actions—from Hiroshima to

Chernobyl—have tended to become a primary objective correlative of such a sentiment.

Like premodern skepticism and Enlightenment optimism, the romantic way of being-with technology can thus be characterized by a pluralism of ideas that constitutes a critical uneasiness: (1) the will to technology is a necessary self-creative act that nevertheless tends to overstep its rightful bounds; (2) technology makes possible a new material freedom but alienates from the decisive strength to exercise it and creates wealth while undermining social affection; (3) scientific knowledge and reason are criticized in the name of imagination; and (4) artifacts are characterized more by process than by structure and invested with a new ambivalence associated with the category of the sublime. The attractive and repulsive interest revealed by the sublime expresses perhaps better than any other the uniqueness of the romantic way of being-with technology.

Coda

As analysis of the romantic being-with technology has especially tended to demonstrate, the ideas associated with the four aspects of technology as volition, as activity, as knowledge, and as object cannot be completely separated. Theology, ethics, epistemology, and metaphysics are ultimately aspects of a way of being in the world. Acknowledging this limitation, it is nevertheless possible to summarize the three ways of life in relation to technology by means of the matrix in table 5.

At the outset, however, the argument of this epilogue indicated a relation to Heidegger's early analysis of technology, although it has taken off in a trajectory not wholly consistent with Heidegger's own analysis or intentions. Yet there remains a final affinity worth noting. In Heidegger's existential analysis there is a paradox that the personal that is revealed through the technical is also undermined thereby. Tools are used with others and in a world of artifacts owned by others, but the others easily become treated as all the same and thus become, as he calls it, a They—mass society. "In utilizing public means of transport and in making use of information services such as the newspaper," Heidegger writes, "every Other [person] is like the next. The Being-with-one-another dissolves one's own Dasein [or existence] completely into the kind of Being of 'the Others,' in such a way, indeed, that the Others, as distinguishable and explicit, vanish more and more" (1927 [trans., 1962, p. 164]).

Table 5. Three Ways of Being-with Technology

Conceptual Elements	Basic Attitudes		
	Ancient Skepticism (suspicious of technology)	Enlightenment Optimism (promotion of technology)	Romantic Uneasiness (ambivalent about technology)
Volition (transcendence)	Will to technology involves tendency to turn away from God or the gods	Will to technology is ordained by God or by nature	Will to technology is an aspect of creativity, which tends to crowd out other aspects
Activity (ethics)	Personal: Technical affluence undermines individual virtue Societal: Technical change weakens political stability	Personal: Technical activities socialize individuals Societal: Technology creates public wealth	Personal: Technology engenders freedom but alienates from affective strength to exercise it Societal: Technology weakens social bonds of affection
Knowledge (epistemology)	Technical information is not true wisdom	Technical engagement with the world yields true knowledge (pragmatism)	Imagination and vision are more crucial than technical knowledge
Objects (metaphysics)	Artifacts are less real than natural objects and thus require external guidance	Nature and artifice operate by the same mechanical principles	Artifacts expand the process of life and reveal the sublime

With regard to the romantic way of being-with technology there is also a paradox. Not only is there a certain ambivalence built into this attitude, but the attitude itself has not been adopted in any whole-hearted way by modern culture. Romanticism is, if you will, uneasy with itself. Indeed, this may be in part why romanticism has so far been unable to demonstrate the kind of practical efficacy exhibited by both premodern skepticism and Enlightenment optimism. The paradox of the romantic way of being-with technology is that, despite an intellectual cogency and expressive power, it has yet to take hold as a truly viable way of life. Given almost two centuries of active articulation, this impotence may well point toward inherent weaknesses. Could it be that romanticism has been adopted, but that it is precisely its internal ambivalences, its bipolar attempt to steer a middle course between premodern skepticism and Enlightenment optimism, that vitiate its power?

NOTES

. . . .

Introduction. Thinking about Technology

1. Len Giovannitti and Fred Freed, *The Decision to Drop the Bomb* (New York: Coward-McCann, 1965), p. 197.

2. "The Atomic Scientists of Chicago," lead editorial explanation, *Bulletin of the Atomic Scientists of Chicago* 1, no. 1 (December 10, 1945): 1.

3. See, e.g., Charles R. Cantor, "Orchestrating the Human Genome Project," *Science* 248 (April 6, 1990): 49–50.

4. Zhores A. Medvedev, *Nuclear Disaster in the Urals,* trans. George Saunders (New York: W. W. Norton, 1979).

5. This program was killed in 1972 (as a result of internal academic politics), but in the course of its existence it produced seven books and eight research reviews on such topics as technology and economics, biomedical innovation, technological change, technology and education, and information technology and politics. For more detail see *Harvard University Program on Technology and Society, 1964–1972: A Final Review* (Cambridge: Harvard University, 1972).

6. See, for example, Alan F. Westin, *Privacy and Freedom* (New York: Atheneum, 1967), pp. 316–321.

7. Quoted from "What Is the Hastings Center?" included on the inside cover of the *Hastings Center Report,* vol. 1 (1971) through vol. 10 (1981), after which there has been a slight modification of the wording.

8. From an undated information flier put out by the Kennedy Institute.

9. For the full story, see Robert M. Anderson, Robert Perrucci, Dan E. Schendel, and Leon E. Trachtman, *Divided Loyalties: Whistle-Blowing at BART* (West Lafayette, Ind.: Purdue University, 1980).

10. For one narrative history of this event, see Clifford Grobstein, *A Double Image of the Double Helix: The Recombinant-DNA Debate* (San Francisco: W. H. Freeman, 1979).

11. Two case studies: Martin Curd and Larry May, *Professional Responsibility for Harmful Actions* (Dubuque, Iowa: Kendall/Hunt, 1984), and John H. Fielder

and Douglas Birsch, *The DC-10 Case: A Study in Applied Ethics, Technology, and Society* (Albany, N.Y.: State University of New York Press, 1993).

12. See V. C. Marshall, *Disaster at Flixborough* (Oxford: Pergamon, 1979). See also *The Flixborough Disaster: Report of the Court of Inquiry* (London: Her Majesty's Stationery Office, 1975), and John Grayson, *The Flixborough Disaster: The Lessons for the British Labour Movement*, Institute for Worker's Control Pamphlet 41 (Nottingham: IWC, n.d.).

13. This sometimes thought "mythical" event is documented in Alan Prendergast, "Launch on Warning: Doomsday Is a Phone Call away at NORAD's Missile Warning Center," *Rocky Mountain Magazine*, May–June 1981, pp. 29–34. See also "Nuclear War by Accident—Is It Impossible?" interview with Gen. James Hartinger, *U.S. News and World Report* 95 (December 19, 1983): 27.

14. Janice R. Long, "Charges against Nuclear Industry Investigated," *Chemical and Engineering News* 54, no. 9 (March 1, 1976): 11–12.

15. See, e.g., "Thrills and Lax Security Cited in Computer Break-In," *New York Times*, August 14, 1983, p. A30; and Joseph B. Treaster, "Trial and Error by Intruders Led to Entry into Computer," *New York Times*, August 23, 1983, pp. A1 and 14.

16. Brojendra Natn Banerjee, *Environmental Pollution and Bhopal Killings* (Delhi: Gian, 1987), is a rather sensationalist report (compares Bhopal to Hiroshima) but full of data and with an extensive bibliography, which aggressively presents the Indian point of view. For a companion Indian perspective, see Anees Chishti, *Dateline Bhopal: A Newsman's Diary of the Gas Disaster* (New Delhi: Concept, 1986). For comparison, see Paul Shrivastara, *Bhopal: Anatomy of a Crisis* (Cambridge, Mass.: Ballinger, 1987), which again includes much information and an extensive bibliography that overlaps little with Banerjee's.

17. World Commission on Environment and Development, *Our Common Future* (New York: Oxford University Press, 1987).

18. Reagan's lecture "The Spirit of Freedom," which he gave in both England and France, went virtually unreported in the United States. No mention of it occurs in the *New York Times*. But see the *Times* of London, Wednesday, June 14, 1989, p. 2.

19. See John Naisbitt, *Megatrends: Ten New Directions Transforming Our Lives* (New York: Warner Books, 1982), and Richard Lamm, *Megatraumas: America at the Year 2000* (Boston: Houghton Mifflin, 1985).

20. Charles Perrow, *Normal Accidents: Living with High-Risk Technologies* (New York: Basic Books, 1984). See also Barry A. Turner, *Man-Made Disasters* (London: Wykeham, 1978).

21. See Michael Bradie, Thomas W. Attig, and Nicholas Rescher, eds., *The Applied Turn in Contemporary Philosophy*, Bowling Green Studies in Applied Philosophy, vol. 5 (Bowling Green, Ohio: Bowling Green State University, 1983).

22. See, e.g., Frederick Ferré (1988) and Don Ihde (1993), and the shift toward global issues in the latter.

23. This is, for instance, the basic argument of Roger Shinn, *Forced Options:*

Social Decisions for the Twenty-first Century, 3d ed. (Cleveland: Pilgrim Press, 1991; first published 1982): "Some options are avoidable and some are forced. That is, some options can be put off indefinitely or evaded forever; others cannot. . . . On many scientific theories you probably are wiser not to take a stand. . . . But some options are forced. Many a person has faced the decision: "Have surgery, or don't." Air fighter pilots, when planes go out of control, must decide—in a hurry—to bail out or stay with the plane. . . . In such cases a refusal to decide, a failure to decide, is itself a decision" (pp. 3–4). It is remarkable the degree to which forced options are associated with modern technology.

24. See, e.g., Stephen H. Unger, *Controlling Technology: Ethics and the Responsible Engineer,* 2d ed. (New York: John Wiley, 1994); Mike W. Martin and Roland Schinzinger, *Ethics in Engineering,* 2d ed. (New York: McGraw-Hill, 1989).

25. This is continuous with a position expressed in the preface to Mitcham and Mackey, eds., *Philosophy and Technology* (1972). "Unlike those who now symbolize . . . the character of our generation," the editors wrote, "we chose, with hesitation, the path of reflection rather than political action to try to come to terms with the brute facts of our existence" (p. v).

26. For the contents of this volume, see Mitcham and Mackey, *Bibliography of the Philosophy of Technology* (1973), p. 6.

27. For supplementary evaluation, see Mitcham and Mackey, *Bibliography of the Philosophy of Technology* (1973), p. 18.

28. For contents, see Mitcham and Mackey, *Bibliography of the Philosophy of Technology* (1973), pp. 13–14.

29. Despite some fluctuation in name from "world" to "international" and back again, this is one series of meetings.

30. For the relevant contents of the proceedings of these congresses, see Mitcham and Mackey, *Bibliography of the Philosophy of Technology* (1973), pp. 12–13.

31. For the contents of the proceedings of this congress, see Mitcham and Grote, "Current Bibliography in the Philosophy of Technology: 1973–1974" (1978), pp. 325–326.

32. See Mitcham and Mackey, *Bibliography of the Philosophy of Technology* (1973), pp. 8–10.

33. See Mitcham and Mackey, *Bibliography of the Philosophy of Technology* (1973), pp. 2–3.

34. Melvin Kranzberg, "Toward a Philosophy of Technology: Prefatory Note," *Technology and Culture* 7, no. 3 (summer 1966): 301.

35. Illustrating the uneasiness philosophers continued to feel in talking about "technology," the first was initially titled "Scientists and Social Responsibility," with a subtitle "Philosophers Look at Technology" eventually added. The second of the set, however, eschewed this hybrid and was forthrightly called a "Philosophy and Technology" conference.

36. See *Research in Philosophy and Technology,* vol. 1 (1978). In the 1977 meeting the shift was even more pronounced; in the proceedings (*Research in Philosophy*

and Technology 3 [1980]: 5–130) all papers are grouped under three headings: "The Citizen and Technological Decision Making," "Ethics and Biomedical Research," and "Reviews of Recent Books" (Bernard Gendron's *Technology and the Human Condition* [1977] and Langdon Winner's *Autonomous Technology: Technics-out-of-Control as a Theme in Political Thought* [1977]).

37. Paul T. Durbin, "Introduction to the Series," *Research in Philosophy and Technology* 1 (1978): 3. The italics are Durbin's.

38. Edited for the first ten years by Paul Durbin (with Carl Mitcham as coeditor during 1984), but taken on by Edmund Byrne (Indiana University) with vol. 10, no. 2 (February 1985), by John Fielder (Villanova University) with vol. 15, no. 1 (September 1989), and then by Candy Torres (NASA Johnson Space Center) with vol. 19, no. 1 (February 1994).

39. Edited by Paul Durbin and Carl Mitcham from 1978 to 1985. After vol. 8 (1985) the Society for Philosophy of Technology withdrew official support from this series and initiated another, *Philosophy and Technology,* with a different publisher. At that time Durbin became editor of the new series and Frederick Ferré (University of Georgia) took over *RPT.* Mitcham edited two transition volumes, a supplement and vol. 9 (both dated, because of delays, 1989). In 1993 Ferré passed the general editorship on to Mitcham for vols. 15 (1995) and following.

40. For a description of INVESCIT, see José Sanmartín and Manuel Medina, "A New Role for Philosophy and Technology Studies in Spain," *Technology in Society* 11, no. 4 (1989): 447–455.

Chapter One. Engineering Philosophy of Technology

1. George Berkeley, *A Treatise concerning the Principles of Human Knowledge* (1710), sec. 151.

2. Two useful studies of this period are Marie Boas Hall, *The Mechanical Philosophy* (New York: Arno Press, 1981), first published as "The Establishment of the Mechanical Philosophy," *Osiris* 10 (1952): 412–541, and Richard Westfall, *The Construction of Modern Science: Mechanisms and Mechanics* (New York: John Wiley, 1971). Westfall includes a good bibliographic essay.

3. Karl Marx, *Das Kapital,* vol. 1 (1867), pt. 4, chap. 15, sec. 4. See also pt. 3, chap. 11, where Ure is called "the philosopher of industry."

4. Adam Smith, *An Inquiry into the Nature and Causes of the Wealth of Nations* (1776), begins the modern economic analysis of machinery, and thus introduces some basic conceptual distinctions. Charles Babbage's *On the Economy of Machinery and Manufactures* (1832) exercised a more direct influence on Ure.

5. Hans-Martin Sass, "Man and His Environment: Ernst Kapp's Pioneering Experience and His Philosophy of Technology and Environment," in *German Culture in Texas,* ed. Glen E. Lich and Dona B. Reeves (Boston: Twayne, 1980), pp. 82–99, with notes pp. 269–271. See also the introduction by Sass to the 1978 reprint of *Grundlinien einer Philosophie der Technik.*

6. This letter is quoted in an obituary that appeared in *Deutsche Rundschau für Geographie und Statistik* 20 (1898): 40–43. The quotation is from Faust's last

speech, *Faust*, part 2, 5.6.11577. As another sidelight on the Marxist comparison, Ludwig Feuerbach also once seriously considered emigrating to the New
World and corresponded with Kapp about this possibility, but he ultimately
decided, like Marx, to remain in Europe.

7. See Aristotle *Eudemian Ethics* 7.9.1241b24, and Ralph Waldo Emerson,
"Wealth," in *English Traits* (Boston, 1860), p. 169: "Man is a shrewd inventor,
and is ever taking the hint of a new machine from his own structure, adapting
some secret of his own anatomy in iron, wood, and leather, to some required
function of the work of the world."

8. See also, e.g., J. D. Bernal, *The Extension of Man: A History of Physics before
1900* (London: Weidenfeld and Nicolson, 1972), pp. 15–16, where he defines
physics as "a limited subject [concerned with] heat, light, sound, electricity
and magnetism, with a small allowance of atomic physics. . . . The reason why
we can block out this particular part of knowledge and experience and call it
physics is because it deals primarily with what might be called *the extension of
the human sensory-motor arrangement.*"

9. For this and related information, see Vitaly G. Gorokhov, "Russkii inzhen-
ermekhanik i filosof tekhniki Petr Klementevich Engelmeier" [Russian mechanical engineer and philosopher of technology Peter Klementevich Engel-
meier], *Voprosy Istorii Estestvoenaniia i Tekhniki*, no. 4 (1990): 51–60. This article
is taken from an unpublished book-length manuscript by Gorokhov, who has
supplied other materials as well.

10. For more detail, see Kendall E. Bailes, *Technology and Society under Lenin
and Stalin: Origins of the Soviet Technical Intelligentsia, 1917–1941* (Princeton:
Princeton University Press, 1978), especially chap. 4, "The Industrial Party Affair," pp. 95–121. For a general account of the parallel conflict in the United
States, see Edwin T. Layton Jr., *The Revolt of the Engineers: Social Responsibility
and the American Engineering Profession* (Baltimore: Johns Hopkins University
Press, 1986).

11. For the story of Palchinsky, see Loren R. Graham, *The Ghost of the Executed
Engineer: Technology and the Fall of the Soviet Empire* (Cambridge: Harvard University Press, 1993). Graham mentions Engelmeier in passing (pp. 43–44) but
only on the basis of what is contained in Kendall Bailes.

12. According to the research of Vitaly Gorokhov, Engelmeier's last letter is
dated July 26, 1939. T. I. Rainow, in a letter to his wife dated April 14, 1943,
recalls the "late" P. K. Engelmeier.

13. Cf. Hannah Arendt, *The Human Condition* (1958), "Prologue": "In 1957,
an earth-born object made by man was launched into the universe, where for
some weeks it circled the earth according to the same laws of gravitation that
swing and keep in motion the celestial bodies. . . . The immediate reaction,
expressed on the spur of the moment, was relief about the first 'step toward
escape from men's imprisonment to the earth.' And this strange statement, far
from being the accidental slip of some American reporter, unwittingly echoed
[feelings which had] been commonplace for some time. . . . The banality of the
statement should not make us overlook how extraordinary in fact it was; for

although Christians have spoken of the earth as a vale of tears and philosophers have looked upon their body [*sic*] as a prison of mind or soul, nobody in the history of mankind has ever conceived of the earth as a prison for men's bodies or shown such eagerness to go literally from here to the moon. . . . It is the same desire to escape from imprisonment to the earth that is manifest in the attempt to create life in the test tube, in the desire to mix 'frozen germ plasm from people of demonstrated ability under the microscope to produce superior human beings' and 'to alter [their] size, shape and function'; and the wish to escape the human condition, I suspect, also underlies the hope to extend man's life-span far beyond the hundred-year limit." Cf. also Hannah Arendt, *Between Past and Future* (New York: Viking, 1968), chap. 8, "The Conquest of Space and the Stature of Man."

14. "VEIFA" is an acronym for Vereinigte Elektrotechnische Institute Frankfurt-Aschaffenburg (United electrotechnical institutes of Frankfurt-Aschaffenburg).

15. Dissertation title: "Über einen neuen Hochspannungstransformator und seine Anwendung zur Erzeugung durchdringungsfähiger Röntgenstrahlen [Concerning a new high voltage transformer and its production of penetrating X rays]. Institute for Applied Physics, University of Frankfurt, 1917.

16. See, e.g., Bernhard Bavink, "Philosophie der Technik," in *Ergebnisse und Probleme der Naturwissenschaften*, 5th ed. (Leipzig: Hirzel, 1933); Andrew G. Van Melsen, *Science and Technology* (1961); and Alwin Diemer, "Philosophie der Technik," in *Grundriß der Philosophie*, vol. 2, *Die philosophischen Sonderdisziplinen* (Meissenheim: Anton Hain, 1964).

17. Max Eyth, *Lebendige Kräfte: Sieben Vorträge aus dem Gebiete der Technik* [Living powers: Seven lectures from the domain of technology]. (Berlin: Julius Springer, 1903). Alard DuBois-Reymond, *Erfindung und Erfinder* [Inventing and inventor] (Berlin: Julius Springer, 1906).

18. See Peter K. Engelmeier, "Allgemeinen Fragen der Technik," installment 8, *Dinglers Polytechnisches Journal* 312, no. 9 (June 3, 1899), esp. p. 130.

19. Tadeusz Kotarbinski, *Praxiology: An Introduction to the Sciences of Efficient Action* (New York: Pergamon, 1965). First published in Polish in 1955. Cf. also Jan Ostrowski, R. Pichon, and R. Durand-Auzis, *Alfred Espinas, précurseur de la praxéologie, ses antécédents et ses successeurs* (Paris: Librairie Général de Droit de Jurisprudence, 1973).

20. That Simondon sees himself as extending the phenomenological tradition is suggested by the dedication to Maurice Merleau-Ponty of his second book, *L'individu et sa genèse physico-biologique* (Paris: Presses Universitaires de France, 1964). For a general study of Simondon's thought, see Gilbert Hottois, *Simondon et la philosophie de la "culture technique"* (Brussels: De Boeck, 1993).

21. See Karl R. Popper, *The Open Society and Its Enemies* (1945). The discussion of social engineering first occurs in vol. 1, chap. 3, sec. 4 but is then returned to throughout the book, with a distinction being drawn between utopian (bad) and piecemeal (good) social engineering.

22. See, for example, John Dewey, *Individualism, Old and New* (1929), especially chap. 8, "Individuality in Our Day," included in *John Dewey: The Later Works, 1925–1953*, vol. 5, *1929–1930* (Carbondale: Southern Illinois University Press, 1984). Compare also Joseph Agassi, *Technology: Philosophical and Social Aspects* (1985).

23. Mike W. Martin and Roland Schinzinger, *Ethics in Engineering*, 2d ed. (New York: McGraw-Hill, 1989), chap. 3, "Engineering as Social Experimentation."

Chapter Two. Humanities Philosophy of Technology
1. See, for example, Mircea Eliade, *The Forge and the Crucible* (1971).
2. Aristotle, *Nicomachean Ethics* 1.1.1094a26-b5.
3. Jean-Jacques Rousseau, *Discours sur les sciences et les arts*, vol. 3 of *Oeuvres complètes*, Pléiade edition, pp. 9 and 17.
4. Rousseau, *Discours sur les sciences et les arts*, p. 19.
5. On Mumford's life, consult Lewis Mumford, *Sketches from Life: The Early Years* (New York: Dial Press, 1982), and Donald L. Miller, *Lewis Mumford: A Life* (Pittsburgh: University of Pittsburgh Press, 1989). The best collection of studies is Thomas P. Hughes and Agatha C. Hughes, eds., *Lewis Mumford: Public Intellectual* (New York: Oxford University Press, 1990).
6. Lewis Mumford, "The Drama of the Machines," *Scribner's Magazine* 88 (August 1930): 150–161.
7. See also John Michael Krois, "Ernst Cassirer's Theory of Technology and Its Import for Social Philosophy," *Research in Philosophy and Technology* 5 (1982): 209–222.
8. Lewis Mumford, *Man as Interpreter* (New York: Harcourt Brace, 1950), p. 2.
9. Mumford, *Man as Interpreter*, pp. 8–9.
10. Lewis Mumford, "Technics and the Future of Western Civilization," in *In the Name of Sanity* (New York: Harcourt Brace, 1954), p. 39.
11. Lewis Mumford, *The Culture of Cities* (New York: Harcourt Brace, 1938), pp. 409–461.
12. Lewis Mumford, *Art and Technics* (1952), p. 35.
13. All page references for "Meditación de la técnica" are to *Obras completas*, vol. 5.
14. All page references for "El mito del hombre allende la técnica" are to *Obras completas*, vol. 9.
15. This logic is analyzed in Albert Camus's *Le Mythe de Sisyphe* (Paris: Gallimard, 1942), in the section titled simply "Kirilov."
16. All page references for Heidegger are to *Die Technik und die Kehre* (1962).
17. See, e.g., "What Are Poets For?" (first published 1950) and "The Thing" (1954), in *Poetry, Language, Thought*, trans. and ed. Albert Hofstadter (New York: Harper and Row, 1971), pp. 116–166.
18. Patrick H. Dust, "Freedom, Power and Culture in Ortega y Gasset's Philosophy of Technology," *Research in Philosophy and Technology* 11 (1991): 126. For two other comparisons, see Juan Vayá Menéndez, "La cuestion de la técnica

en una doble ‹meditación›: Ortega y Heidegger," *Convivium* (Barcelona), nos. 9–10 (1961): 64–91, and 11–12 (1961): 75–97; and Jorge Acevedo, "Introducción a 'La pregunta por la técnica,'" in Martin Heidegger, *Ciencia y técnica* (Santiago de Chile: Editorial Universitaria, 1984), esp. pp. 62–68.

19. Ortega y Gasset, *Obras completas,* 5: 301–302, and 7: 85–86. Cf. José Ortega y Gasset, *Man and People,* trans. Willard R. Trask (New York: W. W. Norton, 1957), p. 20.

20. Ortega y Gasset, *Obras completas,* 5: 302, and 7: 86. *Man and People* (1957), pp. 20–21.

21. This definition, as added in a "Note to the Reader" in the "American edition" (which is not simply a translation) of *La Technique,* reads in English: "The term *technique,* as I use it, does not mean machines, technology, or this or that procedure for attaining an end. In our technological society, *technique* is the *totality of methods rationally arrived at and having absolute efficiency* (for a given stage of development) in *every* field of human activity" (*Technological Society,* p. xxv). Unfortunately, the French original of this definition has not been published. The use of the word "having" in this context is slightly misleading, however; in light of related discussions it seems reasonable to substitute "obtaining as a result" or "aspiring to" (see, e.g., *The Technological Society,* pp. 24 ff. and 79 ff., and *The Technological System,* chap. 1, "Technology as a Concept").

22. Ellul, *A Temps et à contretemps* (1981a), pp. 155–156. Ellul's tendency to overlook details is well illustrated by this passage that, in the original, incorrectly cites the titles of two of his own books. The bracketed dates have been added.

23. Karl Marx and Friedrich Engels, *Deutsche Ideologie* (1845), pt. 1, chap. 1, sec. (a).

24. Trevor J. Pinch and Wiebe E. Bijker, "The Social Construction of Facts and Artifacts, or How the Sociology of Science and the Sociology of Technology Might Benefit Each Other," in Wiebe E. Bijker, Thomas P. Hughes, and Trevor J. Pinch, eds., *The Social Construction of Technological Systems* (1987), pp. 17–50.

25. For more on the argument between technological determinism and social constructivism, see the introductory essay in MacKenzie and Wajcman, eds., *The Social Shaping of Technology* (1985), pp. 2–25; Steven Woolgar, "The Turn to Technology in Social Studies of Science," *Science, Technology, and Human Values* 16, no. 1 (winter 1991): 20–50; Langdon Winner, "Upon Opening the Black Box and Finding It Empty: Social Constructivism and the Philosophy of Technology" (1991); and Wiebe Bijker, "Do Not Despair: There Is Life after Constructionism," *Science, Technology, and Human Values* 18, no. 1 (winter 1993): 113–138.

Chapter Three. From Engineering to Humanities Philosophy of Technology

1. Subsequently (with Heidegger, Gadamer, and Ricoeur), hermeneutics has been put forth as the general form of all human experience—including scientific-technical explanation. For an overview of six definitions of herme-

neutics, see Richard E. Palmer, *Hermeneutics: Interpretation Theory in Schleier-macher, Dilthey, Heidegger, and Gadamer* (Evanston, Ill.: Northwestern University Press, 1969), chap. 3, pp. 33–45. Utilizing the classic contrast, Palmer maintains that "the task of interpretation today . . . is to break out of scientific objectivity and the scientist's way of seeing. . . . So overcome with the perspective of technological thinking are we that only in scattered moments does our historicality come into view at all" (p. 252). In a subsequent update, Palmer questions whether the expansion of interpretation to include explanation does "not rob hermeneutics of its most characteristic elements"; see his "Hermeneutics," in *Contemporary Philosophy: A New Survey*, vol. 2, *Philosophy of Science*, ed. Guttorm Fløistad (The Hague: Martinus Nijhof, 1982), pp. 453–505, esp. 457 ff.

2. C. P. Snow, *The Two Cultures: And a Second Look* (New York: New American Library, 1964), p. 20. This is an expanded version of *The Two Cultures and the Scientific Revolution*, first published 1959.

3. Albert Speer, *Inside the Third Reich*, trans. Richard and Clara Winston (New York: Macmillan, 1970), p. 619.

4. For proceedings, see "Über die Verantwortung des Ingenieurs," *VDI-Zeitschrift* 92 (1950): 589–627.

5. For proceedings, see "Mensch und Arbeit im technischen Zeitalter," *VDI-Zeitschrift* 93 (1951): 655–663, 766–773.

6. For proceedings, see "Die Wandlungen des Menschen durch die Technik," *VDI-Zeitschrift* 96 (1954): 113–159.

7. For proceedings, see "Der Mensch im Kraftfeld der Technik," *VDI-Zeitschrift* 97 (1955): 897–933, subsequently published in book form (Düsseldorf: VDI-Verlag, 1955).

8. Rainer Berger, "Institution, Diskurs und System: Eine Analyse der bürgerlichen Gesellschaftstheorien, ihrer ausgebliebenen Technikkritik und der gesellschaftstheoretischen Voraussetzungen der Technikbewertung" (Ph.D. diss., Political Science, Free University Berlin, 1989), pp. 344 ff.

9. Simon Moser (1973; first published 1958); English trans., p. 124.

10. See, e.g., *Mensch und Technik: Veröffentlichungen* [Humanity and technology: publications] (Düsseldorf: VDI Verlag, 1963).

11. VDI, *Technik und Gesellschaft* (Düsseldorf: VDI Verlag, 1968), p. 252. Subtitled "Wechselwirkungen-Einflüsse-Tendenzen" [Interactions-influences-tendencies], this volume included contributions on technical education, socio-technical organization, and economics.

12. *Wirtschaftliche und gesellschaftliche Auswirkungen des technische Fortschritts* [Economic and social consequences of technical progress] (Düsseldorf: VDI Verlag, 1971), p. 248.

13. For more on this, see Alois Huning and Carl Mitcham, "The Historical and Philosophical Development of Engineering Ethics in Germany," *Technology in Society* 15, no. 4 (1993): 427–439.

14. For a review, see Carl Mitcham, "Philosophy of Technology in Germany," *Research in Philosophy and Technology* 13 (1993): 329–381.

15. Alois Huning, "Philosophy of Technology and the Verein Deutscher In-

genieure," *Research in Philosophy and Technology* 2 (1979): 267. My account of the "Mensch und Technik" committee relies heavily on Huning's work.

16. John Dewey, *Problems of Men* (New York: Philosophical Library, 1946), p. 291; quoted in Hickman, *John Dewey's Pragmatic Technology* (1990), p. 58.

17. Ewart Edmund Turner, "A Textbook for Christianity," *Christian Century* 47 (January 8, 1930): 48. Cited in Harriet Furst Simon, "Textual Commentary," *John Dewey: The Later Works, 1925–1953* vol. 4, *The Quest for Certainty* (Carbondale: Southern Illinois University Press, 1984), p. 271.

18. John Dewey, *John Dewey: The Later Works*, vol. 10, *Art as Experience*, p. 64. Quoted to a slightly different extent in Hickman, *John Dewey's Pragmatic Technology* (1990), pp. 43–44.

19. Alexander S. Kohanski's *Philosophy and Technology: Toward a New Orientation in Modern Thinking* (New York: Philosophical Library, 1977), whose discussions of a host of philosophers from Newton and Kant to Sartre and Wittgenstein remain on an unfortunately superficial level. This oddly exasperating work—many times it seems just on the verge of realizing its promise, only to dissipate itself in intellectual platitudes—concludes with a homily about the philosopher-technologist who would represent "a new orientation in modern thinking" because he would (1) treat all nature, and not just other men, as an end in itself, (2) recognize that truth cannot be pursued for its own sake but only for truly human ends, (3) interpret "universality and necessity . . . in relation to the purposes of man and nature, which it is to serve," (4) give up the pursuit of certainty in favor of the pursuit of truth, and (5) accept the limitations of the human condition (pp. 178–183).

20. See Frederick Engels, *Socialism: Utopian and Scientific* (1883), trans. Edward Aveling (New York: International Publishers, 1935), especially pp. 36–38.

21. Henri de Saint-Simon, "Mémoire sur la science de l'homme" (1813), in *Oeuvres complètes de Saint-Simon et Enfantin* (Paris, 1865–1876), 40: 236, quoted from Felix Markham, ed. and trans., *Henri de Saint-Simon: Social Organization, the Science of Man and Other Writings* (New York: Harper and Row, 1964), p. 23.

22. Henri de Saint-Simon, *Nouveau Christianisme* (1825), in *Oeuvres complètes*, 23: 96–192, quoted from Markham, ed. and trans., *Henri de Saint-Simon* (1964), p. 105.

23. "Er erfaßt die *Arbeit* als das *Wesen* . . . des Menschen. . . . Die Arbeit, welche Hegel allein kennt und anerkennt, ist die *abstrakt geistige*" (*Öknomisch-philosophische Manuskripte aus dem Jahre 1844*, in Karl Marx and Friedrich Engels, *Werke, Ergänzungsband*, pt. 1 [Berlin: Dietz, 1973], "Kritik der Hegelschen Dialektik und Philosophie überhaupt," p. 574). Cf. English version, Karl Marx, *The Economic and Philosophic Manuscripts of 1844*, ed. Dirk J. Struik, trans. Martin Milligan (New York: International Publishers, 1964), p. 177, where *Arbeit* is translated as "labor."

24. Karl Marx, *Grundriße der Kritik der politischen Ökonomie* is the title given to seven notebooks from the winter of 1857–1858 that were first published beginning in 1939 (Moscow: Foreign Language Publishing House, 1939–1941; reprinted, Berlin: Dietz, 1953). English version, *Grundrisse: Foundations of the Cri-

tique of Political Economy, trans. Martin Nicolaus (New York: Vintage, 1973). *Zur Kritik der politischen Ökonomie* (1859) is available in Marx and Engels, *Werke,* vol. 13 (Berlin: Dietz, 1974), pp. 3–160. English version, *Contribution to a Critique of Political Economy,* ed. Maurice Dobb and trans. S. W. Ryazanskaya (New York: International Publishers, 1970). (The opening chapters of *Das Kapital* [1867] review the argument of *Zur Kritik der politischen Ökonomie.* Later chapters pick up and develop the analyses of technology that are present in the *Grundriße* manuscripts but not included in the *Kritik;* cf., e.g., *Grundrisse* [trans. 1973], pp. 690 ff., with *Das Kapital,* vol. 1, pt. 4, chap. 13 [trans. 1967, pp. 371 ff.]).

25. Karl Marx, *Das Kapital,* vol. 1, afterword to the second German edition (trans., 1967, p. 20). This self-description is regularly misquoted, and Marx is regularly described as standing Hegel on his head when in fact Marx argues that he found Hegel standing on his head and desired to place him upright on his feet. See also Karl Marx and Frederick Engels, *The Holy Family, or Critique of Critical Critique* (1845), chap. 7, sec. 4, trans. R. Dixon (Moscow, Foreign Languages Publishing House, 1956), p. 254; and Frederick Engels, *Socialism: Utopian and Scientific* (1883), trans. Edward Aveling (New York: International Publishers, 1935), p. 49.

26. "Der Richtum der Gesellschaften, in welchen kapitalistische Produktionsweise herrscht, erscheint als eine 'ungeheure Warensammlung,' die einzelne Ware als seine *Elementarform"* (*Das Kapital,* vol. 1 [1967], pt. 1, chap. 1, sec. 1). The phrase "ungeheure Warensammlung" is even referenced in a note to *Zur Kritik der politischen Ökonomie* (1859), p. 3.

27. This translation is based on *The Grundrisse,* ed. and trans. David McLellan (New York: Harper and Row, 1971), p. 27, and *Grundrisse,* trans. Martin Nicolaus (New York: Vintage, 1973), p. 94.

28. Karl Marx and Friedrick Engels, *Die deutsche Ideologie: Kritik der neusten deutschen Philosophie in ihren Repräsentanten Feuerbach, B. Bauer und Stirner, und des deutschen Sozialismus in seinen verschiedenen Propheten* (1845), in Marx and Engels, *Werke,* vol. 3 (Berlin: Dietz, 1973), p. 3. Quoted from *The German Ideology,* pts. 1 and 3, ed. R. Pascal (New York: International Publishers, 1947), p. 22. Cf. also the analysis of decreasing specialization of labor in capitalism in *Das Kapital,* vol. 1, pt. 4, chap. 13, sec. 9 [trans. 1967, pp. 487f.] and Marx's brief remarks on labor in communist society in *Critique of the Gotha Program* (1875), n. 3, near the end.

29. Julian M. Cooper, "The Scientific and Technical Revolution in Soviet Theory," in Fleron, ed. *Technology and Communist Culture* (1977), pp. 146–179. See also remarks in Ronald Amann, Julian Cooper, and R. W. Daview, eds., with assistance of Hugh Jenkin, *The Technological Level of Soviet Industry* (New Haven: Yale University Press, 1977), pp. 229 and 234.

30. See, e.g., Günter Wettstädt, *Ideologie im Zwielicht: Zum Einfluß bürgerlicher Technikphilosophie auf die imperialistische Bildungsideologie* [Ideology at twilight: On the influence of bourgeois philosophy of technology in imperialist ideology education], Zur Kritik der bürgerlichen Ideologie, vol. 37 (Berlin: Akademie-Verlag, 1974).

31. Hermann Ley, *Dämon Technik?* [Demon technology?] (Berlin: Deutscher Verlag Wissenschaften, 1961). For annotations on this and related works by Ley and others, see Mitcham and Mackey, *Bibliography of the Philosophy of Technology* (1973), pp. 33–34, and passim. Bernd M. Lindenberg, *Das Technikverständnis in der Philosophie der DDR* [The place of technology in philosophy in the German Democratic Republic], European University Papers 46 (Frankfurt: Peter D. Lang, 1979), is a study of this period.

32. "Die Marxistisch-Leninistische Philosophie und die technische Revolution," *Deutsche Zietschrift für Philosophie*, special issue (1965). For contents, see Mitcham and Mackey, *Bibliography of the Philosophy of Technology* (1973), pp. 9–10.

33. Erwin Herlitzius, "Technik und Philosophie," *Informationsdienst Geschichte der Technik* (Dresden) 5, no. 5 (1965): 1–36.

34. Exemplifying such discussions, see the entries on "Technik" and "technische Revolution" (by Klaus-Dieter Wüstneck) and on "Technizismus" and "Technokratie" (by Günther Hoppe) in Georg Klaus and Manfred Buhr, eds., *Philosophisches Wörterbuch*, 6th revised and expanded edition (Berlin: Europäische Buch, 1969); and even more, Gizella Kovács and Siegfried Wollgast, eds., *Technikphilosophie in Vergangenheit und Gegenwart* [Philosophy of technology in past and present] (Berlin: Akademie-Verlag, 1984) and Hellmuth Lange's monograph-length study "Technikphilosophie" (1988). I am grateful to Käthe Friedrich for calling some of this material to my attention.

35. The analysis and program of *Civilization at the Crossroads* influenced the reform program of the Alexander Dubček government during the "Prague Spring" of 1968 and no doubt subsequent developments as well.

36. S. L. Sobolev, A. I. Kitov, and A. A. Liapunov, "Osnounye cherti kibernetiki" [Basic features of cybernetics], *Voprosy filosofi*, no. 5 (1955): 136–148; and Ernst Kolman, "Chto takoe kibernetika?" [What is cybernetics?], *Voprosy filosofi*, no. 9 (1955): 148–159.

37. For good general analyses of the discussion of cybernetics in the Soviet Union, see Peter Paul Kirschenmann, *Information and Reflection: On Some Problems of Cybernetics and How Contemporary Dialectical Materialism Copes with Them* (Boston: D. Reidel; New York: Humanities Press, 1970); and Loren R. Graham, *Science and Philosophy in the Soviet Union* (New York: Knopf, 1972), esp. chap. 9, "Cybernetics." For related references with annotations, see Mitcham and Mackey, *Bibliography of the Philosophy of Technology* (1973), pp. 105 ff.

38. For a related book on this theme, see Robert Daglish, ed. and trans. *The Scientific and Technological Revolution: Social Effects and Prospects* (Moscow: Progress Publishers, 1972).

39. Mikhail Gorbachev's *perestroika*, 1985–1990, was clearly intended to constitute this social revolution.

40. Max Horkheimer and Theodor Adorno, *Dialectic of Enlightenment* (1972), p. xv.

41. Max Horkheimer, *Critique of Instrumental Reason* (1974), p. vii.

42. For Habermas's theory, see *The Theory of Communicative Action*, trans. Thomas McCarthy, 2 vols. (Boston: Beacon Press, 1984, 1987). Vol. 1, chapter 4, section 2, "The Critique of Instrumental Reason," is an extended analysis of the weaknesses of the ideas of Horkheimer and Adorno. For a short summary of Habermas's overall project, see his *Communication and the Evolution of Society*, trans. Thomas McCarthy (Boston: Beacon Press, 1979).

43. Henry Ford, in an interview with Charles N. Wheeler, *Chicago Tribune*, (May 25, 1916).

44. For a complementary analysis of the making/doing distinction, see *Nicomachean Ethics* 6.4.

Chapter Four. The Philosophical Questioning of Technology

1. Ernest Nagel, *The Structure of Science: Problems in the Logic of Scientific Explanation* (New York: Harcourt, Brace and World, 1961), p. viii.

2. Thomas Kuhn, *The Structure of Scientific Revolutions* (Chicago: University of Chicago Press, 1962), p. 161.

3. Peter Galison, "Bubble Chambers and the Experimental Workplace," in *Observation, Experiment, and Hypothesis in Modern Physical Science*, ed. Peter Achinstein and Owen Hannaway (Cambridge: MIT Press, 1985), pp. 304–373; and idem, *How Experiments End* (Chicago: University of Chicago Press, 1987).

4. Examples: Tom L. Beauchamp, *Principles of Biomedical Ethics* (New York: Oxford University Press, 1979; 2d ed. 1983, 3d ed. 1989), exemplifies the attempt to apply existing ethical principles to biomedical technologies. H. Tristram Engelhardt, *The Foundations of Bioethics* (New York: Oxford University Press, 1986), pursues the development of a distinctive deontology for high-tech biomedical practice.

5. John Ladd, "Physicians and Society: Tribulations of Power and Responsibility," in *The Law-Medicine Relation: A Philosophical Exploration*, ed. S. F. Spicker, J. M. Healey, and H. T. Engelhardt (Boston: D. Reidel, 1981), pp. 33–52.

6. See Otto Mayr, *Authority, Liberty, and Automatic Machinery in Early Modern Europe* (Baltimore: Johns Hopkins University Press, 1986), pp. xviii, 265.

7. See the article of this title included in Langdon Winner, *The Whale and the Reactor* (1986), pp. 19–39.

8. Carolyn Merchant, *The Death of Nature: Women, Ecology, and the Scientific Revolution*, 2d ed. (San Francisco: Harper and Row, 1990). For two complementary studies on relations between gender and technology see Joan Rothschild, *Teaching Technology from a Feminist Perspective: A Practical Guide* (New York: Pergamon Press, 1988), and Judy Wajcman, *Feminism Confronts Technology* (University Park: Pennsylvania State University Press, 1991).

9. See e.g., Max Weber, *The Protestant Ethic and the Spirit of Capitalism*, trans. Talcott Parsons (New York: Scribner, 1930 [1905]); and R. H. Tawney, *Religion and the Rise of Capitalism* (London: Murray, 1926).

10. Mitcham and Grote, eds., *Theology and Technology* (1984), examines this spectrum of theologies.

11. Leo Strauss and Alexandre Kojève, *De la tyrannie* (Paris: Gallimard, 1954), pp. 343–344. See also Leo Strauss, *On Tyranny*, rev. and expanded version including the Strauss-Kojève correspondence, ed. Victor Gourevitch and Michael S. Roth (New York: Free Press, 1991), p. 212.

Chapter Five. Philosophical Questions about *Techne*

1. Charles Singer, E. J. Holmyard, A. R. Hall, Trevor I. Williams, et al., eds. *A History of Technology*, 5 vols. (New York: Oxford University Press, 1955–1959). These five volumes, which cover up through the nineteenth century, have been condensed to one volume in T. K. Derry and Trevor I. Williams, *A Short History of Technology* (New York: Oxford University Press, 1960); they have also since been supplemented by two volumes, *The Twentieth Century* (New York: Oxford University Press, 1984), on technology up through 1950. A better one-volume overview in the internalist tradition is Ian McNeil, ed., *An Encyclopedia of the History of Technology* (London: Routledge, 1990).

2. This summary history is indebted to Robert P. Multhauf, "Some Observations on the State of the History of Technology," *Technology and Culture* 15, no. 1 (January 1974): 1–12, and Reinhard Rürup, "Historians and Modern Technology: Reflections on the Development and Current Problems of the History of Technology," *Technology and Culture* 15, no. 2 (April 1974): 161–193.

3. Maurice Daumas, ed., *Histoire générale des techniques* (Paris: Presses Universitaires de France, 1962–1979): vol. 1, Chita de La Calle et al., *Les origines de la civilisation technique* (1962); vol. 2, Margarite Dubuisson et al., *Les premières étapes du machinisme* (1965); vol. 3, Margarite Dubuisson et al., *L'expansion du machinisme* (1968); vol. 4, Maurice Daumas et al., *Les techniques de la civilisation industrielle, energie et materiaux* (1978); vol. 5, M. Perrot et al., *Les techniques de la civilisation industrielle, transformation, communication, facteur humain* (1979). English version of vols. 1–3, *A History of Technology and Invention: Progress through the Ages*, trans. Eileen B. Hennessey (New York: Crown, 1969–1970): vol. 1, *The Origins of Technological Civilization* (1969); vol. 2, *The First Stages of Mechanization* (1969); vol. 3, *The Expansion of Mechanization* (1979).

4. Melvin Kranzberg and Carroll Pursell Jr., eds., *Technology in Western Civilization*, 2 vols. (New York: Oxford University Press, 1967).

5. The Gille volumes, which make the most philosophical arguments of these histories, and as such are referred to later, are included in the references.

6. Singer et al., eds., *A History of Technology*, 1: vii. See also Gordon Child's contribution to Singer, "Early Forms of Society," where he restates the definition of technology as "the study of those activities, directed to the satisfaction of human needs, which produce alterations in the material world" but then immediately extends "the meaning of the term . . . to include the results of those activities" (1: 38). With Child's further commentary on the concept of human need as historically, socially, and even individually determined, his definition becomes vacuous.

7. Melvin Kranzberg and Carroll Pursell Jr., "The Importance of Technology

in Human Affairs," in Kranzberg and Pursell, eds., *Technology in Western Civilization*, 1: 4–5.

8. Kranzberg and Pursell, "Importance of Technology in Human Affairs," p. 5.

9. Kranzberg and Pursell, "Importance of Technology in Human Affairs," p. 11.

10. Maurice Daumas, "Preface," in *A History of Technology and Invention*, vol. 1, *The Origins of Technological Civilization*, ed. Maurice Daumas, trans. Eileen B. Hennessy (New York: Crown, 1969), p. 7.

11. Daumas, "Preface," p. 7.

12. See Singer et al., eds., *History of Technology*, 1: v.

13. Basic etymological information is drawn from Liddell and Scott, *Greek-English Lexicon*, the *Oxford English Dictionary*, and Eric Partridge, *Origins*, among others, although wherever possible original sources have been consulted. So far the only works specifically on this subject are in German. See especially Wilfried Seibieke, *Versuch einer Geschichte der Wortfamilie um τέχνη in Deutschland vom 16. Jahrhundert bis etwa 1830* (Düsseldorf: VDI-Verlag, 1968); Johannes Erich Heyde, "Zur Geschichte des Wortes 'Technik,'" *Humanismus und Technik* 9, no. 1 (1963): 25–43; and Wolfgang Schadewaldt, "Die Begriffe 'Natur' und 'Technik' bei den Griechen," in *Natur, Technik, Kunst* (Göttingen: Musterschmidt, 1960), pp. 35–53. Seibieke's chapter 3, "*Technologia*—Von der philosophischen 'Kunstlehre' zur 'Handwerkswissenschaft,'" is especially relevant to parts of the argument toward the end of this chapter. English version of Schadewaldt's essay, "The Concepts of *Nature* and *Technique* according to the Greeks," trans. William Carroll, Carl Mitcham, and Robert Mackey, *Research in Philosophy and Technology* 2 (1979): 159–171.

14. For other commentary on Plato's and Aristotle's understandings of *techne*, cf. Simon Moser, "Toward a Metaphysics of Technology," *Philosophy Today* 15, no. 2 (summer 1971): 129–156, with the supplement available from the editor-translator. Also, for a counterbalance to the strictly textual and conceptual analysis, one should compare the provocative work of Plato scholar Robert S. Brumbaugh, who in *Ancient Greek Gadgets and Machines* (New York: Thomas Crowell, 1966), uses archaeological evidence in seeking to correct the notion that Greeks were devoid of an interest in and involvement with technology. "The truth is," he writes, "that invention and gadget-design gradually blossomed to a nearly contemporary fruition by the second century A.D." (p. 131).

15. Jacques Maritain, *Art and Scholasticism and the Frontiers of Poetry*, trans. Joseph W. Evans (New York: Scribner's, 1962), pp. 46–47. This is a translation of *Art et scolastique*, 3d rev. ed. (Paris: L. Rouart, 1935); first edition, 1920.

16. Maritain, *Art and Scholasticism and the Frontiers of Poetry*, pp. 47–48.

17. Jacques Maritain, *Creative Intuition in Art and Poetry* (New York: Pantheon, 1953), p. 48.

18. Maritain, *Creative Intuition in Art and Poetry*, pp. 56 and 57.

19. Maritain, *Creative Intuition in Art and Poetry*, pp. 57–58.

20. Maritain, *Creative Intuition in Art and Poetry*, pp. 58–59.

21. Along this line compare Abraham Maslow's distinction between nomothetic and idiographic knowing—the one concerned with laws and classes, the other with unique individuals, especially persons—in his *The Psychology of Science: Reconnaissance* (New York: Harper and Row, 1966), pp. 8–11. Cf. also John Julian Ryan, *The Humanization of Man* (New York: Newman Press, 1972), esp. p. 22, and two articles by Cyril Stanley Smith: "Matter versus Materials: A Historical View," *Science* 162 (November 8, 1966): 637–644, and "Metallurgical Footnotes to the History of Art," *Proceedings of the American Philosophical Society* 116, no. 2 (April 17, 1972): 97–135.

22. Cf. C. S. Lewis, *The Four Loves* (New York: Harcourt Brace, 1960).

23. See, e.g., Joseph Tusiani, ed. and trans., *The Complete Poems of Michelangelo* (New York: Noonday Press, 1960), poems 83 and 84, pp. 76–77.

24. The notion is more prevalent in the crafts, however. Particularly in discussions of Asian crafts, one can often find this theme quite consciously formulated. See, e.g., Bernard Leach, *A Potter's Book* (New York: Transatlantic, 1965), as well as other works by the same author, and D. T. Suzuki, *Zen and Japanese Culture* (Princeton: Princeton University Press, 1959), especially his remarks on the insufficiency of technique in the art of swordsmanship, pp. 14, 113, 173, 212. One relevant quotation is the story of a woodcarver in *Chuang Tzu* 19.10. Asked by the king how he created a perfect bell stand, the artisan replied:

> When I began to think about the work you commanded
> I guarded my spirit, did not expend it
> On trifles, that were not to the point.
> I fasted in order to set
> My heart at rest.
> After three days fasting,
> I had forgotten gain and success.
> After five days
> I had forgotten praise or criticism.
> After seven days
> I had forgotten my body
> With all its limbs.
>
> By this time all thought of your Highness
> And of the court had faded away.
> All that might distract me from the work
> Had vanished.
> I was collected in the single thought
> Of the bell stand.
>
> Then I went to the forest
> To see the trees in their own natural state.

When the right tree appeared before my eyes,
The bell stand also appeared in it, clearly, beyond doubt.
All I had to do was to put forth my hand
And begin.

If I had not met this particular tree
There would have been
No bell stand at all.

What happened?
My own collected thought
Encountered the hidden potential in the wood;
From this live encounter came the work.

From the "imitations" by Thomas Merton in *The Way of Chuang Tzu* (New York: New Directions, 1965), pp. 110–111. Cf. also *Chuang Tzu* 3.2 and 13.10.

25. Walter J. Ong, *Ramus, Method, and the Decay of Dialogue: From the Art of Discourse to the Art of Reason* (Cambridge: Harvard University Press, 1958), p. 197.

26. See William Ames, *Technometry*, trans. with introduction and commentary by Lee W. Gibbs (Philadelphia: University of Pennsylvania Press, 1979). For commentary see Keith L. Sprunger, "Technometria: A Prologue to Puritan Theology," *Journal of the History of Ideas* 29, no. 1 (March 1968): 115–122.

27. With regard to alchemy—that hermetic science in which the Aristotelian imagination is transmogrified—two works are especially helpful to reflection on technology: Mircea Eliade, *The Forge and the Crucible* (1971), and Titus Burckhardt, *Alchemy: Science of the Cosmos, Science of the Soul* (Baltimore: Penguin, 1971). It is interesting, moreover, that in an alchemical work attributed to Saint Thomas Aquinas matter is raised to the status of a second principle and, along with form, is given divine significance. See *Aurora Consurgens* (New York: Pantheon, 1966), the "Commentary" by Marie-Louise von Franz, p. 385. As this Jungian interpreter observes in a footnote (p. 341), the importance of matter in the alchemical tradition is attested by the figure of Hyle, a mother goddess found in *Asclepius Latinus*, as well as by the teachings of Hermogenes and Numenius. Perhaps it would not be too bold to suggest the reason is that alchemy is more a practical than a theoretical science, one concerned with the actual process of generation—whether it be of gold or of the psychic reality for which this metal is a common symbol.

If I may be even bolder and venture a comparison, in the theory of knowledge the scholastic principle *quodquod recipitur, recipitur per modum recipiens* (whatever is received must be received in accordance with the manner of the recipient) can be left as an abstract principle in a way that allows considerations of form to dominate an epistemological discussion; in educational practice, however, the principle requires an attention to the particular student that was not always appreciated by scholastics. In practice, matter becomes of virtually equal importance. More succinctly, from the perspective of thought there is but one principle, form (matter is only a relationship to form); but from the perspective of practice there are two principles, form and matter (although the exigencies of practice may at times make form seem to be no more than a relationship to matter). This interpretation is borne out by Giordano Bruno's description of matter as "the divine and excellent progenitor, generator and mother of natural things" (see *Concerning the Cause, Principle, and One*, Fourth Dialogue, in Sidney Greenberg, *The Infinite in Giordano Bruno* [New York: King's Crown Press, 1950], p. 156). However, the nature of Renaissance alchemy is ambiguous to say the least.

The relation of alchemy to the transition from ancient to modern conceptions of matter should not be investigated without reference to Frances A. Yates, *Giordano Bruno and the Hermetic Tradition* (New York: Random House, 1964) and Paolo Rossi, *Francis Bacon: From Magic to Science*, trans. Sacha Rabinovitch (London: Routledge and Kegan Paul, 1968). According to Yates, "the basic difference between the attitude of the magician to the world and the attitude of the scientist to the world is that the former wants to draw the world into himself, whilst the scientist does just the opposite" (p. 454). Donald Brinkmann in *Mensch und Technik* (1946) and William Leiss in *The Domination of Nature* (1972) both consider the relation between alchemy and modern technology from a somewhat different angle. For the accepted interpretation by technological history see R. J. Forbes, *Studies in Ancient Technology*, vol. 1 (Leiden: E. J. Brill, 1955), pp. 121–144.

28. Galileo's famous metaphor is to be found in *The Assayer* (1623): "Philosophy is written in this grand book, the universe, which stands continually open to our gaze. But the book cannot be understood unless one first learns to comprehend the language and read the letters in which it is composed. It is written in the language of mathematics, and its characters are triangles, circles, and other geometric figures without which it is humanly impossible to understand a single word of it; without these, one wanders about in a dark labyrinth" (*Discoveries and Opinions of Galileo*, trans. Stillman Drake [Garden City, N.Y.: Doubleday, 1957], pp. 237–238). Two decades before Galileo, Francis Bacon in *The Advancement of Learning* (1605) articulated a distinction between "the book of God's word" and "the book of God's works" (I.1.3) without specifying the language of the latter. Approximately the same amount of time after Galileo, Thomas Browne in *Religio Medici* (1643) makes a similar reference to "nature, that universal and publick manuscript, that lies expansed unto the eyes of all" and concludes by denying the fundamental Greek distinction between nature and artifice: "All things are artificial; for nature is the art of God" (1.16). Lest it too readily be assumed that this follows simply from the Christian doctrine of creation ex nihilo, and the monastic idea of the "nature of created things" as a spiritual book (see, e.g., Evagrius Ponticus, *Praktikos* 92), note that Augustine explicitly qualifies the metaphor. In a sermon from the *Ennarationes in Psalmos* (Psalm 103, pt. 1) Augustine, with obvious reference to Rom. 1:20, describes

nature as a visible image of the invisible God, but he then contrasts the word of Scripture with the silence of nature and argues for a need to read natural descriptions in Scripture and in nature itself in an allegorical manner. Applying his own principle, in *Epistle 55* (8.13) he writes that "we make use of parables, formulated with reverent devotion, to illustrate our religion, drawing freely in our speech on the whole creation, the winds, the sea, the earth, birds, fishes, flocks, trees, men; just as, in the administration of the sacraments, we use with Christian liberty, but sparingly, water, wheat, wine, oil" (*Fathers of the Church*, vol. 12 [Washington, D.C.: Catholic University of America Press, 1950], p. 271). For Augustine the book of nature can be interpreted only through Scripture. From Bacon, Galileo, and Browne on, the two books are to be separated. "Man is the Interpreter of Nature," as William Whewell writes, "and Science is the right Interpretation" (*Philosophy of the Inductive Sciences*, 2d ed. [1847], vol. 1, chap. 2, sec. 9). (According to Whewell the empiricist, in contrast to Galileo, "the letters and symbols which are presented to the Interpreter are really objects of sensation.") For further discussion, see A. R. Peacocke, *Creation and the World of Science* (Oxford: Clarendon Press, 1979), especially part 1, "The Two Books"; and Hans Blumenberg's study, *Die Lesbarkeit der Welt* (Frankfurt: Suhrkamp, 1981). Although not the primary intention of Joseph Pitt (1992), in analyzing "how Galileo thought we could use mathematics to read the Book of Nature" (p. 1) he also shows how this leads to an alliance between science and technology.

29. For some collaborative confirmation of the speculative thesis regarding ancient versus modern ontologies of matter, compare R. G. Collingwood, *The Idea of Nature* (New York: Oxford University Press, 1945), pp. 111–112:

> For the early Greeks quite simply, and with some qualification for all Greeks whatever, nature was a vast living organism, consisting of a material body spread out in space and permeated by movements in time; the whole body was endowed with life, so that all its movements were vital movements; and all these movements were purposive, directed by intellect. This living and thinking body was homogeneous throughout in the sense that it was all alive, all endowed with soul and with reason; it was non-homogeneous in the sense that different parts of it were made of different substances each having its own specialized qualitative nature and mode of acting. The problems which so profoundly exercise modern thought, the problem of the relation between dead matter and living matter, and the problem of the relation between matter and mind, did not exist. . . . There was no material world devoid of mind, and no mental world devoid of materiality; matter was simply that of which everything was made, in itself formless and indeterminate, and mind was simply the activity by which everything apprehended the final cause of its own changes.

For the seventeenth century all this was changed. Science had discovered a material world in a quite special sense: a world of dead matter, infinite in extent and permeated by movement throughout, but utterly devoid of ultimate qualitative differences and moved by uniform and purely quantitative forces. The word "matter" had acquired a new sense: it was no longer the formless stuff of which everything is made by the imposition upon it of form, it was the quantitatively organized totality of moving things.

Based on what has been argued in the main body of the chapter, Collingwood's idea of the ancient ontology of matter leaves something to be desired. But as Whitehead has remarked in one of his own studies on this theme, "The history of the doctrine of matter has yet to be written" (*The Concept of Nature* [Cambridge: Cambridge University Press, 1920], p. 16). But cf. also Carolyn Merchant, *The Death of Nature: Women, Ecology and the Scientific Revolution*, 2d ed. (San Francisco: Harper and Row, 1990).

30. Cf. Gary Gutting, "Paradigms, Revolutions, and Technology," in Laudan, ed., *The Nature of Technological Knowledge* (1984), pp. 47–66.

Chapter Six. From Philosophy to Technology

1. According to James Boswell's *Life of Johnson* (1791), Samuel Johnson attributes this definition to Benjamin Franklin in a morning conversation, April 7, 1778. No citation is given, and the statement does not occur in Franklin's written works, but it is now universally credited to him. In a complementary remark a half century later, Thomas Carlyle's *Sartor Resartus* (1833), bk. 1, chap. 5, credits the protagonist Herr Teufelsdröckh with the definition, "Man is a Tool-using Animal."

2. Lord Richie-Calder, "Knowing How and Knowing Why," introduction to pt. 7, Technology, in *New Encyclopaedia Britannica, Propaedia: Outline of Knowledge and Guide to the Britannica* (Chicago: Encyclopaedia Britannica, 1975), p. 434.

3. Bernard P. Dauenhauer, *Silence: The Phenomenon and Its Ontological Significance* (Bloomington: Indiana University Press, 1980), especially pp. 33–49.

4. The writings of engineer Henry Petroski, while not quite as philosophically inclined as Florman's, are predicated on a related attempt to see the humanities in engineering terms. See, e.g., *To Engineer Is Human* (1985) and *Beyond Engineering* (1986), esp. chap. 1, "Introduction: Writing as Bridge-Building," which sees the two activities as fundamentally similar.

5. With regard to the identification between engineer and soldier: In *Troilus and Cressida*, for instance, Thersites refers to Achilles as "a rare enginer" (2.3.8; in Shakespeare see also "the enginer / Hoist with his own petar" [*Hamlet* 2.4.206] and the more ambiguous "tire the ingener" [*Othello* 2.1.65]). In *Paradise Lost* (6.553) Milton describes "the foe / Approaching, gross and huge, in hollow cube / Training his devilish enginery [= artillery]." In his *Tractate of Educa-*

tion (1644), Milton also refers to "engineers" as just another source for that "real tincture of natural knowledge" appropriate to a proper education. See Oliver Morley Ainsworth, ed., *Milton on Education* (New Haven: Yale University Press, 1928), p. 58. The first professional organization of engineers took place in Milton's time, with the formation in the French army of the Corps du Génie in 1672, leading to an associated Ecole du Génie in 1749. (Notice the distinctive relation between the French words *génie* and *ingénieur,* the science of engineering and a person who knows this science, and their common reflection of the Latin.) Indeed, the first schools to grant engineering degrees were all associated with the military: the Ecole des Ponts et Chaussées (1746), which grew out of the semimilitary Corps des Ponts et Chaussées (formed 1716); the Ecole Polytechnique (1794, placed by Napoleon under the direction of the Ministry of the Armed Forces); and the United States Military Academy at West Point (1802). The Corps of Engineers remains an important branch of the United States Army and all modern armies. This material on the military associations of early formal engineering education can be found elaborated in George S. Emmerson, *Engineering Education: A Social History* (Newton Abbot, England: David and Charles, 1973), esp. chap. 2, and John Hubbel Weiss, *The Making of Technological Man: The Social Origins of French Engineering Education* (Cambridge: MIT Press, 1982), esp. chap. 1.

6. See, e.g., Mannyngs's *Chronicle of England* (1338). Richard Sibbes's commentary in Charles Haddon Spurgeon's *The Treasury of Psalms* (1635), Psalm 9, line 15, refers to "that great engineer, Satan." And of course Milton's Lucifer is the engineer of Pandemonium, the city built to rival God.

7. For material on Smeaton, see A. W. Skempton, ed., *John Smeaton, FRS* (London: Thomas Telford, 1981); for civil engineering more generally, see Garth Watson, *The Civils: The Story of the Institution of Civil Engineers* (London: Thomas Telford, 1988).

8. The term "mechanic" continues to share, especially in Great Britain, many of the connotations of "technician." "The man who fixes the gas cooker or wires the house for electricity is not an engineer. He is a mechanic or an electrician." William T. O'Dea, *The Meaning of Engineering* (London: Scientific Book Club, 1961), p. 11.

9. See, e.g., L. T. C. Rolt, *The Mechanicals: Progress of a Profession* (London: Heinmann, 1967), especially chaps. 1–3.

10. The wording has not varied from the first edition, 1974.

11. Ralph J. Smith, Blaine R. Butler, and William K. LeBold, *Engineering as a Career,* 4th ed. (New York: McGraw-Hill, 1983), p. 9. For a sample of definitions running from that of the British architect and civil engineer Thomas Tredgold (1788–1829) to the Accreditation Board for Engineering and Technology (1982), merge Smith, 3d ed. (1969), pp. 8–9, with Smith, 4th ed. (1983), p. 8. The first three editions (1956, 1962, 1969) were all written by Smith alone. Although he credits the fourth edition as "primarily the work of coauthors," a brief comparison reveals nothing new in the quoted passages; hence the accreditation to

Smith alone in the text. Smith also wrote the article "Engineering" in the *New Encyclopaedia Britannica* (1975).

12. Smith et al., *Engineering as a Career,* 4th ed. (1983), p. 10.

13. Smith et al., *Engineering as a Career,* 4th ed. (1983), p. 12.

14. Smith et al., *Engineering as a Career,* 4th ed. (1983), p. 160. Thomas T. Woodson's often-cited *Introduction to Engineering Design* (New York: McGraw-Hill, 1966), while noting that not all engineering functions involve design in the strict sense, also identifies "engineering design" as "the essential activity of professional engineering" (p. 8). Variations on this thesis can be found in most general engineering texts and are too numerous for further citation here, but they will be considered subsequently in greater detail.

15. For sample illustration, see Douglas M. Considine, ed., *Chemical and Process Technology* (New York: McGraw-Hill, 1974). According to the preface of this once standard handbook, it "is *not* a compilation of generalities, but rather is packed with detailed information" (pp. xxii). As specified earlier on p. xxi, this detailed information concerns "the traditional spheres of interest in industrial chemistry and chemical technology as reflected by the petroleum, petrochemical, chemical, paper, textile, and other long-established process industries" as well as "more recent applications of an advancing and broadening chemical technology, including, as examples, the materials and processes now required by the electronics, optics, and aerospace industries." A pie diagram on p. xxviii gives a conceptual overview of the kinds and percentages of information included: roughly one-third is devoted to data on raw materials, one-third to equipment specifications, and one-third to descriptions of consumer products. Further examples of this use of "technology" referring to empirical data on the raw materials of industrial processes, their equipment, and their products can be found in any number of such handbooks. See, e.g., Alexander S. Craig, *Dictionary of Rubber Technology* (New York: Philosophical Library, 1969), and Alan Gilpin, *Dictionary of Fuel Technology* (New York: Philosophical Library, 1970).

16. For more on this term, see Carroll W. Pursell Jr., "Technological Sciences," in *New Encyclopaedia Britannica, Macropaedia* (1975), 18: 19–21.

17. Smith, *Engineering as a Career,* 3d ed. (1969), pp. 210–211. The earlier edition is cited because in the fourth, in order to avoid the sexist implications of referring to the engineer as a "man" and using "him" or "his," the terms are always either "engineers" or "engineering" and "their," and the prose has become too contorted for easy quotation.

18. Philip Sporn, *Foundations of Engineering* (Oxford: Pergamon Press, 1964), pp. 18–19. A late 1960s proposal to compensate for the increasingly theoretical cast of post–World War II engineering education with the creation of a new bachelor of engineering technology (B.E.T.) degree reflects Sporn's ideas. As John Dustin Kemper describes it in his widely used textbook, "The intention of the creators of the B.E.T. degree is that its holders would occupy a middle ground between the craftsman and the engineer and would use the title 'engi-

neering technologist,' rather than 'engineer.' In the words of the Advisory Committee for the Engineering Technology Education Study: 'The technologist should be a master of detail; the engineer, of the total system. . . . The development of methods or new applications is the mark of the engineer. Effective use of established methods is the mark of the technologist" (John Dustin Kemper, *Engineers and Their Profession*, 3d ed. [New York: Holt, Rinehart and Winston, 1982], p. 277). For further documentation see James J. Duderstadt, Glenn F. Knoll, and George S. Springer, *Principles of Engineering* (New York: John Wiley, 1982), pp. 7–11, and George C. Beakley, *Careers in Engineering and Technology*, 3d ed. (New York: Macmillan, 1984), pp. 29 ff.

19. Note once more the subtlety of usage being so summarily appealed to here. "Medical technician" and "medical technologist" can be the same person (one proficient in the use of technical medical hardware without the nurse or physician's comprehension of how it contributes to a patient's health) or different ones (the medical technician as the medical technologist's assistant).

20. Jay Weinstein, *Sociology-Technology: The Foundations of Postacademic Social Science* (Brunswick, N.J.: Transaction Books, 1982), pp. 35–36.

21. Robert S. Sherwood and Harold B. Maynard, "Technology," in *McGraw-Hill Encyclopedia of Science and Technology*, 6th ed. (New York: McGraw-Hill, 1982), 18: 142–146. This article has not changed from the first edition of 1960.

22. Tom Burns, "Technology," in *A Dictionary of the Social Sciences*, ed. Julius Gould and William L. Kolb (New York: Free Press, 1964), pp. 716–717. (This reference work, unlike its just-cited counterpart, does not even have an entry for "engineering.")

23. Robert S. Merrill, "The Study of Technology," in *International Encyclopedia of the Social Sciences* (New York: Macmillan and Free Press, 1968), 15: 576–577. This "classic" social science conception of technology is reiterated in, e.g., Nicholas Abercrombie, Stephen Hill, and Bryan S. Turner, *The Penguin Dictionary of Sociology*, 2d ed. (London: Penguin, 1988), pp. 251–252, and David Jary and Julia Jary, *The HarperCollins Dictionary of Sociology* (New York: HarperCollins, 1991), pp. 515–516.

24. Economists are the most prominent. "Technology means the systematic application of scientific or other organized knowledge to practical tasks" (John Kenneth Galbraith, *The New Industrial State*, 2d rev. ed. [Boston: Houghton Mifflin, 1971], p. 12).

25. Mumford is the best-known illustration of this latter case, although he is not always consistent. For more consistency, see D. S. L. Cardwell, *Turning Points in Western Technology: A Study of Technology, Science and History* (New York: Neale Watson Academic, 1972), who states his theme as "the development of technics and its evolution into *technology* [his italics], which we shall briefly define for the present as technics based on science" (p. 2). Cf. also Arnold Pacey, *The Maze of Ingenuity: Ideas and Idealism in the Development of Technology* (Cambridge: MIT Press, 1976), p. 19: "Dictionaries tend to define technology in terms of 'systematic knowledge' of practical subjects, and clearly such

knowledge is essential to the modern discipline of technology. But clearly also, systematic knowledge of this kind did not exist before 1600 or 1650 at the earliest, so technology, defined in this way, did not exist then either. For this reason, terms such as *the practical arts,* or *the mechanical and chemical arts* are used in this book rather than *technology* to refer to the technical skills of earlier historical periods." But as Pacey himself immediately admits, these latter terms did not occur before the seventeenth century either, so it is not clear why they should be preferred for any earlier period. For a vigorous critique of all such approaches, see Michael Fores, *"Technik,* or Mumford Reconsidered," *History of Technology* 6 (1981): 121–137.

26. Peter F. Drucker, "Work and Tools," *Technology and Culture* 1, no. 1 (winter 1959): 28–37. This important article has been reprinted in Drucker's *Technology, Management and Society* (New York: Harper and Row, 1970) and in Melvin Kranzberg and William H. Davenport, eds., *Technology and Culture: An Anthology* (New York: Schocken, 1972).

27. Nathan Rosenberg, *Perspectives on Technology* (New York: Cambridge University Press, 1976), pp. 1–2. The same view is repeated in Rosenberg, *Inside the Black Box* (1982).

28. Michael Fores, "Some Terms in the Discussion of Technology and Innovation," *Technology and Society* 6, no. 2 (October 1970): 56–63.

29. Cf. Paul T. Durbin, *Dictionary of Concepts in the Philosophy of Science* (New York: Greenwood Press, 1988), entry on "Technology," p. 315: "In short, today the term 'technology' has many meanings, and almost all but the most technical are controversial." It is not clear why Durbin excludes the technical ones.

30. Cf. also Daniel Bell's related definition in *The Winding Passage: Essays and Sociological Journeys, 1960–1980* (New York: Basic Books, 1980), p. 20: "Technology is the instrumental ordering of human experience within a logic of efficient means."

31. John Kenneth Galbraith, *The New Industrial State,* 2d rev. ed. (Boston: Houghton Mifflin, 1971), p. 12; and Nathan Rosenberg, *Inside the Black Box: Technology and Economics* (New York: Cambridge University Press, 1982), p. 145. The second definition is repeated in Nathan Rosenberg and Claudio Frischtak, eds., *International Technology Transfer: Concepts, Measures, and Comparisons* (New York: Praeger, 1985), p. 4. Cf. the closely related definition by another economist: Emmanuel G. Mesthene, *Technological Change: Its Impact on Man and Society* (Cambridge: Harvard University Press, 1970), p. 25.

32. Robert E. McGinn, "What is Technology?" (1978), p. 180. The idea of a "characterology" of technology can be traced back at least to Jacques Ellul, *La Technique* (1954), chap. 2.

33. For discussion of these problems among historians, see the critical symposium "The Historiography of Technology," *Technology and Culture* 15, no. 1 (January 1974): 1–48.

34. *New Encyclopaedia Britannica, Propaedia: Outline of Knowledge and Guide to the Britannica* (1975).

35. The three-volume *Encyclopaedia Britannica,* 1st ed. (1768–1771), is more a

glorified dictionary than an encyclopedia; the word "technology" is conspicuous by its complete absence. The classic eleventh edition (1910–1911) does not have an entry on technology either, but it does have a substantial article, "Technical Education," by Philip Magnus, 26: 487–498. This article is continued and revised up through the fourteenth edition. This situation with other encyclopedias is only slightly less surprising.

36. Hugh of St. Victor, *Didascalicon* (ca. 1125), identifies these as weaving, weapons forging, navigation, agriculture, hunting, medicine, and acting. For commentary see Ivan Illich, "Research by People," in Illich, *Shadow Work* (Boston: Marian Boyers, 1981), pp. 77–95.

37. Jacob Bigelow, *The Useful Arts*, 2 vols. (Boston: Thomas Webb, 1842).

38. André Leroi-Gourhan, *Evolution et techniques*, vol. 1, *L'Homme et la matière* (Paris: Michel, 1943), and vol. 2, *Milieu et techniques* (Paris: Michel, 1945). This classification scheme provides the organizing framework for Maurice Daumas, ed., *Histoire générale des techniques* (Paris: Presses Universitaires de France, 1962–), a work to which Leroi-Gourhan is a major contributor. For an outline in English of the Leroi-Gourhan framework see the appendix to Ronald Bruzina, "Art and Architecture: Ancient and Modern," *Research in Philosophy and Technology* 5 (1982): 184–187.

39. Leo Marx, *The Machine in the Garden: Technology and the Pastoral Ideal in America* (New York: Oxford University Press, 1964).

40. Donald W. Shriver Jr., "Man and His Machines: Four Angles of Vision," *Technology and Culture* 13, no. 4 (October 1972): 531–555.

41. Daniel Callahan, "Modes and Manifestations of Technology," chap. 3 in *The Tyranny of Survival* (New York: Macmillan, 1973), pp. 55–84.

42. See H. J. Eisenman, "Technology, Society, and Values in Twentieth Century America: The UCLA 1973 Summer Seminar," *Technology and Culture* 16, no. 2 (April 1975): 182–188, which reports on a seminar taught by Burke.

43. The influence of Schuurman's analysis is clear in Steven V. Monsma, ed., *Responsible Technology: A Christian Perspective* (Grand Rapids, Mich.: Wm. B. Erdmans, 1986), esp. chaps. 2, 9, and 10.

44. D. Teichmann, "On the Classification of the Technological Sciences," in Friedrich Rapp, ed., *Contributions to a Philosophy of Technology* (1974), p. 138.

45. Stephen J. Kline, "What is Technology?" *Bulletin of Science, Technology and Society* 5, no. 3 (1985): 215–218.

46. Philosopher of science Karl Popper, in his distinction between three worlds—world 1 of living and nonliving material objects, world 2 of subjective mental states, and world 3 of human products—can be interpreted as anticipating aspects of this new division. Cf. Karl R. Popper, *Objective Knowledge: An Evolutionary Approach*, rev. ed. (New York: Oxford University Press, 1979), esp. chaps. 3 and 4.

Chapter Seven. Types of Technology as Object

1. See also Lewis Mumford, "Machines," *Encyclopedia Americana*, International edition (New York: Americana, 1972), 18: 57–62.

2. Note the equivocation on "work" here and in reference to tools. Tools perform work and produce works, whereas machines operate and produce products. When machines are spoken of as performing work, the definition of work becomes that of physics—force times displacement in the line of force.

3. Ludwig Noiré, *Das Werkzeug und seine Bedeutung für die Entwickelungsgechichte der Menschheit* (Mainz: I Diemer, 1880). Quoted from Paul Carus, "The Philosophy of the Tool," *Open Court* 7, no. 29, issue 308 (July 20, 1893), p. 3736. Carus is reviewing Noiré's book.

4. This last point is attested by the way tools tend to impose certain forms of operating on their environment, even when their user might desire otherwise. "If the only tool you have is a hammer, you tend to treat everything as if it were a nail" (Abraham Maslow, *Psychology of Science* [New York: Harper and Row, 1966], pp. 15–16).

5. John Donne, *Eighty Sermons* (1640), no. 37, pp. 253–254.

6. The simple machines of antiquity are sometimes numbered as five, at other times as six. When five are listed, either the inclined plane or the wedge is omitted as an application of the other. In modern mechanics the gear drive and hydraulic press are often considered additional simple machines.

7. Vitruvius *De architectura* 10.1.3.

8. Note the two senses of "drive": in one sense the motor "drives the car," in another the "driver" does.

9. See, e.g., Vincent Edward Smith, "Toward a Philosophy of Physical Instruments," *Thomist* 10, no. 3 (July 1947): 307–333, and F. V. Lazarev and M. K. Trifonova, "The Role of Apparatus in Cognition and Its Classification," in Friedrich Rapp, ed., *Contributions to a Philosophy of Technology* (1977), pp. 197–209.

10. F. M. Feldhaus, "Machines and Tools: Ancient, Medieval and Early Modern," in *Encyclopedia of the Social Sciences* (New York: Macmillan, 1933), 10: 14.

11. In this sense all modern "mechanical" devices (if not electronic, chemical, etc., ones) are made up of simple machines as elements, some of which will invariably be tools—e.g., the switch.

12. For Hegel's most fully developed discussion of tools and machines and their relation to work, see the two sets of lecture notes from his Jena period: *Jenenser Realphilosophie I*, ed. J. Hoffmeister (Leipzig, 1932)—notes from 1803–1804; and *Jenenser Realphilosophie II*, ed. J. Hoffmeister (Leipzig, 1931)—notes from 1805–1806. *Jenenser Realphilosophie II* has been republished as *Jenaer Realphilosophie* (Hamburg: Felix Meiner, 1967). See especially *Realphilosophie I*, pp. 23 ff.; and *Realphilosophie II*, pp. 197 ff. Other relevant comments can be found in Hegel's *Encyclopaedia*, sec. 526, and in the *Philosophy of Right*, sec. 198 along with the addition to sec. 203. (Note in passing, however, that in all of Hegel's system—which includes virtually every other type of consciousness—there is no clear place for technology or engineering as such.) See also Marx's discussion of tools and machines in *Das Kapital* (1867), vol. 1, chap. 15, "Machinery and Modern Industry," sec. 1, "The Development of Machinery." Cf. also Aris-

totle's description of the tool (*organon*) as an inanimate servant in *Eudemian Ethics* 7.9.1241b24 and the converse allusion to the slave as a living tool in *Politics* 1.4.1253b23–1254a18. Finally, Edmond Barbotin, *The Humanity of Man* (Maryknoll, N.Y.: Orbis, 1975), pp. 197–200, gives a brief phenomenological description of the tool-machine relationship that parallels the present analysis.

13. "Technology . . . is characterized by the fact that it itself is the product of technological means. Technology was not created by man with his bare hands and capacity; it is one technology that produces another" (Nathan Rotenstreich, "Technology and Politics," *International Philosophical Quarterly* 7, no. 2 [June 1967]: 98).

14. The locus classicus for this argument is Henry Adams (1838–1918), "The Dynamo and the Virgin," chap. 25 in *The Education of Henry Adams* (Boston: Houghton Mifflin, 1918). But cf. Lynn White Jr., "Dynamo and Virgin Reconsidered," in *Machina ex Deo* (Cambridge: MIT Press, 1968), pp. 57–73; and Harvey Cox, "The Virgin and the Dynamo Revisited: An Essay on the Symbolism of Technology," *Soundings* 44, no. 2 (summer 1971): 125–146.

15. Norris W. Clarke, "System: A New Category of Being," *Proceedings of the Twenty-third Annual Convention of the Jesuit Philosophical Association* (Woodstock, Md.: Woodstock College Press, 1961), pp. 5–17. See also H. Romback's monumental *Substance System Struktur: Die Ontologie des Funktionalismus und der philosophische Hintergrund der modernen Wissenschaften*, 2 vols. (Freiburg: Alber, 1966).

16. Classical (or Newtonian) mechanics, which deals with material bodies of a size appreciable by the unaided human senses moving at speeds small in comparison with the speed of light, is to be distinguished from quantum mechanics, which describes the behavior of atoms and subatomic particles, and from relativity mechanics, which deals with speeds approaching the speed of light. Celestial mechanics is an astronomical application of classical mechanics; fluid mechanics applies classical mechanics to nonrigid bodies.

17. "This dynamics is entirely a modern science. The mechanical speculations of the ancients, particularly of the Greeks, related wholly to statics. Only in mostly unsuccessful paths does their thinking extend into dynamics" (Ernst Mach, *The Science of Mechanics: A Critical and Historical Account of Its Development* [LaSalle, Ill.: Open Court, 1960; [original 1883], p. 151).

18. *McGraw-Hill Dictionary of Science and Technology* and *McGraw-Hill Encyclopedia of Science and Technology*. Cf. Alexander Cowie, "Machines and Machine Components," in *New Encyclopaedia Britannica, Macropaedia* (1975), 11: 231: a machine is "a device consisting of two or more resistant, relatively constrained parts that may serve to transmit and modify force and motion in order to do work"; and George H. Martin, *Kinematics and Dynamics of Machines* (New York: McGraw-Hill, 1969), p. 3: "A machine is a device for transforming or transferring energy."

19. A mechanism is "that part of a machine which contains two or more pieces so arranged that the motion of one compels the motion of the others"

(*McGraw-Hill Dictionary of Science and Technology*). See also *McGraw-Hill Encyclopedia of Science and Technology,* under "machine." Joseph Edward Shigley, *Theory of Machines* (New York: McGraw-Hill, 1961), p. 3: "A machine is a device which uses power to accomplish a physical effect. But a mechanism is a combination of machine elements to achieve a certain motion."

20. A slightly different conceptualization of the machine/mechanism distinction is contained in Anthony Esposito, *Kinematics for Technology* (Columbus, Ohio: Charles Merrill, 1973): "A mechanism is a device, consisting of two or more members, which accepts an input and modifies it in some way to produce a desired output. Inputs and outputs are generally considered to be either forces or motions" (p. 3). "A machine consists of one or more mechanisms and, as such, usually performs various functions" (p. 6). Esposito uses "mechanism" to denote any component of a machine complex.

21. Donna J. Haraway, *Simians, Cyborgs, and Women: The Reinvention of Nature* (New York: Routledge, 1991), p. 152. The *Oxford English Dictionary,* 2d ed. (1992), cites the first occurrences of "cyborg" as *New York Times* (May 22, 1960), p. 31, and an article by Clynes and Kline in, *Astronautics* (September 1960), p. 27—thus showing its simultaneous appearance in both popular and scientific press. Cf. also John Todd's brief for "living machines"—that is, ecologies structured for specific purposes such as purifying water in symbiosis with urban environments—which might be called cyborg-landscapes (John Todd, "Living Machines," *Annals of Earth* 10, no. 3 [1992], p. 3).

22. Donella H. Meadows, Dennis L. Meadows, Jørgen Randers, and William W. Behrens III, *Limits to Growth: A Report for the Club of Rome's Project on the Predicament of Mankind* (New York: Universe Books, 1972). The Worldwatch Institute, founded by Lester Brown in 1974, has since 1984 issued each January 200-page-plus *State of the World* reports on "Progress toward a Sustainable Society" (New York: W. W. Norton, 1984–). See also Norman Myers, with Uma Ram Nath and Melvin Westlake, *Gaia: An Atlas of Planet Management* (Garden City, N.Y.: Doubleday Anchor, 1984; 2d ed. 1993); Walter H. Corson, ed., for the Global Tomorrow Coalition, *The Global Ecology Handbook: What You Can Do about the Environmental Crisis* (Boston: Beacon Press, 1990). For restoration ecology, see William R. Jordan III, Michael E. Gilpin, and John D. Aber, eds., *Restoration Ecology: A Synthetic Approach to Ecological Research* (New York: Cambridge University Press, 1987).

23. Although no one uses this term, it loosely characterizes the views of the following spectrum of critics: Robert Elliot, "Faking Nature," *Inquiry* 25, no. 1 (March 1982): 81–93; Wolfgang Sachs, "The Gospel of Global Efficiency: On Worldwatch and Other Reports on the State of the World," *IFDA* (International Foundation for Development Alternatives, 4 place du Marché, 1260 Nyon, Switzerland) *Dossier,* no. 68 (November–December 1988): 33–39; Wolfgang Sachs and Ivan Illich, "A Critique of Ecology: The Virtue of Enoughness," *NPQ: New Perspectives Quarterly* 6, no. 1 (spring 1989): 16–19; and Wolfgang Sachs,

ed., *Global Ecology: A New Area of Political Conflict* (London: Zed Books, 1993); Eric Higgs, "A Quality of Engaging Work to Be Done: Ecological Restoration and Morality in a Technological Culture," *Restoration and Management Notes* 9, no. 2 (winter 1991): 97–104. See also Higgs, "Musings at the Confluence of the Rivers *Techne* and *Oikos*," *Research in Philosophy and Technology* 12 (1992): 243–258; David Strong, "The Technological Subversion of Environmental Ethics," *Research in Philosophy and Technology* 12 (1992): 33–66; Eric Katz, "The Big Lie: Human Restoration of Nature," *Research in Philosophy and Technology* 12 (1992): 231–241.

24. This point has become the basic thesis of restoration ecology and is frequently expounded in the pages of *Restoration and Management Notes*. See, e.g., William R. Jordan III, "On the Imitation of Nature," and Donald Worster, "The Anti-management Revolt," *Restoration and Management Notes* 3, no. 1 (summer 1985): 2–3 and 4–5; William R. Jordan III, "Thoughts on the Acid Test," and the exchange of letters between Reed Noss and Lynn Starnes on "Restoration" vs. "Habitat Management," *Restoration and Management Notes* 3 no. 2 (winter 1985): 58 and 59–60; William R. Jordan III, "Restoration and the Reentry of Nature," *Restoration and Management Notes* 4, no. 1 (summer 1986): 2; Peter Losin, "Faking Nature—a Review," *Restoration and Management Notes* 4, no. 2 (winter 1986): 55; James Sayen, "Notes toward a Restoration Ethic," *Restoration and Management Notes* 7, no. 2 (winter 1989): 57–59; and Steve Packard, "No End to Nature" (a reply to Bill McKibben's *The End of Nature*), *Restoration and Management Notes* 8, no. 2 (winter 1990): 72. The reflections of Wendell Berry also deserve serious consideration in this regard; see, e.g., his essays on "Getting Along with Nature" and "Preserving Wilderness" in *Home Economics* (San Francisco: North Point Press, 1987).

25. Cf., e.g., Brian Cotterell and Johan Kamminga, *Mechanics of Pre-industrial Technology: An Introduction to the Mechanics of Ancient and Traditional Material Culture* (New York: Cambridge University Press, 1990), which covers machines, structures, stone tools, projectiles (or weapons), land transport, water transport, and musical instruments, in that order. See also Henry Hodges, *Artifacts: An Introduction to Primitive Technology* (New York: Praeger, 1964); and David Earl Young and Robson Bonnichsen, *Understanding Stone Tools: A Cognitive Approach* (Orono, Maine: Center for the Study of Early Man, 1984).

26. Mumford, *Technics and Civilization* (1934), p. 80, himself citing Reuleaux.

27. Lynn White Jr., *Medieval Technology and Social Change* (New York: Oxford University Press, 1962), p. 115. Cf. also Siegfried Giedion's remarks on the natural movements of the hand in his *Mechanization Takes Command* (New York: W. W. Norton, 1969 [1948]), pp. 46–47.

28. Although the analysis here has focused on the term "extension," one could also consider the notion of "projection," making use of its various psychological meanings.

29. Sōetsu Yanagi, *The Unknown Craftsman: A Japanese Insight into Beauty,*

adapted by Bernard Leach (New York: Kodansha International, 1972), p. 108.

30. See, e.g. Jerome S. Bruner, *Beyond the Information Given* (New York: W. W. Norton, 1973), especially pp. 252 and 300–301.

31. For an impressionistic account of this distinction, see the section "Soft Technology" in Stewart Brand, ed., *Whole Earth Epilog* (Baltimore: Point/Penguin Books, 1974). But see also E. F. Schumacher, *Small Is Beautiful* (New York: Harper and Row, 1973); David Dickson, *Alternative Technology and the Politics of Technical Change* (New York: Universe Books, 1975); and R. Clark, "The Pressing Need for Alternative Technology," *Impact of Science on Society* 23, no. 4 (October–December 1973). Another set of terms sometimes used for this distinction is high versus low technology. Since its origins in the 1970s the idea of alternative technologies has spawned a continuing discussion, usually of a more practical bent, that it is not necessary to survey here.

32. Octavio Paz, "Use and Contemplation," in *In Praise of Hands* (Greenwich, Conn.: New York Graphic Society, 1974), p. 21. Cf. also: "It was a long room of agreeable shape. The thick clay walls had been finished on the inside by the deft palms of Indian women, and had that irregular and intimate quality of things made entirely by the human hand. There was a reassuring solidity and depth about those walls, rounded at door-sills and window-sill, rounded in wide wings about the corner fire-place" (Willa Cather, *Death Comes for the Archbishop* [New York: Knopf, 1927], pp. 32–33).

33. The social constructivist school in the sociology and history of technology can, of course, be read as meeting the first of these requirements. See, e.g., MacKenzie and Wajcman, eds., *The Social Shaping of Technology* (1985); Bijker, Hughes, and Pinch, eds., *The Social Construction of Technological Systems* (1987); and Bijker and Law, eds., *Shaping Technology / Building Society* (1992). The first essay in the 1985 collection, Pinch and Bijker's "The Social Construction of Facts and Artifacts, or How the Sociology of Science and the Sociology of Technology Might Benefit Each Other," includes a history of the bicycle that exemplifies well the social constructivist approach. Of related but generally unrecognized interest are material culture studies. See, e.g., Thomas J. Schlereth, ed., *Material Culture Studies in America* (Nashville, Tenn.: American Association for State and Local History, 1982).

34. An aside on terminology: The practical proposals of socialism are based on sociological theory. If, to avoid ad hominem characterizations, it is preferable to use "socialist" and "socialism" in place of "Marxist" and "Marxism," then some less personal terms should be found for that position commonly referred as "Luddite" or "Luddism." For theoretical studies on which Luddite practice could be based, one possible candidate is the term "mechanology," used by Lafitte and Simondon to refer to a phenomenology of machines, taking machines as a generic term that includes tools. But the Lafitte-Simondon theoretical project confines itself to the inner evolution of mechanical development and fails to address issues dealing with the social implications of alternative inner structures among artifacts. Moreover, insofar as theoretical study

leads to political program, the terms "mechanist" and "mechanism" would have exactly the wrong connotations. What the antisocialist (*not* antitechnologist) school promotes is a phenomenology of artifacts or "artifactology," from which one can formulate a political program that might be termed "artifactism."

35. Richard Weaver, "Humanism in an Age of Science," ed. Robert Hamlin, *Intercollegiate Review* 7, nos. 1–2 (fall 1970): 130–135. Reprinted in Mitcham and Mackey, eds., *Philosophy and Technology* (1972), pp. 136–142.

36. For an appraisal of the necessity and beneficence of artifice, see Jacques Ellul, "Technique and the Opening Chapters of Genesis," in Mitcham and Grote, eds., *Theology and Technology* (1984), pp. 123–138.

37. See, e.g., Marshall McLuhan and Quentin Fiore, with Jerome Agel, *The Medium Is the Massage* (New York: Bantam, 1967).

38. Extending this analysis, which obviously adapts terms from Aristotle, one could describe the tool itself as signated matter. Then one could say that the more technologically advanced the tool, the more signated its matter, and thus the more predetermined its motion.

39. Shoshana Zuboff, *In the Age of the Smart Machine: The Future of Work and Power* (New York: Basic Books, 1988), p. 9.

40. For a broad-brush historical study of relations between different technologies and the forms of work they imply that necessarily stops short of Zuboff's analysis, see Melvin Kranzberg and Joseph Gies, *By the Sweat of Thy Brow: Work in the Western World* (New York: G. P. Putnam, 1975).

41. Mihaly Csikszentmihalyi and Eugene Rochberg-Halton, *The Meaning of Things: Domestic Symbols and the Self* (New York: Cambridge University Press, 1981).

42. Weston, "Ivan Illich and the Radical Critique of Tools" (1989), p. 182, n. 6. There are some minor citation mistakes in Weston.

43. Csikszentmihalyi and Rochberg-Halton, *Meaning of Things*, p. 16.

44. Baudrillard, *Le Système des objets*, from the partial translation in Mark Poster, ed., *Jean Baudrillard: Selected Writings* (Stanford: Stanford University Press, 1988), p. 24. Italics in the original.

45. Poster, ed., *Baudrillard: Selected Writings*, p. 25.

46. Borgmann, *Technology and the Character of Contemporary Life* (1984), pp. 167–168.

Chapter Eight. Types of Technology as Knowledge

1. Charles Augustin Coulomb (1736–1806), who served for thirty years in the Corps du Génie, although best known for his work in electrical theory, actually began his studies in applied mechanics and only later in life moved on to physics.

2. The engineering sciences, as defined in the authoritative James H. Potter, ed., *Handbook of the Engineering Sciences* (Princeton, N.J.: Van Nostrand, 1967), include what are called the basic engineering sciences (mathematics, physics,

chemistry, graphics, statistics, theory of experiments, and mechanics) and the applied engineering sciences (thermal phenomena, heat and mass transfer, chemical energy conversion, turbomachinery, nuclear reactor engineering, aeronautics and astronautics, field theory, electromechanical energy conversion, physical electronics, electronic circuits, system dynamics, materials science, machine elements, control systems, operations research, information retrieval, preparation of reports, and computers).

3. For a contemporary statement of this argument in a different context (which also contains a criticism of Piaget's tendency to subordinate earlier to later stages of cognitive development), see Gareth Matthews, *Philosophy and the Young Child* (Cambridge: Harvard University Press, 1980), chap. 4, "Piaget," pp. 37–55.

4. Patricia Benner, *From Novice to Expert: Excellence and Power in Clinical Nursing Practice* (Reading, Mass.: Addison-Wesley, 1984).

5. Mario Bunge, "Philosophical Inputs and Outputs of Technology" (1979b), pp. 266–267. Bunge is not, however, always completely consistent in his technic/technology distinction. Cf. "Technology: From Engineering to Decision Theory" (1985), p. 220: "Our concept of technology is wide as well: it includes all the crafts, such as farming and plumbing, and all the science-based technologies, from mechanical to social engineering. Following Mumford we call *technics* the body of prescientific technical knowledge, and *technology* the body of science-based technical knowledge." Obviously technology cannot at once include and be something other than technics.

6. Bunge, "Technology," table on p. 238. But see also Michael Polanyi, *Pure and Applied Science and Their Appropriate Forms of Organization* (1953); James K. Feibleman, "Pure Science, Applied Science, and Technology: An attempt at Definitions" (1961); and Henryk Skolimowski, "Problems of Truth in Technology" (1970–71).

7. For more discussion of the distinction between scientific and technological experiments, see Simon Moser, "Toward a Metaphysics of Technology" (1971), pp. 137–138. Cf. also Hilbert van Nydeck Schenck Jr., *Theories of Engineering Experimentation*, 2d ed. (New York: McGraw-Hill, 1968).

8. See, e.g., Joseph Agassi, "The Confusion between Science and Technology in the Standard Philosophies of Science" (1966), and Henryk Skolimowski, "The Structure of Thinking in Technology" (1966).

9. Thomas M. Smith, "Project Whirlwind: An Unorthodox Development Project," *Technology and Culture* 17, no. 3 (July 1976): 462. (The "unorthodox" character of this project has nothing to do with the point at issue here.)

10. This is the classic wording of George Cayley, quoted from C. H. Gibbs-Smith, *Sir George Cayley's Aeronautics, 1796–1855* (London, 1962), as cited in Vincenti (1990), p. 208.

11. Ronald Kline, "Science and Engineering Theory in the Invention and Development of the Induction Motor, 1880–1900," *Technology and Culture* 28, no. 2 (April 1987): 283–313.

12. Bruce E. Seely, "The Scientific Mystique in Engineering: Highway Research at the Bureau of Public Roads, 1918–1940," *Technology and Culture* 25, no. 4 (October 1984): 799.

13. See, e.g., Sidney M. Edelstein, "The Allerley Matkel (1532): Facsimile Text, Translation, and Critical Study of the Earliest Printed Book on Spot Removing and Dyeing," *Technology and Culture* 5, no. 3 (summer 1964): 297–321; Lon R. Shelby, "Setting out the Keystones of Pointed Arches: A Note on Medieval 'Baugeometrie,'" *Technology and Culture* 10, no. 4 (October 1969): 537–548; and Barton C. Hacker, "Greek Catapults and Catapult Technology: Science, Technology and War in the Ancient World," *Technology and Culture* 9, no. 1 (January 1968): 34–50.

14. On the distinction between work and labor see, e.g., Hannah Arendt, *The Human Condition* (1958), pts. 3 and 4.

15. Harold I. Brown, *Perception, Theory and Commitment: The New Philosophy of Science* (Chicago: University of Chicago Press, 1979), p. 10. My italics.

16. See Don Ihde, *Technics and Praxis* (1979), and *Existential Technics* (1983), and Peter Galison, *How Experiments End* (Chicago: University of Chicago Press, 1987).

17. Patrick A. Heelan, *Space-Perception and the Philosophy of Science* (1983), p. 251. Heelan adapts the phrase from studies in the psychology of perception.

18. See especially the essays collected in Cyril Stanley Smith, *A Search for Structure: Selected Essays on Science, Art, and History* (Cambridge: MIT Press, 1981).

19. Hans Jonas, "The Practical Uses of Theory" (1966), pp. 189–190. On this issue see also Nicholas Lobkowicz, *Theory and Practice: History of a Concept from Aristotle to Marx* (Notre Dame, Ind.: University of Notre Dame Press, 1967), especially pt. 1.

Chapter Nine. Types of Technology as Activity

1. These two concepts, action and process, are strongly linked with two distinct schools of philosophy. In the first, the linguistic-analytic philosophy of action stresses the distinction between (nonvoluntary) behavior that simply happens and (voluntary) action that has results in ways that are in accordance with the intuitive associations presented here. One suggestive distinction in this school is that between such linguistic acts as announcing, persuading, proposing, encouraging, promising, and such. According to J. L. Austin in *How to Do Things with Words* (New York: Oxford University Press, 1962), this diversity can be separated into perlocutionary and illocutionary linguistic acts. The former essentially involve the production of some effect (making?) and do not always require work (tools?); the latter do not necessarily produce any effect (using?) but do require a locutionary base (tools?). Furthermore, illocutionary acts (using?) can be a means to perlocutionary acts (making?), but not vice versa (at least not in the same way). Similar suggestive parallels can be drawn with analytic discussions of the relations between choosing (inventing?), decid-

ing (designing?), and doing (making?). For more detailed introductions to the linguistic-analytic philosophy of action, see Alvin I. Goldman, *A Theory of Human Action* (Englewood Cliffs, N.J.: Prentice-Hall, 1970), and Lawrence Davis, *Theory of Action* (Englewood Cliffs, N.J.: Prentice-Hall, 1979).

In the second, the school of process philosophy views process, following Alfred North Whitehead and others, as a fundamental but overlooked metaphysical category. For general introduction, see George R. Lucas Jr., *The Genesis of Modern Process Thought: A Historical Outline with Bibliography* (Metuchen, N.J.: Scarecrow Press and American Theological Library Association, 1983), and Ernest Wolf-Gazo, ed., *Process in Context: Essays in Post-Whiteheadian Perspectives* (New York: Peter Lang, 1988). Dorothy Emmet's *The Passage of Nature* (Philadelphia: Temple University Press, 1992) defines process as "a continuant with an internal order and a direction of change" (p. 6) and distinguishes three main types: natural, as in physiological changes; social, as in human political activities; and artificial, as in the building of a house. Although Emmet thus defines the artificial as "creative processes," by stressing the ordered continuity of this creativity over an extended period of time, she also seems to at least include using. A creative process could be said to be distinct from a creative action in its extention and inclusion of using.

2. The introduction of a distinction between "making action" and "using process" is not unlike that found in, say, moral philosophy, when the contrasting pairs of right/wrong (applied to the inner character of actions) and good/bad (applied to external consequences) are proposed as technical expressions according to a more fluid but nonetheless allusive linguistic practice.

3. Claude Lévi-Strauss, *The Savage Mind* (Chicago: University of Chicago Press, 1966), p. 17.

4. Lévi-Strauss, *The Savage Mind*, p. 19.

5. Bernard Rudofsky, *Architecture without Architects: A Short Introduction to Non-pedigreed Architecture* (New York: Doubleday, 1964), opposite pl. 4. (This is the exhibition catalog; pages are unnumbered.).

6. "It is the function of the engineering and development department of a modern corporation to take equipment which it has been decided by management to manufacture, and to do the detailed study of the design and manufacturing process which is necessary if the device is to be produced cheaply and in volume, and if it is to be free from minor defects under field operation" (Francis Russel Bichowsky, *Industrial Research* [New York: Arno Press, 1972, 1942], p. 26). Bichowsky also points out that such design "is never a fixed thing" because of changes in demand, experience, the availability of materials, and other variables.

7. For a discussion of these and other aspects of engineering from the viewpoint of the engineer, see Ralph J. Smith, *Engineering as a Career*, 4th ed. (New York: McGraw-Hill, 1983), chap. 6, "Functions of Engineering," pp. 118–166. See also A. W. Futrell Jr., *Orientation to Engineering* (Columbus, Ohio: Charles E. Merrill, 1961), especially chap. 16, "Functions of Engineering."

8. A second example along this line: In 1862 Alphonse Beau de Rochas wrote a pamphlet on improving the efficiency of locomotives in which he clearly and in detail conceived the four-stroke-cycle internal combustion engine and gave a correct theoretical explanation of its working principles. But it was not until fourteen years later that Nicolaus August Otto invented the four-stroke-cycle engine—based on an incorrect theory. Notice how these examples point up the weakness of, e.g., R. J. Forbes's definition of invention as "a mental process in which various discoveries and observations are combined and guided by experience into some new tool or operation" ("The Beginnings of Technology and Man," in *Technology in Western Civilization*, ed. M. Kranzberg and C. Pursell [New York: Oxford University Press, 1967], 1: 14).

9. Francis Bacon (1561–1626) is the first to argue explicitly and at length for the need to promote inventing. In so doing he distinguishes between those inventions that have been based on an appropriate understanding of nature and those that have been virtually independent of scientific knowledge—and, it could be added, method. The former are what today would be called science-based inventions; the latter are more traditional or evolutionary inventions. See, e.g., Bacon's "Thoughts and Conclusions," trans. from the Latin in Benjamin Farrington, *The Philosophy of Francis Bacon* (Chicago: University of Chicago Press, 1966), pp. 90 ff.

10. See, e.g., S. C. Gifillan, *The Sociology of Invention* (Cambridge: MIT Press, 1963 [1935]). Two other primary sources for philosophical reflection on the nature and meaning of inventing are John Jewkes, David Sawers, and Richard Stillerman, *The Sources of Invention*, 2d ed. (New York: W. W. Norton, 1969), and H. Stafford Hatfield, *The Inventor and His World* (New York: Dutton, 1933). The most philosophical discussion is René Boirel, *Théorie générale de l'invention* (Paris: Presses Universitaires de France, 1961).

11. Alfred North Whitehead, *Science and the Modern World* (New York: Free Press, 1967 [1925]), p. 96.

12. See John B. Rae, "The Invention of Invention," in Kranzberg and Pursell, eds., *Technology in Western Civilization* (1967), 1: 325–336, and Daniel J. Boorstin, *The Americans: The Democratic Experience* (New York: Random House, 1973), chap. 56, "The Social Inventor: Inventing the Market," and chap. 57, "Communities of Inventors: Solutions in Search of Problems."

13. See especially the English translation "Technology in Its Proper Sphere" [from *Philosophie der Technik* (1927)], in Mitcham and Mackey, eds., *Philosophy and Technology* (1972), pp. 317–334.

14. David Pye, *The Nature of Design* (New York: Reinhold, 1964), p. 19.

15. Some representative general engineering design texts from the past few decades include Morris Asimov, *Introduction to Design* (Englewood Cliffs, N.J.: Prentice-Hall, 1962); Edward V. Krick, *An Introduction to Engineering and Engineering Design* (New York: John Wiley, 1965; 2d ed., 1969; reprint, Huntington, N.Y.: Krieger, 1976); John R. Dixon, *Design Engineering: Inventiveness, Analysis, and Decision Making* (New York: McGraw-Hill, 1966); Thomas T. Woodson, *In-*

troduction to Engineering Design (New York: McGraw-Hill, 1966); D. Henry Edel Jr., ed., *Introduction to Creative Design* (Englewood Cliffs, N.J.: Prentice-Hall, 1967); Duncan Morrison, *Engineering Design: The Choice of Favourable Systems* (New York: McGraw-Hill, 1968); William H. Middendorf, *Engineering Design* (Boston: Allyn and Bacon, 1969); Joseph P. Vidosic, *Elements of Design* (New York: Ronald Press, 1969); Michael French, *Engineering Design: The Conceptual State* (London: Heinemann, 1971), 2d ed., *Conceptual Design for Engineers* (London: Design Council; and New York: Springer-Verlag, 1985); G. Pitts, *Tech niques in Engineering Design* (New York: John Wiley, 1973); N. L. Svensson, *Introduction to Engineering Design* (Randwich: New South Wales University Press, 1974; London: Pittman, 1976; 2d rev. ed., Kensington: New South Wales University Press, 1981); W. H. Mayall, *Principles of Design* (New York: Van Nostrand Reinhold, 1979); Percy H. Hill, *The Science of Engineering Design* (New York: Holt, Rinehart and Winston, 1970); J. Christopher Jones, *Design Methods: Seeds of Human Futures* (New York: John Wiley, 1970; 2d enlarged ed. without subtitle, 1992); Keith Sherwin, *Engineering Design for Performance* (Chichester, England: Ellis Horwood, and New York: John Wiley, 1982); George E. Dieter, *Engineering Design: A Materials and Processing Approach* (New York: McGraw-Hill, 1983; 2d ed. 1990); G. Pahl and W. Beitz, *Engineering Design,* ed. Ken Wallace (London: Design Council; and New York: Springer-Verlag, 1984); and Nigel Cross, *Engineering Design Methods* (New York: John Wiley, 1989). Also worth mention is the tetralogy by Gordon L. Glegg: *The Design of Design* (New York: Cambridge University Press, 1969); *The Selection of Design* (New York: Cambridge University Press, 1972); *The Science of Design* (New York: Cambridge University Press, 1973); and *The Development of Design* (New York: Cambridge University Press, 1981). Continuity in basic ideas over this period is indicated by reprints and consistency of mutual citations. The typical structure of such books is to begin with a general definition of the engineering design activity, to proceed to a description of the designing process or method, and then to take up in varied order topics related to this process, such as modeling, detailing, graphic representation, standards, testing, reliability, evaluation and optimization, and economics. Over the past four decades the main changes in engineering design textbooks has been to augment such topics with material on computer graphics and creativity.

16. "Report on Engineering Design" (MIT Committee on Engineering Design), *Journal of Engineering Education* 51, no. 8 (April 1961): 647.

17. Woodson, *Introduction to Engineering Design* (1966), p. 3. In the *McGraw-Hill Dictionary of Science and Technology* (1984) design is defined as "the act of conceiving and planning the structure and parameter values of a system, device, process, or work of art."

18. Pahl and Beitz, *Engineering Design,* p. 1.

19. See José Ortega y Gasset, "Thoughts on Technology," in Mitcham and Mackey, eds., *Philosophy and Technology* (1972), pp. 295 ff. Ortega, *Obras completas,* 5: 331 ff.

20. Joseph Edward Shigley, *Theory of Machines* (New York: McGraw-Hill, 1961), pp. v and vi. Over the course of almost thirty years, and despite the increasing influence of the use of computers in both computational and graphic analysis, Shigley has continued to stress this point. In an expanded *Theory of Machines and Mechanisms*, written with John Joseph Uicker Jr. (New York: McGraw-Hill, 1980), Shigley remains "firmly of the opinion that graphical computation is . . . basic" (p. xi). Joseph Edward Shigley and Charles R. Mischke, *Mechanical Engineering Design* (New York: McGraw-Hill, 1989), continues to emphasize the centrality of graphics in the design process.

21. E. F. O'Doherty, "Psychological Aspects of the Creative Act," in J. Christopher Jones and D. G. Thornley, *Conference on Design Methods* (1963), pp. 197–203.

22. "Report on Engineering Design," *Journal of Engineering Education* 51, no. 8 (April 1961): 647–648.

23. For a related discussion of phantasmal thinking, see L. R. Rogers, "Sculptural Thinking—I" and "Sculptural Thinking—III" (part 2 is a commentary by Donald Brook), in *Aesthetics in the Modern World*, ed. Harold Osborne (New York: Weybright and Talley, 1968). For an empirical account of the high correlation between drawing talent and general engineering abilities, see Steve M. Slaby and Arthur L. Bigelow, "Engineering Graphics—a Predictor for Academic Performance in Engineering," *Journal of Engineering Education* 51, no. 7 (March 1961): 581–587. At the conclusion of their statistical presentation, the authors suggest that "graphics is part of the thinking of an engineer" (p. 586), because "the ability to 'think' space is a necessary condition if we are to define engineering correctly" (p. 587).

24. These factors are usually enumerated as four: materials, interrelation of parts, methods of construction, and effect of the whole upon those who will become involved with it. Only the first three, however, can be analyzed in the drawing itself; the fourth denotes a social context that is not amenable to quantification and is, in fact, a stumbling block and source of frustration to many engineers. For example, technically speaking a long, flat bridge can be constructed that is completely safe but that, because of the curvature of the earth, will appear to an approaching motorist to be sagging; as a result the public will be afraid to use it. This compels the engineer to arch the bridge in a way that is not required by any of the first three factors. A second example: floors in concrete buildings have to have almost twice as much concrete in them as they really need to support a designated load in order to keep them from vibrating in ways that pose no structural danger but would make the occupants nervous.

25. For an elementary discussion of engineering modeling that unintentionally brings out its inherent character as miniature construction, see *The Man-Made World: Engineering Concepts Curriculum Project* (New York: McGraw-Hill, 1971), chap. 4, pp. 139–178. "Models are used, not only to describe a set of ideas, but also to evaluate and to predict the behavior of systems before they

are built. This procedure can save enormous amounts of time and money. It can avoid expensive failures and permit the best design to be found without the need for construction of many versions of the real thing. Models evolve, and it is customary to go through a process of making successive refinements to find a more suitable model" (p. 177).

26. Joseph Edward Shigley and Charles R. Mischke, *Mechanical Engineering Design*, 5th ed. (New York: McGraw-Hill, 1989), p. 5; italics in the original.

27. The standard history of engineering drawing in English is Peter J. Booker, *A History of Engineering Drawing* (London: Chatto and Windus, 1963). This history deserves to be related to more general and extended studies of the relations between instrumentation and vision in the modern period such as Jonathan Crary's *Techniques of the Observer: On Vision and Modernity in the Nineteenth Century* (Cambridge, Mass.: MIT Press, 1990), Don Gifford's *The Father Shore: A Natural History of Perception* (New York: Atlantic Monthly Press, 1990), and Martin Kemp's *The Science of Art: Optical Themes in Western Art from Brunelleschi to Seurat* (New Haven: Yale University Press, 1990). One philosophical essay in this direction is Edward Wachtel, "The Invention of Glass: The Discovery of the Western World View," in Pitt and Lugo, eds., *The Technology of Discovery and the Discovery of Technology* (1991), pp. 193–207. Also of relevance is Harold Belofsky, "Engineering Drawing—a Universal Language in Two Dialects," *Technology and Culture* 32, no. 1 (January 1991): 23–46, with the follow-up "Comment and Response on 'Engineering Drawing—a Universal Language in Two Dialects'," by Sadahiko Mori and Harold Belofsky, *Technology and Culture* 33, no. 4 (October 1992): 853–857.

28. Jean-François Lyotard, *The Postmodern Condition: A Report on Knowledge*, trans. Geoff Bennington and Brian Massumi (Minneapolis: University of Minnesota Press, 1984), p. 44.

29. See, e.g., Allen Buchanan, *Ethics, Efficiency, and the Market* (Totowa, N.J.: Rowman and Allanheld, 1985), and Sumner H. Slichter, "Efficiency," *Encyclopedia of the Social Sciences* (New York: Macmillan, 1937), vol. 3, esp. p. 439: "There is no such thing as efficiency in general or efficiency as such—there are simply a multitude of particular kinds of efficiency. Actions and procedures which are efficient when measured with one measuring stick may be inefficient when measured with a different measuring stick."

30. It is interesting, however, that the original *Oxford English Dictionary* (1933) did not include this meaning, which had to await the 1972 supplement for proper recognition. On John Smeaton's contribution to the development of this concept (if not the actual term), see Arnold Pacey, *The Maze of Ingenuity: Ideas and Idealism in the Development of Technology* (Cambridge: MIT Press, 1974), pp. 206 ff.

31. "Among the problems here are those concerning the lever. Indeed, it is incredible that a larger weight can be moved by a weak power, even when more weight is applied; for the same weight that a human cannot move with-

out a lever, one quickly moves by applying the weight of the lever" (Pseudo-Aristotle "Mechanical Problems" 847b1–847b15).

32. See, e.g., Salomon de Caus, *Les Raisons des forces mouvantes* (1615).

33. See, e.g., William E. Akin, *Technocracy and the American Dream: The Technocrat Movement, 1900–1941* (Berkeley and Los Angeles: University of California Press, 1977), and Howard P. Segal, "The Technological Utopians," in *Imagining Tomorrow: History, Technology, and the American Future*, ed. Joseph J. Corn (Cambridge: MIT Press, 1986), pp. 119–136.

34. See Henryk Skolimowski, "The Structure of Thinking in Technology" (1966). Skolimowski was a student of the Polish praxiologist Tadeusz Kotarbinski; see Kotarbinski's *Praxiology: An Introduction to the Sciences of Efficient Action* (Oxford: Pergamon Press, 1965).

35. Cf. Jacques Ellul, *The Technological Society* (1964), p. xxv.

36. For a good update on the spectrum of research regarding such input-output forms of rationality, see Jon Elster, ed., *Rational Choice* (New York: New York University Press, 1986).

37. Middendorf, *Engineering Design,* p. 184.

38. Vidosic, *Elements of Design Engineering,* pp. 222–223.

39. Cross, *Engineering Design Methods,* p. 17.

40. I. C. Jarvie, "The Social Character of Technological Problems: Comments on Skolimowski's Paper" (1966), objects that what the engineer strives for is really determined by the social definition of the problem. For instance, in warfare there are times when civil engineers are called upon to design a bridge for speed of construction rather than durability. But Skolimowski's point is that within such historically and socially set parameters as materials, cost, and time, a civil engineer qua civil engineer will always strive for as much durability as feasible. In fact Jarvie's own example tells against him, because it is the military rather than the civil engineer who would be called upon to design a pontoon bridge for maximum military efficiency (that is mobility and resistance to damage by firepower).

41. Although beyond consideration of the present remarks, this representation process may be related to the creation of writing and what Bruno Latour calls "inscription devices." See Bruno Latour, "Visualization and Cognition: Thinking with Eyes and Hands," *Knowledge and Society: Studies in the Sociology of Culture Past and Present* 6 (1986): 1–40. See also Ivan Illich and Barry Sanders, *ABC: The Alphabetization of the Popular Mind* (1988).

42. The following truncated analysis is implicitly in dialogue with two general volumes on design methodology included in the references: J. Christopher Jones and D. G. Thornley, eds., *Conference on Design Methods* (1963), and M. J. de Vries, Nigel Cross, and D. P. Grant, eds., *Design Methodology and Relationships with Science* (1993). Referenced articles by Luis L. Bucciarelli and by Lambert J. Van Poolen are equally important. Complementing Bucciarelli's ethnographic study (1988) is Crispin Hales's dissertation, *Analysis of the Engineering*

Design Process in an Industrial Context (Eastleigh, Hampshire, England: Grants Hill Publications, 1987). Beyond these and other works already cited one could also consult Gerald Nadler, "An Investigation of Design Methodology," *Management Science* 13, no. 10 (June 1967): B-642–B-655; Bohdan Walentyowicz, "On Methodology of Engineering Design," in *Proceedings of the XIV International Congress of Philosophy*, Vienna, September 2–9, 1968 (Vienna: Herder, 1968), 2: 586–590; Vladimir Hubka, ed., *Review of Design Methodology*, Proceedings of the International Conference on Engineering Design in Rome 1981 (Zurich: Heurista, 1981); Barrie Evans, James A. Powell, and Reg Talbot, eds., *Changing Design* (New York: John Wiley, 1982); Vladimir Hubka, "Attempts and Possibilities for Rationalisation of Engineering Design," in *Design and Synthesis*, ed. H. Yoshikawa, Proceedings of the International Symposium on Design and Synthesis, Tokyo, July 11–13, 1984 (Amsterdam: North-Holland, 1985), pp. 133–138; Patrick Whitney, ed., *Design in the Information Environment: How Computing Is Changing the Problems, Processes and Theories of Design* (Carbondale: Southern Illinois University Press, 1985); W. Ernst Eder, ed., *Proceedings of the 1987 International Conference on Engineering Design* (New York: American Society of Mechanical Engineers, 1987); Vladimir Hubka and W. Ernst Eder, *Theory of Technical Systems: A Total Concept Theory for Engineering Design* (New York: Springer-Verlag, 1988); and Ladislav Tondl, "Changes in Cognitive and Value Orientations in System Design," *Philosophy and Technology* 7 (1990): 87–98. Friedrich Rapp, ed., *Contributions to a Philosophy of Technology* (1974), also includes two articles on design methodology by Morris Asimov and R. J. McCrory.

43. Art and architecture books on the subject of design invariably concentrate on issues of form. A book on roof design, for example, provides an inventory of various ways to build roofs—not ways as actions, but ways as forms, patterns, shapes; one on lighting design contains pictures and drawings of various alternative formal solutions to lighting design problems. A work whose subtitle aptly illustrates this approach is Kurt Hoffmann, Helga Friese, and Walter Meyer-Bohe, *Designing Architectural Facades: An Ideas File for Architects* (New York: Whitney Library of Design, 1975). For some comprehensive architectural discussions of design that approach the philosophical, see Paul J. Grillo, *What Is Design?* (Chicago: P. Theobald, 1960); Christopher Alexander, *Notes on the Synthesis of Form* (Cambridge: Harvard University Press, 1964); Pye, *The Nature of Design*; Bryan Lawson, *How Designers Think: The Design Process Demystified* (London: Architectural Press, 1980); and Peter G. Rowe, *Design Thinking* (Cambridge: MIT Press, 1987). For further generalization, see Victor Margolin, ed., *Design Discourse: History, Theory, Criticism* (Chicago: University of Chicago Press, 1989).

44. Industrial design, especially the Bauhaus school of industrial design, is an attempt to bridge this gap between art and engineering and either to include aesthetic formal properties in the design process or to find aesthetic value in purely functional designs. In fact, however, the attempt has led to the triumph of engineering efficiency as influenced by economic pressures. For a

brief overview, see the article "Industrial Design," in *New Encyclopaedia Britannica, Macropaedia* (1975), 9: 512–520. For more critical analyses in and of the industrial design tradition, see Victor Papanek, *Design for the Real World* (New York: Van Nostrand, 1971), 2d ed., rev., *Design for the Real World: Human Ecology and Social Change* (New York: Van Nostrand, 1984); Victor Papanek, *Design for Human Scale* (New York: Van Nostrand, 1983); and Jonathan M. Woodham, *The Industrial Designer and the Public* (London: Pembridge Press, 1983).

45. Although the artificial character of its artifacts is most apparent in the plastic arts (sculpture, painting, etc.), to some extent this remains true even in poetry and music.

46. Edward S. Casey, *Imagining: A Phenomenological Study* (Bloomington: Indiana University Press, 1976), provides some foundations for a more extended comparison of engineering and artistic imagination.

47. Two exceptions that supplement the succeeding analysis are Yves R. Simon, "Pursuit of Happiness and Lust for Power in Technological Society," in Mitcham and Mackey, eds., *Philosophy and Technology* (1972, 1983), especially pp. 173–175, and Hans Jonas, "The Practical Uses of Theory," ibid., especially pp. 339–341.

48. This discussion draws on a parallel analysis in James Russell Woodruff, "The Question of the Neutrality of Technology" (Ph.D. diss., University of Rochester, 1993).

49. See, e.g., James K. Feibleman, "Technology as Skills" (1966). See also Michael Polanyi, *Personal Knowledge* (1958), chap. 4, "Skills."

50. See Max Weber's comments on the universality of technique in human activities in *Economy and Society*, trans. Guenther Roth and Claus Wittich (New York: Bedminster Press, 1968 [1922]), 1: 65.

51. This tension between heuristics (i.e., problem-solving strategies that propose solutions without testing all possible outcomes) and algorithms (problem-solving methods that analyze all possible outcomes and thus guarantee the best solutions) is intimately involved with the ultimate nature of matter and energy, their knowability, and the dream of a complete technology. Laplace's scientific postulate that if given a complete description of matter and motion at some point in time, he could deduce the remainder of the world may be restated technologically as: Given a complete description of matter and energy at some point in time, humans can intervene to produce whatever they desire. Heisenberg's uncertainty principle and various other aspects of quantum mechanics, as well as chaos theory, raise fundamental questions about the foundations of Laplace's hypothesis and its technological correlate.

52. See, e.g., Paul A. Samuelson and William D. Nordhaus, *Economics*, 14th ed. (New York: McGraw-Hill, 1992), p. 553.

53. Material technology = the efficient production of goods (with efficiency judged in terms of matter and energy); social technology = the efficient organization of society (with efficiency judged either in terms of technological productivity or psychological stress). Social technology is closely related to B. F. Skin-

ner's conception of a "technology of behavior," not to mention his technology of teaching. See B. F. Skinner, *Beyond Freedom and Dignity* (New York: Knopf, 1971).

54. Joseph A. Schumpeter, *Theorie der wirtschaftlichen Entwicklung* (Leipzig: Dunker and Humbolt, 1912); English version, *Theory of Economic Development*, trans. Redvers Opie (Cambridge: Harvard University Press, 1934). For a marginally independent development of Schumpeter's analysis see Randall Collins, *Weberian Sociological Theory* (New York: Cambridge University Press, 1986), chap. 4, "A Theory of Technology," pp. 77 ff.

55. H. G. Barnett, *Innovation* (New York: McGraw-Hill, 1953), p. 7. But cf. W. F. Ogburn, *On Culture and Social Change: Selected Papers*, ed. Otis Dudley Duncan (Chicago: University of Chicago Press, 1964), p. 23: "Invention is defined as a combination of existing and known elements of culture, material and/or nonmaterial, or a modification of one to form a new one."

56. The very terms "invention" and "innovation" are conspicuous by their absence in most scientific and technical dictionaries and encyclopedias. One exception, Christopher Morris, ed., *Academic Press Dictionary of Science and Technology* (San Diego: Academic Press, 1992), defines "innovation" under "innovative behavior" as "any action that occurs spontaneously in a new situation, rather than as the result of trial and error learning" (p. 1113) and defines "to invent" as "to create . . . a previously unknown device or physical process" (p. 1136).

57. Stephen Toulmin, "Innovation and the Problem of Utilization," in *Factors in the Transfer of Technology*, ed. William H. Gruber and Donald G. Marquis (Cambridge: MIT Press, 1969), p. 25. See also Toulmin's *Human Understanding*, vol. 1 (Princeton: Princeton University Press, 1972), pp. 364–378.

Bernard Lonergan's discussion of what he calls common sense and its merging with science to produce technology is also relevant to this discussion, although Toulmin's focus is more limited than Lonergan's. See Lonergan, *Insight* (New York: Philosophical Library, 1956), pp. 207 ff. Lonergan is developing a point first made by Alexander Koyré in a review of Mumford's *Technics and Civilization* titled "Du monde de l'à peu près à l'universe de la précision," *Critique* 4, issue 28 (September 1948): 806–823.

For a broad but rich discussion of related issues, see George Basalla, *The Evolution of Technology* (New York: Cambridge University Press, 1988).

58. Toulmin, "Innovation and the Problem of Utilization," p. 27.

59. A similar biological model of the utilization-innovation process—but one that emphasizes more the interrelations of a feedback-based ecological system—was developed by the Georgia Tech Innovation Project. See the executive summary, *Technological Innovation: A Critical Review of Current Knowledge* (1975), available from the Advanced Technology and Sciences Studies Group, Georgia Tech.

60. Harold Koontz, "The Management Theory Jungle Revisited," *Academy of Management Review* 5, no. 2 (1980): 175–187. The original argument that this one revisits was Koontz's "The Management Theory Jungle," *Academy of Management Journal* 4, no. 3 (1961): 174–188. The distinctions argued in these ar-

ticles have played a role not only in *Management*, the widely used textbook by Koontz (and Heinz Weihrich), but in numerous other textbooks as well.

61. See e.g., Alan L. Porter, Frederick A Rossini, Stanley R. Carpenter, and A. T. Roper, with Ronal W. Larson and Jeffrey S. Tiller, *A Guidebook for Technology Assessment and Impact Analysis* (New York: North Holland, 1980).

62. For the latter, a classic analysis is Max Weber's *The Protestant Ethic and the Spirit of Capitalism* (1905). But the most comprehensive general study is Herbert Applebaum's *The Concept of Work: Ancient, Medieval, and Modern* (Albany: State University of New York Press, 1992). The most accessible short history is Melvin Kranzberg and Joseph Gies, *By the Sweat of Thy Brow: Work in the Western World* (New York: Putnam, 1975).

63. Karl Polanyi, "Aristotle Discovers the Economy," in *Trade and Market in Early Empires: Economies in History and Theory*, ed. Karl Polanyi, Conrad M. Arensberg, and Harry W. Pearson (Glencoe, Ill.: Free Press, 1957), pp. 64–94.

64. For a complementary discussion of the distinction between substantive and functional divisions of labor, see Ursula Franklin's distinction between holistic and prescriptive technologies in *The Real World of Technology* (Toronto: Anansi, 1990), pp. 18 ff.

65. The most sustained articulation of the work/labor distinction is provided by Hannah Arendt, *The Human Condition* (1958). But see also, e.g., Ivan Illich, *Tools for Conviviality* (1973), p. 32, where it is assumed.

66. Paul Edwards, ed., *The Encyclopedia of Philosophy*, 8 vols. (New York: Macmillan, 1967), contains no entry on "work." Neither does the *Encyclopedia of the Social Sciences* (1935) or the revised *International Encyclopedia of the Social Sciences* (1968), which under "work" refers readers to "labor force," "labor relations," "occupations and careers," "professions," and "workers." The only English texts using "philosphy" and either "work" or "labor" in the titles are C. Delisle Burns's *The Philosophy of Labour* (London: Allen and Unwin, 1925), Etienne Borne and François Henry's *A Philosophy of Work*, trans. Francis Jackson (New York: Sheed and Ward, 1938), Frank Tannenbaum's *A Philosophy of Labor* (New York: Knopf, 1951), Remigius C. Kwant's *Philosophy of Labor* (Pittsburgh: Duquesne University Press, 1960), and Edmund F. Byrne's *Work, Inc.: A Philosophical Inquiry* (Philadelphia: Temple University Press, 1990). The first and third are labor-movement statements; the second and third are based in Catholic moral theology. For the latter, see also Marie Dominique Chenu, *The Theology of Work*, trans. Lilian Soiron (Chicago: Regnery, 1966). Byrne's book, as its title suggests, is more concerned with workers in relation to community and corporation than with work itself, and constitutes a non-Marxist, non-Catholic social philosophy of work.

67. Karl Marx, "Estranged Labor," in *Economic and Philosophic Manuscripts of 1844*, ed. Dirk J. Struik, trans. Martin Milligan (New York: International Publishers, 1964), p. 108.

68. Robert Blauner's *Alienation and Freedom: The Factor Worker and His Industry* (Chicago: University of Chicago Press, 1964), for instance, offers a systematic

and empirical study that attempts to relate many of these aspects of the problem. For Blauner there are four dimensions of alienation: powerlessness, meaninglessness, isolation, and self-estrangement—the most visible of which is powerlessness, the opposite of freedom and control. Alienation and freedom are then conceived as two poles of the experience of technology as a production process. By means of empirical studies of a printshop, textile mill, automobile assembly line, and automated chemical plant, Blauner argues that alienation and freedom are differentially affected by different kinds of technologies. Richard Schacht's *Alienation* (Garden City, N.Y.: Doubleday, 1970) offers a more theoretical overview from Hegel through existential psychology.

69. Harry Braverman, *Labor and Monopoly Capital: The Degradation of Work in the Twentieth Century* (New York: Monthly Review Press, 1974), pp. 444–445. This is also essentially the view argued in David Noble's *America by Design: Science, Technology, and the Rise of Corporate Capitalism* (New York: Knopf, 1977), and *Forces of Production: A Social History of Industrial Automation* (New York: Oxford University Press, 1986).

70. Michael Goldhaber, *Reinventing Technology: Policies for Democratic Values* (New York: Routledge and Kegan Paul, 1986), especially pp. 30–31.

71. On this topic, see E. P. Thompson's *The Making of the English Working Class* (New York: Pantheon, 1963).

Chapter Ten. Types of Technology as Volition

1. Joseph Schumpeter, *Theorie der wirtschaftlichen Entwicklung* (Leipzig: Dunker and Humbolt, 1912); English version, *Theory of Economic Development*, trans. Redvers Opie (Cambridge: Harvard University Press, 1934). Max Weber, *Die protestantische Ethik und der Geist des Kapitalismus* (1920–1921); English version, *The Protestant Ethic and the Spirit of Capitalism*, trans. Talcott Parsons (Cambridge: Harvard University Press, 1930). Friedrich von Gottl-Ottlilienfeld, *Wirtschaft und Technik* [Economics and technology], *Die naturlichen und technischen Beziehungen der Wirtschaft*, pt. 2 (Tubingen: J. C. B. Mohr, 1923), like Schumpeter, defines technology as means for economic ends. What Daniel Bell considers *The Cultural Contradictions of Capitalism*, paperback ed. (New York: Basic Books, 1978) is based in part on the dependence of an expanding democratic economy on technological change. What in Nathan Rosenberg, Ralph Landau, and David C. Mowery, eds., *Technology and the Wealth of Nations* (Stanford: Stanford University Press, 1992) is termed "innovative performance" is another version of the entrepreneurial spirit.

2. See, e.g., in Mitcham ed., *Philosophy of Technology in Spanish Speaking Countries* (1993), José Gaos's mention of a technological "yearning" (p. 112) and Miguel Angel Quintanilla's remark on "desire" (p. 189).

3. Kitaro Nishida, *An Inquiry into the Good*, trans. Masao Abe and Christopher Ives (New Haven: Yale University Press, 1987), p. 23.

4. Alexander Pfänder, *Phenomenology of Willing and Motivation*, trans. Herbert Spiegelberg (Evanston, Ill.: Northwestern University Press, 1967). Pfänder's

highly condensed description of the act of willing is adopted here because it leaves open the question of philosophical interpretation of the foundations of this phenomenon. Not to mention other resources, a full account of volition would include consideration of William James's analysis titled "Will," chap. 26 of *The Principles of Psychology* (1891); Austin Farrer's *The Freedom of the Will* (London: Adam and Charles Black, 1958); Roberto Assagioli's *The Act of the Will* (Baltimore: Penguin, 1973); and Brian O'Shaughnessy, *The Will: A Dual Aspect Theory*, 2 vols. (New York: Cambridge University Press, 1980). For a useful outline of various philosophical theories of the will see Vernon J. Bourke, *Will in Western Thought: An Historico-critical Survey* (New York: Sheed and Ward, 1964); James N. Lapsley, ed., *The Concept of Willing* (Nashville, Tenn.: Abington, 1967), is also helpful.

5. Pfänder, *Phenomenology of Willing*, pp. 22–23.

6. Pfänder, *Phenomenology of Willing*, p. 26.

7. Hannah Arendt, *The Life of the Mind*, vol. 2, *Willing* (New York: Harcourt Brace Jovanovich, 1978), pp. 13–14.

8. Arendt, *Willing*, p. 15.

9. Arendt, *Willing*, p. 18.

10. Paul Ricoeur's comprehensive *La Philosophie de la volunté* is composed of vol. 1, *Le Volontaire et l'involontaire* (1980), English version *Freedom and Nature: The Voluntary and the Involuntary*, trans. Erazim Kohák (Evanston, Ill.: Northwestern University Press, 1966); vol. 2, *Finitude et culpabilité*, pt. 1, *L'Homme fallible* (1960), English version, *Fallible Man*, trans. Charles Kelbley (Chicago: Henry Regnery, 1965), and pt. 2, *La Symbolique du mal* (1960), English version *The Symbolism of Evil*, trans. Emerson Buchanan (New York: Harper and Row, 1967). Vol. 2, pt. 2, and vol. 3 have never appeared.

11. All quotations from *Being and Time* are from the Macquarrie translation cited in the references.

12. George F. Will, *The Morning After: American Successes and Excesses, 1981–1986* (New York: Free Press, 1986), p. 268.

13. Quotations adapted from Augustine, *On Free Choice of the Will*, trans. Anna S. Benjamin and L. H. Hackstaff (Indianapolis: Bobbs-Merrill, 1964).

Conclusion. Continuing to Think about Technology

1. Alfred North Whitehead, *Science and the Modern World* (New York: Free Press, [1925]), p. 3.

2. Recent literature on the history and character of STS teaching and research includes the basic textbooks by Rudi Volti, *Society and Technological Change* (New York: St. Martin's Press, 1988; 2d ed. 1992), Robert E. McGinn, *Science, Technology, and Society* (1991), and Ron Westrum, *Technologies and Society: The Shaping of People and Things* (Belmont, Calif.: Wadsworth, 1991). Of a more reflective analytic bent are Andrew Webster, *Science, Technology, and Society: New Directions* (New Brunswick, N.J.: Rutgers University Press, 1991), and Steve Fuller, *Philosophy, Rhetoric, and the End of Knowledge: The Coming of Science and*

Technology Studies (Madison: University of Wisconsin Press, 1993). Necessary background is Ina Spiegel-Rösing and Derek de Solla Price, *Science, Technology, and Society: A Cross-Disciplinary Perspective* (Beverly Hills, Calif.: Sage, 1977).

Epilogue. Three Ways of Being-with Technology

1. Quoted in *Time*, September 12, 1960, p. 74.

2. One locus classicus of such celebration is Sophocles *Antigone*, lines 332 ff.

3. For an interpretation of the specifically religious dimensions of this negative mythology, see Carl Mitcham, "The Love of Technology Is the Root of All Evil," *Epiphany* 8, no. 1 (1985): 17–28. (Correct title: "On the Saying: 'The Love of Technology Is the Root of All Evil.'")

4. For Plato, see especially the *Laws* 5.743d, where agriculture is described as keeping production within proper limits and as helping to focus attention on the care of the soul and the body. Cf. also *Laws* 8.842d–e and 10.889d.

5. For Aristotle, see especially the *Politics* 1.8–9 and the distinction between two ways of acquiring goods, agriculture and business, the former said to be "by nature" (1258a38), the latter "not by nature" (1258b41). In the *Politics* 6.2, agrarian-based democracy is described as both "oldest" and "best" (1318b7–8).

6. Following Aristotle, Thomas Aquinas's commentary on the *Politics* terms farming "natural," "necessary," and "praiseworthy" (*Sententia libri politicorum* 1, lect. 8), and again in *De regimine principum* 2.3, Thomas identifies farming as "better" than commercial activities for providing for material welfare. For Thomas, however, farming tends to be spoken of in relation to all manual labor, and in consequence of the doctrine of the Fall it takes on a certain ambiguity not found in Aristotle. For instance, in the *Summa theologiae* 2.2, qu. 187, art. 3, "Whether Religious Are Bound to Manual Labor," it is argued that all human beings must work with their hands for four reasons: to obtain food (as proof texts Thomas cites Gen. 3:19 and Ps. 128:2), to avoid idleness (Sir. 33:27), to restrain concupiscence by mortifying the body (2 Cor. 6:4–6), and to enable one to give alms (Eph. 4:28). Note that there is a subtle difference between the first two reasons (which cite the Hebrew Scriptures) and the second two (which cite the Greek Scriptures). For a relevant interpretation of Thomas's thought that nevertheless fails to recognize the tensions alluded to here, see George H. Speltz, *The Importance of Rural Life according to the Philosophy of St. Thomas Aquinas* (Washington, D.C.: Catholic University of America Press, 1945). Cf. also Philo "De agricultura," a commentary on Noah as farmer.

7. "Those who labour in the earth are the chosen people of God, if ever he had a chosen people, whose breasts he has made his peculiar deposit for substantial and genuine virtue. . . . Corruption of morals in the mass of cultivators is a phaenomenon of which no age nor nation has furnished an example" (Thomas Jefferson, *Notes on the State of Virginia* [1782], qu. 19, "Manufactures"). See also a letter to John Jay, August 23, 1785: "Cultivators of the earth are the most valuable citizens. They are the most vigorous, the most independent, the most virtuous, and they are tied to their country, and wedded to its liberty

and interest, by the most lasting bonds. As long, therefore, as they can find employment in this line, I would not convert them into mariners, artisans or anything else."

8. See also Cicero *Academica* 1.4.15.

9. On the inadequacy of human knowledge, see the Book of Job, Prov. 1:7, Isa. 44:25, and Col. 2:8. Power over the world, Satan says in the Gospel of Luke, has been given to him (Luke 4:6). The prince of this world, according to the Gospel of John, is to be cast out (John 12:31).

10. In his study of Milton (in *The Lives of the Poets*, 1: 99–100, pars. 39–41), Samuel Johnson criticizes a program of education that would concentrate on natural philosophy: "The truth is, that the knowledge of external nature, and the sciences which that knowledge requires or includes, are not the great or the frequent business of the human mind. Whether we provide for action of conversation, . . . the first requisite is the religious and moral knowledge of right and wrong. . . . Physiological learning is of such rare emergence, that one may know another half his life, without being able to estimate his skill in hydrostatics or astronomy; but his moral and prudential character immediately appears. [And] if I have Milton against me, I have Socrates on my side. It was his labour to turn philosophy from the study of nature to speculations upon life; but the innovators whom I oppose . . . seem to think, that we are placed here to watch the growth of plants, or the motions of stars. Socrates was rather of the opinion that what he had to learn was, how to do good, and avoid evil." Cf. also *The Rambler*, no. 24 (Saturday, June 9, 1750).

11. Norbert Wiener, "A Scientist Rebels" (1947): 46.

12. John Wesley, *Works* (Grand Rapids, Mich.: Zondervan, n.d. [photomechanical reprint of the edition published by the Wesleyan Conference, London, 1872]), 7: 289.

13. Martin Heidegger, "The Thing," in *Poetry, Language, Thought* (1971), p. 166. See also Heidegger's essay on Rilke, "What Are Poets For?" in the same volume, esp. pp. 112–117.

14. According to the Talmud, "As God fills the entire universe, so does the soul fill the whole body" (Berakhot 10a). According to the teachings of Jesus, "Love your enemies and pray for those who persecute you, so that you may be sons of your Father who is heaven; for he makes his sun rise on the evil and on the good, and sends rain on the just and on the unjust" (Matt. 5:44–45).

15. Jean Le Rond d'Alembert, *Preliminary Discourse to the Encyclopedia of Diderot*, trans. Richard N. Schwab and Walter E. Rex (Indianapolis: Bobbs-Merrill, 1963), p. 42.

16. Immanuel Kant, "Idea for a Universal History from a Cosmopolitan Point of View" (1784), 3d thesis. Quoted from Immanuel Kant, *On History*, trans. Lewis White Beck (Indianapolis: Bobbs-Merrill, 1963), p. 13.

17. Immanuel Kant, "What Is Enlightenment?" (1784), opening sentence. Quoted from Immanuel Kant, *On History*, trans. Lewis White Beck (Indianapolis: Bobbs-Merrill, 1963), p. 3.

18. See Aristotle *Politics* 1268b25–1269a25, and Thomas Aquinas *Summa theologiae* 1–2, qu. 97, art. 2.

19. Charles-François de Saint-Lambert, "Luxury," in *Encyclopedia,* opening paragraphs. Quoted, with minor revisions, from *Encyclopedia: Selections,* trans. Nelly S. Hoyt and Thomas Cassirer (Indianapolis: Bobbs-Merrill, 1965), p. 204.

20. Saint-Lambert, "Luxury." Quoted from *Encyclopedia: Selections,* p. 231.

21. David Hume, *Essays* (London: Oxford University Press, 1963), p. 262.

22. Hume, *Essays,* pp. 276 and 277.

23. Hume, *Essays,* pp. 277–278.

24. See Francis Bacon, *The Great Instauration,* "The Plan of the Work."

25. D'Alembert, *Preliminary Discourse,* p. 75.

26. D'Alembert, *Preliminary Discourse,* p. 122.

27. Denis Diderot, "Art," in *Encyclopedia.* Quoted from *Encyclopedia: Selections,* p. 5.

28. *Ibid.,* p. 4.

29. For discussion of this contrast, see Nicholas Lobkowicz, *Theory and Practice: History of a Concept from Aristotle to Marx* (Notre Dame, Ind.: University of Notre Dame Press, 1967).

30. On this interesting topic, see K. J. H. Berland, "Bringing Philosophy down from the Heavens: Socrates and the New Science," *Journal of the History of Ideas* 47, no. 2 (April–June 1986): 299–308, a commentary on Amyas Busche's *Socrates: A Dramatic Poem* (1758). One point Berland does not consider is the extent to which this view of Socrates, which is also found in Aristophanes' *The Clouds* as well as other sources, might be legitimate; see, e.g., Leo Strauss, *Socrates and Aristophanes* (New York: Basic, 1966).

31. Alexander Pope, *An Essay on Man,* 1.289.

32. Julien Offroy de La Mettrie, *L'Homme-machine* (1748), near the end. Quoted from Julien Offroy de La Mettrie, *Man a Machine* (La Salle, Ill.: Open Court, 1912), p. 148.

33. This is vividly demonstrated by the vicissitudes of development taking place throughout the world. Geographic advantage, scientific knowledge, imported hardware, political or economic decisions, piecemeal optimism, and envious desire cannot by themselves or even in concert effect industrialization. Despite the ideological rhetoric of Maoist China and Islamicist Iran, modern technology does not seem able to be adopted independent of certain key elements of Western culture. The westernization of Japan confirms the argument from the other side of the divide.

34. For one collection of texts that does begin to point in this direction, see Humphrey Jennings, *Pandaemonium: The Coming of the Machine as Seen by Contemporary Observers, 1660–1886,* ed. Mary-Lou Jennings and Charles Madge (New York: Free Press, 1985).

35. Cf. Friedrich Nietzsche, *The Gay Science* (1882), vol. 1, sec. 12. For a mundane philosophy of gothic pathos, see Jean-Paul Sartre, *Being and Nothingness* (1943); the last sentence of the last chapter declares that "Man is a useless passion."

36. See Francis Bacon, "The Masculine Birth of Time," trans. in Benjamin Farrington, *The Philosophy of Francis Bacon* (Chicago: University of Chicago Press, 1966). For a complementary interpretation of "Steamboats, Viaducts, and Railways," see Don Gifford, *The Farther Shore: A Natural History of Perceptions, 1798–1984* (New York: Vintage, 1991), pp. 69, 86, and 118.

37. The note is to *The Excursion*, 8.112, at the beginning of a passage describing the industrial transformation of the English landscape so that "where not a habitation stood before, / Abodes of men" are now "irregularly massed / Like trees in forests" (lines 122–124) and as a "triumph that proclaims / How much the mild Directress of the plough / Owes to alliance with these newborn arts!" (lines 130–132). "In treating of this subject," Wordsworth writes in his note, "it was impossible not to recollect, with gratitude, the pleasing picture . . . Dyer has given of the influences of manufacturing industry upon the face of this Island. He wrote at a time when machinery was first beginning to be introduced, and his benevolent heart prompted him to augur from it nothing but good." Wordsworth, as much as Sophocles (*Antigone*, lines 331 ff.), is capable of appreciating the benefits of technology. But, he adds, now "Truth has compelled me to dwell upon the baneful effects arising out of an ill-regulated and excessive application of powers so admirable in themselves."

38. Jean-Jacques Rousseau, *Discours sur les sciences et les arts*, in *Oeuvres complètes*, Pléiade edition, 3: 22.

39. Ibid., p. 29.

40. Ibid., pp. 17, 10–11, and 25.

41. Cf., in this same regard, Niccolò Machiavelli's use of *virtú* as power in *The Prince* (1512).

42. Rousseau, *Discours sur les sciences et les arts*, p. 5.

43. William Wordsworth, letter to Charles James Fox, January 14, 1801. In this commentary on his presentation of "domestic affections" in the poems "The Brothers" and "Michael," Wordsworth further remarks that "The evil [of the destruction of domestic affections] would be the less to be regretted, if these institutions [of industrialization] were regarded only as palliatives to a disease [in a manner not unlike that associated with ancient skepticism]; but the vanity and pride of their promoters are so subtly interwoven with them, that they are deemed great discoveries and blessings to humanity [as per Enlightenment optimism]."

44. See John Milton, *Paradise Lost*, 1.670 ff. Milton also associates Satan's legions with engines and engineering at 1.750 and 6.553.

45. Samuel Taylor Coleridge, *Biographia Literaria* (1817), ed. George Watson (New York: Dutton, 1956), chap. 13, p. 167.

46. William Blake, *Jerusalem*, pt. 4, "To the Christians," introduction.

47. Edmund Burke, *A Philosophical Enquiry into the Origin of Our Ideas of the Sublime and Beautiful* (1757), pt. 1, sec. 7, first sentence.

REFERENCES

· · · ·

As indicated by the citations policy statement in the prefatory notes, the following references focus on the major authors and works in the philosophy of technology with which this book is in dialogue. Other works, except for classic or marginal texts and standard reference volumes that require no note, are cited in full by endnotes the first time they occur in each chapter. In the endnote citations, special attention may be called to the standard works in the history of technology that are mentioned initially in chapter 5. Philosophy of technology depends on a broad knowledge provided by such historical studies, of which only the most philosophical are included in the general references.

For some authors supplementary works are included by means of a note so the reference list can serve on its own as a general introduction to theoretical literature in the philosophy of technology.

Agassi, Joseph. 1985. *Technology: Philosophical and Social Aspects.* Boston: D. Reidel. xix, 272 pp.

Anders, Günther. 1961. *Off Limits für das Gewissen.* Reibek bei Hamburg: Rowohlt. 150 pp. English version, apparently translated by Anders's wife, *Burning Conscience: The Case of the Hiroshima Pilot, Claude Eatherly, Told in His Letters to Günther Anders* (New York: Monthly Review Press, 1961), xxiii, 135 pp. From this exchange of letters an essay, "Commandments in the Atomic Age," is included in Mitcham and Mackey, eds., *Philosophy and Technology* (1972), pp. 130–135. This essay was first published as "Gebote des Atomzeitalters," *Frankfurter Allgemeine Zeitung,* July 14, 1957. *Off Limits für das Gewissen* is also included, along with *Der Mann auf der Brücke: Tagebuch aus Hiroshima und Nagasaki* [The man on the bridge: Diaries from Hiroshima and Nagasaki] (1959) and *Die Toten: Rede über die drei Weltkriegen* [The dead: Speech on the three world wars] (1965), in *Hiroshima is überall* [Hiroshima is everywhere] (Munich: Beck, 1982), xxxvi, 394 pp.

Note on Anders: For other works of interest by Anders, see *Die Atomare Drohung: Radikale Überlegungen* [The nuclear threat: Radical considerations] (Munich: Beck, 1981), xiii, 223 pp.; *Die Antiquiertheit des Menschen: Über die*

351

Seele im Zeitalter der zweitern industriellen Revolution [The antiquitization of man: On the soul in the age of the second industrial revolution] (Munich: Beck, 1956), 353 pp.; republished together with vol. 2, subtitled, *Über die Zerstörung des Lebens im Zeitalter der dritten industrielle Revolution* [On the destruction of life in the age of the third industrial revolution] (Munich: Beck, 1980), 465 pp.; *Endzeit und Zeitenende: Gedanken über die atomare Situation* [Final time and the end of time: Thoughts on the atomic situation] (Munich: Beck, 1972), xiv, 221 pp.

Arendt, Hannah. 1958. *The Human Condition.* Chicago: University of Chicago Press. vi, 333 pp.

"Are There Any Philosophically Interesting Questions in Technology?" 1977. In *PSA 1975,* Proceedings of the 1975 biennial meeting of the Philosophy of Science Association, ed. Patrick Suppe and Peter D. Asquith, 2: 137–201. East Lansing: Philosophy of Science Association, Michigan State University. Contents: Paul T. Durbin, "Are There Interesting Philosophical Issues in Technology as Distinct from Science? An Overview of Philosophy of Technology"; Mario Bunge, "The Philosophical Richness of Technology"; Edwin T. Layton, "Technology and Science, or '*Vive la Petite Difference'* "; Max Black, "Are There Any Philosophically Interesting Questions in Technology?" Ronald N. Giere, "A Dilemma for Philosophers of Science and Technology."

Ashby, W. Ross. 1956. *An Introduction to Cybernetics.* New York: John Wiley. ix, 295 pp.

Baudrillard, Jean. 1968. *Le Système des objets* [The system of objects]. Paris: Gallimard. 288 pp.

Beck, Heinrich. 1979. *Kulturphilosophie der Technik: Perspektiven zu Technik— Menschheit—Zukunft* [Philosophy of culture of technology: Perspectives on technology, humanity, the future]. Trier: Spee-Verlag. 292 pp. This is a 2d rev. ed. of *Philosophie der Technik* (Trier: Spee-Verlag, 1969), 226 pp.

Bergson, Henri. 1907. *L'Evolution créatrice.* Paris: F. Alcan. viii, 403 pp. English version, *Creative Evolution,* trans. Arthur Mitchell (New York: Henry Holt, 1911), 453 pp.

———. 1932. *Les Deux Sources de la morale et de la religion.* Paris: Alcan. 346 pp. English version, *Two Sources of Morality and Religion,* trans. R. Ashley Audra, Cloudesley Brereton, and W. Horsfall Carter (New York: Henry Holt, 1935), viii, 308 pp.

Bijker, Wiebe E., Thomas P. Hughes, and Trevor Pinch, eds. 1987. *The Social Construction of Technological Systems: New Directions in the Sociology and History of Technology.* Cambridge: MIT Press. x, 405 pp. Papers of a workshop held at the University of Twente, the Netherlands, July 1984.

Bijker, Wiebe E., and John Law, eds. 1992. *Shaping Technology/Building Society: Studies in Sociotechnical Change.* Cambridge: MIT Press. vii, 341 pp.

Billington, David P. 1974. "Structures and Machines: The Two Sides of Technology." *Soundings* 57, no. 3 (fall): 275–288.

———. 1986. "In Defense of Engineers." *Wilson Quarterly* 10, no. 1 (New Year's 1986): 86–97.

Note on Billington: Billington's reflection on engineering also include *Robert Maillart's Bridges: The Art of Engineering* (Princeton: Princeton University Press, 1979), xv, 146 pp.; *The Tower and the Bridge: The New Art of Structural Engineering* (New York: Basic Books, 1983), xx, 306 pp.; "The Acts in Technology," *Anglican Theological Review* 65, no. 1 (January 1983): 31–48.

Bon, Fred. 1898. *Über das Sollen und das Gute: Eine begriffsanalytische Untersuchung* [Concerning the right and the good: An essay in conceptual analysis]. Leipzig: W. Engelmann. iv, 188 pp. See also this author's *Grundzuege der wissenschaftlichen und technischen Ethik* (Leipzig: W. Engelmann, 1896).

Borgmann, Albert. 1984. *Technology and the Character of Contemporary Life: A Philosophical Inquiry.* Chicago: University of Chicago Press. ix, 302 pp. Note 1, p. 251, references a series of Borgmann's articles and reviews 1971–1984 that have been important independent of the incorporation of their arguments into this book.

———. 1987. "The Question of Heidegger and Technology: A Critical Review of the Literature." *Philosophy Today* 31, no. 2 (summer): 98–194. A special issue with analysis and bibliography, with the assistance of Carl Mitcham.

Note on Borgmann: See also *Crossing the Postmodern Divide* (Chicago: University of Chicago Press, 1992), vii, 173 pp.

Brinkmann, Donald. 1946. *Mensch und Technik: Grundzüge einer Philosophie der Technik* [Humanity and technology: Foundations of a philosophy of technology]. Bern: A. Franke. 167 pp.

———. 1953. "L'Homme et la technique" [Humanity and technology]. In *Proceedings of the XI International Congress of Philosophy,* Brussels, August 20–26, 1953, 8: 149–150. Amsterdam: North-Holland.

———. 1963. "Die Technik als philosophische Problem." In *Herders Zeitbericht: Enzyklopädische Beschreibung des zwanzigsten Jahrhunderts,* cols. 1235–1246. Freiburg: Herder. English version, "Technology as Philosophic Problem," trans. William Carroll, Carl Mitcham, and Robert Mackey, *Philosophy Today* 15, no. 2 (summer 1971): 122–128.

Brun, Jean. 1961. *Les Conquêtes de l'homme et la séparation ontologique* [The conquests of humanity and ontological separation]. Paris: Presses Universitaires de France. 298 pp.

———. 1969. *Le Retour de Dionysus* [The return of Dionysos]. Paris: Desclée et Cie. 239 pp.

———. 1981. *Les Masques du désir* [The mass of desire]. Paris: Buchet/Chastel. 263 pp.

———. 1992. *Le Rêve et la machine: Technique et existence* [The dream and the machine: Technology and existence]. Paris: Table Ronde. 367 pp.

Bucciarelli, Louis L. 1988. "An Ethnographic Perspective on Engineering Design." *Design Studies* 9, no. 3 (July): 159–168.

Bugliarello, George, and Dean B. Doner, eds. 1979. *The History and Philosophy of Technology*. Urbana: University of Illinois Press. xxxi, 384 pp. Proceedings of an international Symposium on the History and Philosophy of Technology, 1973, at the University of Illinois at Chicago Circle. Part 2, "The Philosophy of Technology," includes Carl Mitcham, "Philosophy and the History of Technology"; Jean-Claude Beaune, "Technology from an Encyclopedic Point of View"; Peter Caws, "*Praxis* and *Techne*"; David Wojick, "The Structure of Technological Revolutions"; Mario Bunge, "Philosophical Inputs and Outputs of Technology"; Werner Koenne, "On the Relationship between Philosophy and Technology in the German-Speaking Countries"; Frances Svensson, "The Technological Challenge to Political Theory"; David Edge, "Technological Metaphor and Social Control"; Henryk Skolimowski, "Philosophy of Technology as a Philosophy of Man."

Bungard, Walter, and Hans Lenk, eds. 1988. *Technikbewertung: Philosophische und psychologische perspektivan* [Technical evaluation: Philosophical and psychological perspectives]. Frankfurt: Suhrkamp. 383 pp.

Bunge, Mario. 1967. "Action." In *Scientific Research*, vol. 2, *The Search for Truth*, pp. 121–150. New York: Springer. This is a revised and expanded version of "Technology as Applied Science," *Technology and Culture* 7, no. 3 (summer 1966): 329–347. The revised version is reprinted, under its original title (which was co-opted as a general title for the *Technology and Culture* symposium where it initially appeared), "Toward a Philosophy of Technology," in Mitcham and Mackey, eds., *Philosophy and Technology* (1972), pp. 62–76.

———. 1975. "Toward a Technoethics." *Philosophic Exchange* 2, no. 1 (summer): 69–79. Reprint, *Monist* 60, no. 1 (January 1977): 96–107. A Spanish version of this essay is included as appendix 1 in the more extended analysis of *Etica y ciencia*, 3d ed. (Buenos Aires: Siglo Veinte, 1976).

———. 1979a. "The Five Buds of Technophilosophy." *Technology in Society* 1, no. 1 (spring): 67–74.

———. 1979b. "Philosophical Inputs and Outputs of Technology." In Bugliarello and Doners, eds., *History and Philosophy of Technology* (1979), pp. 262–281.

———. 1985. "Technology: From Engineering to Decision Theory." In *Treatise on Basic Philosophy*, vol. 7, *Epistemology and Methodology III: Philosophy of Science and Technology*, part 2, "Life Science, Social Science, and Technology," pp. 219–311. Boston: D. Reidel.

Note on Bunge: For commentary on Bunge's philosophy of technology see Paul Weingartner and Georg J. W. Dorn, eds., *Studies on Mario Bunge's "Treatise"* (Amsterdam: Rodopi, 1990), especially the four papers included in chap. 11, "Social Sciences and Technology," and Bunge's response in chap. 20. Bunge's "Instant Autobiography" (chap. 23) is also useful.

Burlingame, Roger. 1949. *Backgrounds of Power: The Human Story of Mass Production*. New York: Scribner's. xi, 372 pp.

Capurro, Rafael. 1986. *Hermeneutik der Fachinformation* [Hermeneutics of scientific information]. Freiburg: Karl Alber. 239 pp.

Note on Capurro: For works in English, see *Epistemology and Information Science* (Stockholm: Royal Institute of Technology Library, 1985), 37 pp.; *Moral Issues in Information Science* (Stockholm: Royal Institute of Technology Library, 1985), 31 pp.; "Informatics and Hermeneutics," in *Software Development and Reality Construction*, ed. Christiane Floyd et al., pp. 363–375 (New York: Springer, 1992); "What Is Information Science For? A Philosophical Reflection," in *Conceptions of Library and Information Science: Historical, Empirical, and Theoretical Perspectives*, ed. Pertti Vakkari and Blaise Cronin, pp. 82–96 (Los Angeles: Taylor Graham, 1992).

Carpenter, Stanley. 1974. "Modes of Knowing and Technological Action." *Philosophy Today* 18, no. 2 (summer): 162–168. This is a condensed version of chapter 3, "The Role of Knowledge in Technological Action," of Carpenter's doctoral dissertation, "The Structure of Technological Action" (Boston University, 1971).

———. 1978. "The Cognitive Dimension of Technological Change." *Research in Philosophy and Technology* 1: 213–228.

———. 1983. "Alternative Technology and the Norm of Efficiency." *Research in Philosophy and Technology* 6: 65–76.

Casey, Timothy, and Lester Embree, eds. 1990. *Lifeworld and Technology*. Lanham, Md.: University Press of America. xii, 313 pp. Contents: Langdon Winner, "Living in Electronic Space"; Robert N. Procter, "Nazi Biomedical Technologies"; Algis Mickunas, "Technology and Liberation"; Ivan Illich, "From Recorded Speech to the Record of Thought"; Albert Borgmann, "Text and Things: Holding on to Reality"; Larry Hickman, "Literacy, Mediacy and Technological Determinism"; Wolfgang Schirmacher, "Media as Lifeworld"; Lester Embree, "A Perspective on the Rationality of Scientific Technology on How to Buy a Car"; Joseph C. Pitt, "Technology and the Objectivity of Values"; Edward W. Constant II, "Patterns of Discovery or Social Construction of Technology: The Invention of the Turbojet"; Ronald Bruzina, "Architectura Architecturans: World(s) in the Making"; Carl Mitcham, "On Going to Church and Technology"; Frederick Ferré, "The Religious Dialectic and Technology"; Timothy Casey, "Designing Excellence: Some Functional and Aesthetic Considerations"; Gayle L. Ormiston, "Translating Technology: From Artifact to Habitat"; Thomas R. Flynn, "Sartre and Technological Being-in-the-World"; John J. McDermott, "The Hidden Life of Technological Artifacts."

Cassirer, Ernst. 1930. "Form und Technik" [Form and technology]. In *Kunst und Technik*, ed. Leo Kestenberg, pp. 15–61. Berlin: Wegweiser. Reprinted in Ernst Cassirer, *Symbol, Technik, Sprache: Aufsätze aus den Jahren 1927–1933*, ed.

Wolfgang Orth, John Michael Krois, and Josef M. Werle, pp. 39–91 (Hamburg: Felix Meiner, 1985). Citations are to the second publication. Cassirer also briefly considers Ernst Kapp's theories in *The Philosophy of Symbolic Forms*, trans. Ralph Manheim, vol. 2, *Mythical Thought* (New Haven: Yale University Press, 1955), pp. 215 ff.

Cérézuelle, Daniel. 1979. "Fear and Insight in French Philosophy of Technology." *Research in Philosophy and Technology* 2: 53–75.

Choe, Wolhee. 1989. *Toward an Aesthetic Criticism of Technology.* Worcester Polytechnic Institute Studies in Science, Technology and Culture, vol. 2. New York: Peter Lang. 208 pp.

Cohen, Joseph W. 1955. "Technology and Philosophy." *Colorado Quarterly* 3, no. 4 (spring): 409–420.

Cooper, Barry. 1991. *Action into Nature: An Essay on the Meaning of Technology.* Notre Dame, Ind.: University of Notre Dame Press. xvi, 291 pp.

Cutcliffe, Stephen H., Steven L. Goldman, Manuel Medina, and José Sanmartín, eds. 1992. *New Worlds, New Technologies, New Issues.* Research in Technology Studies, vol. 6. Bethlehem, Pa.: Lehigh University Press. 233 pp. Contents: Part 1, "New World": George Bugliarello, "Introduction: Philosophy of Technology, or the Quest for a 'Hominis as Hominem ad Machinam Proportio'"; Albert Borgmann, "The Postmodern Economy"; Manuel Medina, "The Culture of Risk: New Technologies and Old Worlds"; José Sanmartín, "The New World of New Technology"; Don Ihde, "New Technologies/Old Cultures." Part 2, "New Technologies and Political Responses to Them": Melvin Kranzberg, "Introduction: Technological and Cultural Change—Past, Present, and Future"; Paul T. Durbin, "Culture and Technical Responsibility"; Margarita M. Peña, "New Technologies and an Old Debate: Implications for Latin America"; Richard Worthington, "The Nature of Global Processes"; Steven L. Goldman, "No Innovation without Representation: Technological Action in a Democratic Society." Part 3, "New Issues": Elena Lugo, "Introduction: New Dimensions for Action"; Michael E. Zimmerman, "Deep Ecology's Mode of 'Technology Assessment'"; Carl Mitcham, "Science, Technology, and the Military"; Miguel A. Quintanilla, "Scientific and Technical Development in a Democratic Society: The Roles of Government and the Media"; Antonio Ten, "Science and the People: Science Museums and Their Context."

Dessauer, Friedrich. 1908. *Technische Kultur? Sechs Essays* [Technological culture? Six essays]. Kempten and Munich: Kosel. 57 pp.

———. 1927. *Philosophie der Technik: Das Problem der Realisierung* [Philosophy of technology: The problem of its realization]. Bonn: F. Cohen. 280 pp. An English version of three chapters from this book can be found under the title "Technology in Its Proper Sphere," trans. William Carroll, in Mitcham and Mackey, eds., *Philosophy and Technology* (1972), pp. 317–334.

———. 1956. *Streit um die Technik* [The controversy concerning technology]. Frankfurt: J. Knecht. 471 pp. Abridged edition, Freiburg: Herder, 1959. 205 pp.

Dessauer, Friedrich, and Xavier von Hornstein. 1945. *Seele im Bannkreis der Technik* [The soul under the influence of technology]. Olten: Otto Walter. 307 pp.

Dreyfus, Hubert L. 1972. *What Computers Can't Do: A Critique of Artificial Reason.* New York: Harper and Row. xxxv, 259 pp. This book grew out of a RAND Corporation research report P-3244, "Alchemy and Artificial Intelligence" (December 1965); 2d rev. ed., with a new subtitle, *The Limits of Artificial Intelligence* (1979), added a second introduction and new pagination, xiv, 354 pp.; 3d ed., *What Computers Still Can't Do: A Critique of Artificial Reason* (Cambridge: MIT Press, 1992), liii, 354 pp.

Dreyfus, Hubert L., and Stuart Dreyfus, with Tom Athanasiou. 1986. *Mind over Machine: The Power of Human Intuition and Expertise in the Era of the Computer.* New York: Free Press. xviii, 231 pp.

Drucker, Peter F. 1970. *Technology, Management and Society.* New York: Harper and Row. x, 209 pp.

Duque, Félix. 1986. *La filosofía de la técnica de la naturaleza* [The philosophy of the technology of nature]. Madrid: Tecnos. 311 pp.

Durbin, Paul T. 1972. "Technology and Values: A Philosopher's Perspective." *Technology and Culture* 13, no. 4 (October): 556–576.

———. 1978. "Toward a Social Philosophy of Technology." *Research in Philosophy and Technology* 1: 67–97.

———. 1992. *Social Responsibility in Science, Technology, and Medicine.* Bethlehem, Pa.: Lehigh University Press. 230 pp.

———, ed. 1991. *Critical Perspectives on Nonacademic Science and Engineering.* Research in Technology Studies, vol. 4. Bethlehem, Pa.: Lehigh University Press. 299 pp. Contents: Paul T. Durbin, "Introduction"; Billy V. Koen, "The Engineering Method"; Edwin T. Layton Jr., "A Historical Definition of Engineering"; Carl Mitcham, "Engineering as Productive Activity: Philosophical Remarks"; Steven L. Goldman, "The Social Captivity of Engineering"; Ronald Laymon, "Idealizations and the Reality of Dimensional Analysis"; Hans Lenk, "Real-World Contexts and Types of Responsibility"; Henryk Skolimowski, "The Eco-Philosophy Approach to Technological Research"; Sheila Jasanoff, "Judicial Construction of New Scientific Evidence"; Richard E. Sclove, "The Nuts and Bolts of Democracy: Toward a Democratic Politics of Technological Design"; Taft H. Broome Jr., "Bridging Gaps in Philosophy and Engineering"; Günter Ropohl, "Deficiencies in Engineering Education."

Note on Durbin: Durbin's extensive contribution to the philosophy of technology includes his work as editor of *Research in Philosophy and Technology* (1978–1985) and *Philosophy of Technology* (1983–present). Two other books of significance are Durbin, ed., *A Guide to the Culture of Science, Technology, and Medicine* (New York: Free Press, 1980), xl, 723 pp.; and *Dictionary of Concepts in the Philosophy of Science*, Reference Sources for the Social Sciences and Humanities 6 (New York: Greenwood Press, 1988), xvi, 362 pp.

Dussel, Enrique. 1984. *Filosofía de la producción* [Philosophy of production]. Bogotá: Nueva América. 242 pp. For an English version of the appendix of this book see "Technology and Basic Needs," trans. Ana Mitcham, James A. Lynch, and Carl Mitcham, in Mitcham, ed., *Philosophy of Technology in Spanish Speaking Countries* (1993), pp. 101–109. Dussel's *Philosophy of Liberation*, trans. Aquilina Martinez and Christine Morkovsky (Maryknoll, N.Y.: Orbis, 1985), chap. 4, "From Nature to Economics," chap. 5, "From Science to Philosophy of Liberation," and an appendix, "Philosophy and Praxis," include arguments that overlap with the analyses of sections 1 and 3 of *Filosofía de la producción*.

Eliade, Mircea. 1971. *The Forge and the Crucible: The Origins and Structures of Alchemy*. Trans. Stephen Corrin. New York: Harper and Row. 230 pp.

Ellul, Jacques. 1954. *La Technique, ou L'Enjeu de siècle* [Technology, or The bet of the century]. Paris: Colin. 401 pp. Reprint, with minor revisions (including a new "Postface" and appendix) done ca. 1960, Paris: Economica, 1990, vi, 423 pp. "American edition," with two additional introductory sections, an extra final chapter (based on the "Postface" of the 1960 revision), and other minor changes, *The Technological Society*, trans. John Wilkinson (New York: Knopf, 1964), xxxvi, 449, xiv pp.

———. 1962. "The Technological Order." Trans. John Wilkinson. *Technology and Culture* 3, no. 4 (fall): 394–421. Also included in Stover, ed., *The Technological Order* (1963), pp. 10–37. Reprinted in Mitcham and Mackey, eds., *Philosophy and Technology* (1972), pp. 86–105. Includes the appendix to the 1960 revision of *La Technique*. Also reprinted without the appendix in Hickman, ed., *Technology as a Human Affair* (1990), pp. 59–72.

———. 1975. *Sans Feu ni lieu: Signification biblique de la Grande Ville*. Paris: Gallimard. 304 pp. English version (which actually appeared first), *The Meaning of the City*, trans. Dennis Pardee (Grand Rapids, Mich.: Eerdmans, 1970). xix, 209 pp.

———. 1977. *Le Système technicien*. Paris: Calmann-Levy. 361 pp. English version, *The Technological System*, trans. Joachim Neugroschel (New York: Continuum, 1980), xi, 362 pp. Updates the first two chapters of *La Technique* (1954).

———. 1980. "The Ethics of Nonpower." Trans. Nada K. Levy. In Melvin Kranzberg, ed., *Ethics in an Age of Pervasive Technology* (Boulder, Colo.: Westview Press, 1980), pp. 204–212.

———. 1981a. *A Temps et à contretemps* (entretiens avec Madelein Garrigou-Lagrange.) Paris: Le Centurion. 210 pp. English version, *In Search, out of Season*, trans. Lani K. Niles (San Francisco: Harper and Row, 1982), xiv, 242 pp.

———. 1981b. *Perspectives on Our Age*. Ed. William H. Vanderburg. Trans. Joachim Neugroschel. New York: Seabury. 111 pp. *A Temp et à contretemps* and *Perspectives on Our Age* are the two best general introductions to Ellul.

———. 1983. "Recherche pour une Ethique dans une société technicienne." In Hottois, ed., "Ethique et Technique" (1983), pp. 7–20. English version, "The Search for Ethics in a Technicist Society," trans. Dominique Gillot and Carl Mitcham, *Research in Philosophy and Technology* 9 (1989): 23–36.

————. 1987. *Le Bluff technologique*. Paris: Hachette. 489 pp. English version, *The Technological Bluff*, trans. Geoffrey W. Bromiley (Grand Rapids, Mich.: Eerdmans, 1990), xvi, 418 pp. A second update of *La Technique* (1954).

Note on Ellul: For complete bibliography of primary and secondary sources on Ellul, see Joyce Hanks, with assistance of Rolf Asal, *Jacques Ellul: A Comprehensive Bibliography*, in *Research in Philosophy and Technology*, suppl. 1 (1984), xiii, 282 pp. Updates to this fundamental reference can be found in *Research in Philosophy and Technology*, vols. 11 (1991) and 15 (forthcoming). For studies of Ellul's thought, see esp. Clifford G. Christians and Jay M. Van Hook, eds., *Jacques Ellul: Interpretive Essays* (Urbana: University of Illinois Press, 1981); Darrell J. Fasching, *The Thought of Jacques Ellul* (Lewiston, N.Y.: Edwin Mellen Press, 1981); Etienne Dravasa, Claude Emeri, and Jean-Louis Seurin, eds., *Religion, société et politique: Mélanges en hommage à Jacques Ellul* (Paris: Presses Universitaires de France, 1983); David Lovekin, *Technique, Discourse, and Consciousness: An Introduction to the Philosophy of Jacques Ellul* (Bethlehem, Pa.: Lehigh University Press, 1991); and Patrick Troude-Chastenet, *Lire Ellul: Introduction a l'oeuvre socio-politique de Jacques Ellul* (Bordeaux: Presses Universitaires de Bordeaux, 1992).

Engelmeier, Peter K. 1894. "Grundriß der Philosophie der Technik" [Foundation of philosophy of technology]. *Kölnische Zeitung*, no. 606 (July 24, 1894): 1–2, and no. 608 (July 25, 1894): 1–2.

————. 1899. "Allgemeinen Fragen der Technik" [General questions of technology]. *Dinglers Polytechnisches Journal* (Berlin-Stuttgart) 311, no. 2 (January 14): 21–22. This long article is continued in the following: 311, no. 5 (February 4): 69–71; no. 7 (February 18): 101–103; no. 9 (March 4): 133–134; no. 10 (March 11): 149–151; 312, no. 1 (April 8): 1–3; no. 5 (May 6): 65–67; no. 7 (May 20): 97–99; no. 9 (June 3): 129–130; no. 10 (June 10): 145–147; 313, no. 2 (July 15): 17–19; and no. 5 (August 5): 65–67.

————. 1900. "Philosophie der Technik, eine neue Forschungsrichtung" [Philosophy of technology, a new research course]. *Prometheus* 11, no. 564: 689–692. Continued, no. 565: 707–710.

————. 1911. "Philosophie der Technik" [Philosophy of technology]. In *Atti del 4. Congrèsso internazionàle di filosofía* (Bologna), vol. 3 (Genoa, 1911), pp. 587–596. See also Engelmeier, "Essai d'une 'heurologie' ou théorie générale de la création humaine," pp. 582–595.

————. 1912. *Filosofia tekhniki* [Philosophy of technology]. Moscow. Vol. 1, *Obshchiy obeor predmeta* [General survey of problems], 69 pp.; Vol. 2, *Sovremennaia filasofiia* [Modern philosophers], 160 pp.; Vol. 3, *Nasha zhizn* [Our life], 94 pp.; Vol. 4. *Tekhnitsizm* [Technicism], 147 pp.

————. 1927. "Vorarbeit zur Philosophie der Technik" [Preliminary to the philosophy of technology]. *Technik und Kultur* (Verband Deutscher Diplomingenieure, Berlin) 18: 85.

————. 1929. "Nuzhna li nam filosofiia tekhniki?" [Is philosophy of technology necessary?] *Inzhenerny Trud*, no. 2: 36–40.

Espinas, Alfred. 1897. *Les Origines de la technologie* [The origins of technology]. Paris: Alcan. 290 pp.

"Ethical Aspects of Experimenting with Human Subject." 1969. Theme issue of *Daedalus* 98, no. 2 (spring): xiv, 219–594. Contents: Paul A. Freund, "Introduction"; Hans Jonas, "Philosophical Reflections on Experimenting with Human Subjects"; Herrman L. Blumgart, "The Medical Framework for Viewing the Problem of Human Experimentation"; Henry K. Beecher, "Scarce Resources and Medical Advancement"; Paul A. Freund, "Legal Frameworks for Human Experimentation"; Talcott Parsons, "Research with Human Subjects and the 'Professional Complex'"; Margaret Mead, "Research with Human Beings"; Guido Calabresi, "Reflections on Medical Experimentation in Humans"; Louis L. Jaffe, "Law as a System of Control"; David F. Caver, "The Legal Control of the Clinical Investigation of Drugs"; Louis Lasagna, "Special Subjects in Human Experimentation"; Geoffrey Edsall, "A Positive Approach to the Problem of Human Experimentation"; Jay Katz, "The Education of the Physician-Investigator"; Francis D. Moore, "Therapeutic Innovation"; David D. Rutstein, "The Ethical Design of Human Experiments"; William J. Curran, "Governmental Regulation of the Use of Human Subjects in Medical Research."

Feenberg, Andrew. 1991. *Critical Theory of Technology*. New York: Oxford University Press. xi, 235 pp.

Feibleman, James K. 1961. "Pure Science, Applied Science, and Technology: An Attempt at Definitions." *Technology and Culture* 2, no. 4 (fall): 305–317. Reprinted in Mitcham and Mackey, eds., *Philosophy and Technology* (1972), pp. 33–41.

————. 1966. "Technology as Skills." *Technology and Culture* 7, no. 3 (summer): 318–328.

Note on Feibleman: Other works of interest by Feibleman: "Artifactualism," *Philosophy and Phenomenological Research* 25, no. 4 (June 1965): 544–559; "The Importance of Technology," *Nature* 209 (January 8, 1966): 122–125; *Understanding Human Nature: A Popular Guide to the Effects of Technology on Man and His Behavior* (New York: Horizon, 1977), 143 pp.; "The Artificial Environment," in *The Built Environment, Environment and Man*, vol. 8, ed. John Lenihan and William W. Fletcher, pp. 145–168 (New York: Academic Press, 1978; *Technology and Reality* (The Hague: Martinus Nijhoff, 1982), xii, 210 pp. The two books are casual but nonetheless sometimes suggestive studies of interactions of technology with both nature and human nature.

Ferguson, Eugene S. 1992. *Engineering and the Mind's Eye*. Cambridge, Mass.: MIT Press. xiv, 241 pp. An expansion of Ferguson's influential article, "The Mind's Eye: Nonverbal Thought in Technology," *Science* 197, issue no. 4306 (August 26, 1977): 827–836.

Ferré, Frederick. 1988. *Philosophy of Technology.* Englewood Cliffs, N.J.: Prentice-Hall. x, 147 pp.

Note on Ferré: Ferré has also served as general editor of *Research in Philosophy and Technology* from 1988 to 1994.

Fleron, Frederic J., Jr., ed. 1977. *Technology and Communist Culture: The Sociocultural Impact of Technology under Socialism.* New York: Praeger. xii, 520 pp. Proceedings for a conference on Technology and Communist Culture, August, 1975. The most philosophically important papers are Fleron's "Introduction" and "Afterword"; Andrew L. Feenberg, "Transition or Convergence: Communism and the Paradox of Development"; William Leiss, "Technology and Instrumental Rationality in Capitalism and Socialism"; and Julian M. Cooper, "The Scientific and Technical Revolution in Soviet Theory."

Florman, Samuel. 1968. *Engineering and the Liberal Arts: A Technologist's Guide to History, Literature, Philosophy, Art, and Music.* New York: McGraw-Hill. x, 278 pp.

———. 1976. *The Existential Pleasures of Engineering.* New York: St. Martin's Press. xi, 160 pp.

———. 1981. *Blaming Technology: The Irrational Search for Scapegoats.* New York: St. Martin's Press. xi, 207 pp.

———. 1987. *The Civilized Engineer.* New York: St. Martin's Press. xii, 258 pp.

Freyer, Hans, Johannes C. Papalekas, and Georg Weippert, eds. 1965. *Technik im technischen Zeitalter: Stellungnahmen zur geschichtlichen Situation* [Technology in the technological age: Attitudes toward the historical situation]. Düsseldorf: Schilling. 414 pp. One contribution, Arnold Gehlen, "Anthropologische Ansicht der Technik," has been translated into English as "A Philosophical-Anthropological Perspective on Technology" (1983).

Fuller, Buckminster R. 1963. *No More Secondhand God.* Carbondale: Southern Illinois University Press. xiv, 163 pp.

———. 1969. *Operating Manual for Spaceship Earth.* Carbondale: Southern Illinois University Press. 143 pp.

Note on Fuller: Fuller is perhaps the most visionary proponent of an engineering philosophy of technology, and his voluminous, eccentric, jargon-laden writings span half a century. See especially *Nine Chains to the Moon* (New York: Lippincott, 1938), xvi, 405 pp.; *Untitled Epic Power on the History of Industrialization* (New York: Simon and Schuster, 1962), xii, 227 pp.; *Ideas and Integrities: A Spontaneous Autobiographical Disclosure,* ed. Robert W. Marks (Englewood Cliffs, N.J.: Prentice-Hall, 1963), 318 pp.; *Utopia or Oblivion: The Prospects for Humanity* (New York: Bantam, 1969), xi, 336 pp.; *Intuition* (Garden City, N.Y.: Doubleday, 1972; 2d rev. ed., 1973), 210 pp.; *Synergetics: Explanations in the Geometry of Thinking* (New York: Macmillan, 1972); *Synergetics Two: Explorations in the Geometry of Thinking* (New York: Macmillan, 1979).

xxiv, 592 pp.; *Critical Path*, with Kiyoshi Kuromiya (New York: St. Martin's Press, 1981). xxxviii, 471 pp.

García Bacca, Juan David. 1968. *Elogio de la técnica* [Praise of technology]. Caracas: Monte Avila Editores. 181 pp. Republished Barcelona: Anthropos, 1987. 154 pp.

―――. 1969. *Curso sistemático de filosofía actual (filosofía, ciencia, historia, dialéctica y sus aplicaciones)* [Systematic course in contemporary philosophy (Philosophy, science, history, dialectics, and their applications)]. Caracas: Universidad Central de Venezuela, 1969. 373 pp.

―――. 1977. *Teoría y metateoría de la ciencia* [Theory and metatheory of science]. 2 vols. Caracas: Ediciones de la Biblioteca de la Universidad Central de Venezuela. 834 pp.

―――. 1981. *Ciencia, técnica, historia y filosofía en la atmosfera cultural de nuestro tiempo*. Caracas: Universidad de Venezuela, Ediciones de la Biblioteca. 50 pp. English version, "Science, Technology, History, and Philosophy in the Cultural Atmosphere of Our Time," trans. Carl Mitcham and Waldemar López Pineiro, in Mitcham, ed., *Philosophy of Technology in Spanish Speaking Countries* (1993), pp. 229–247.

―――. 1989. *De magica a técnica: Ensayo de teatro filosófico-literario-técnico* [From magic to technology: An essay of the philosophical-literary-technical theater]. Barcelona: Anthropos. 223 pp. For a brief English review, see James Lynch, *Research in Philosophy and Technology* 13 (1993).

Gehlen, Arnold. 1957. *Die Seele im technischen Zeitalter: Sozialpsychologische Probleme in der industriellen Gesellschaft*. 132 pp. English version, *Man in the Age of Technology*, trans. Patricia Lipscomb (New York: Columbia University Press, 1980), xvi, 185 pp.

―――. 1983. "A Philosophical-Anthropological Perspective on Technology." Trans. Dorthe Thrane Rogers and Carl Mitcham. *Research in Philosophy and Technology* 6: 205–216. From "Anthropologische Ansicht der Technik," in Freyer, Papalekas, and Weippert, eds., *Technik im technischen Zeitalter* (1965), pp. 101–118.

Gille, Bertrand, André Fel, Jean Parent, and François Russo. 1978. *Histoire des techniques*. Vol. 1, *Techniques et civilisations*. Vol. 2, *Technique et science*. Paris: Gallimard. xiv, 1652 pp. English version, *The History of Techniques* (New York: Gordon and Breach, 1986). Vol. 1, *Techniques and Civilizations*, trans. P. Southgate and T. Williamson. Vol. 2, *Techniques and Sciences*, trans. J. Brainch, K. Butler, A. D. R. Dawes, W. Extavour, S. Romeo, A. Smith, P. Southgate, and T. Williamson, xv, 1410 pp. For both volumes, A. Keller revised technical terminology and Eda F. Kranakis added supplementary bibliographies. (Although Gille is sometimes given as the editor, he would more accurately be described as the senior author of this synthetizing work.)

Goffi, Jean-Yves. 1988. *La Philosophie de la technique* [Philosophy of technology]. Que sais-je? vol. 2405. Paris: Presses Universitaires de France. 127 pp.

Goldman, Steven L. 1984. "The Technē of Philosophy and the Philosophy of Technology." *Research in Philosophy and Technology* 7: 115–144.

———. 1990. "Philosophy, Engineering, and Western Culture." In *Broad and Narrow Interpretations of Philosophy of Technology, Philosophy and Technology,* vol. 7, ed. Paul T. Durbin, pp. 125–152 (Boston: Kluwer).

———. 1992. "No Innovation without Representation: Technological Action in a Democratic Society." In Stephen H. Cutcliffe et al., eds., *New Worlds, New Technologies, New Issues* (1992), pp. 148–160.

Gorokhov, Vitaly. 1990. "Die Methodologie der Technik in der UdSSR: Ergebnisse und Probleme: Eine Literaturübersicht." *Fridericiana: Zeitschrift der Universität Karlsruhe* 45: 27–38. Expanded English version, "Methodological Research and Problems of the Technological Sciences: A Review of the Literature in Russian," *Research in Philosophy and Technology* (forthcoming).

Grant, George Parkin. 1969. *Technology and Empire: Perspectives on North America.* Toronto: House of Anansi. 143 pp.

Note on Grant: Related philosophy of technology books: *Time as History* (Toronto: Canadian Broadcasting Corporation, 1969), 52 pp.; *English-Speaking Justice* (Sackville, N.B., Canada: Mount Allison University, 1974), 112 pp.; *Can We Think outside Technology?* (Sussex, England: *Tract,* vol. 24, n.d.), which is also published as "The Computer Does Not Impose on Us the Ways It Should Be Used," in *Beyond Industrial Growth,* ed. Abraham Rotstein, pp. 117–131 (Toronto: University of Toronto Press, 1976); *George Grant in Process: Essays and Conversations,* ed. Larry Schmidt (Toronto: House of Anansi Press, 1978), x, 223 pp.; *Technology and Justice* (Notre Dame: University of Notre Dame Press, 1986), 113 pp.; A useful study of Grant's thought is Ian H. Angus, *George Grant's Platonic Rejoinder to Heidegger: Contemporary Political Philosophy and the Question of Technology* (Lewiston, N.Y.: Edwin Mellen Press, 1987).

Gruender, C. David. 1971. "On Distinguishing Science and Technology." *Technology and Culture* 12, no. 3 (July): 456–463.

Gunderson, Keith. 1971. *Mentality and Machines.* Garden City, N.Y.: Doubleday Anchor. xviii, 173 pp. 2d ed. Minneapolis: University of Minnesota Press. xxii, 260 pp.

Habermas, Jürgen. 1975. *Technik und Wissenschaft als Ideologie.* Frankfurt: Suhrkamp. 169 pp. English version of title essay and two others included in *Toward a Rational Society: Student Protest, Science, and Politics,* trans. J. J. Shapiro (Boston: Beacon, 1970). ix, 132 pp.

Hannay, N. Bruce, and Robert E. McGinn. 1980. "The Anatomy of Modern Technology: Prolegomenon to an Improved Public Policy for the Social Management of Technology." *Daedalus* 109, no. 1 (winter): 25–53.

Harrison, Andrew. 1978. *Making and Thinking: A Study of Intelligent Activities.* Indianapolis: Hackett. ix, 207 pp.

Heelan, Patrick. 1983. *Space-Perception and the Philosophy of Science*. Berkeley: University of California Press, xiv, 383 pp.

Heidegger, Martin. 1927. *Sein und Zeit*. First Part, *Jahrbuch für Phänomenologie und phänomenologische Forschuung* (Halle), vol. 8. xi, 438 pp. English version, *Being and Time*, trans. John Macquarrie and Edward Robinson (New York: Harper and Row, 1962), 589 pp.

————. 1954. "Die Frage nach der Technik." In *Vorträge und Aufsätze*, pp. 13–44. Pfullingen: Neske. Reprinted in *Die Technik und die Kehre* pp. 5–36 (Pfullingen: Neske, 1962). English version, "The Question concerning Technology," in *The Question concerning Technology and Other Essays*, trans. William Lovitt, pp. 3–35 (San Francisco: Harper and Row, 1977). From lectures first delivered in 1949–1950.

————. 1959. *Gelasssenheit*. Pfullingen: Günther Neske. 73 pp. English version, *Discourse on Thinking*, trans. John M. Anderson and E. Hans Freund (New York: Harper and Row, 1966), 95 pp.

————. 1961. *Nietzsche*. 2 vols. Pfullingen: Neske. Vol. 1, 662 pp.; vol. 2, 493 pp. English version, with supplementary material, *Nietzsche* (San Francisco: Harper and Row): vol. 1, *The Will to Power as Art*, trans. David Farrell Krell (1979), xvi, 263 pp.; vol. 2, *The Eternal Recurrence of the Same*, trans. David Farrell Krell (1984), xii, 290 pp.; vol. 3, *The Will to Power as Knowledge and as Metaphysics*, trans. Joan Stambaugh, David Farrell Krell, and Frank A. Capuzzi (1987), xiii, 288 pp.; vol. 4, *Nihilism*, trans. Frank A. Capuzzi, ed. David Farrell Krell (1982), x, 301 pp.

————. 1962. "Die Kehre." In *Die Technik und Die Kehre*, pp. 37–47. Pfullingen: Neske. English version, "The Turning," in *The Question concerning Technology and Other Essays*, trans. William Lovitt, pp. 36–49 (San Francisco: Harper and Row, 1977).

————. 1971. *Poetry, Language, Thought*. Trans. Albert Hofstadter. New York: Harper and Row. xxv, 229 pp. Essays from *Hotzwege* (Frankfurt: Klostermann, 1950) and *Vorträge und Aufsätze* (Pfullingen: Neske, 1954).

Note on Heidegger: For analysis and bibliography, see Borgmann, "The Question of Heidegger and Technology: A Critical Review of the Literature," (1987). For further analysis consult Zimmerman, *Heidegger's Confrontation with Modernity: Technology, Politics, Art* (1990).

Hickman, Larry A. 1990. *John Dewey's Pragmatic Technology*. Bloomington: Indiana University Press. xv, 234 pp.

————, ed. 1990. *Technology as a Human Affair*. New York: McGraw-Hill. xiv, 495 pp. Thirty-eight readings divided into seven parts, each with its own useful introduction. Of particular importance are part 1, "Toward a Philosophy of Technology" (with articles by Robert E. McGinn, Alan R. Drengson, Hans Jonas, and Jacques Ellul); part 3, "Technology as Embodiment" (Don Ihde, Maurice Merleau-Ponty, Shoshana Zuboff, and John J. McDermott); and part

4, "The Phenomenology of Everyday Affairs" (Douglas Browning, Glen Jeansonne, Robert Linhart, George Berbner, Edmund Carpenter, Lewis Mumford, Daniel Boorstin, John J. McDermott, Paul B. Thompson, and Ruth Schwartz Cowan). Earlier versions of this anthology are Larry A. Hickman and Azizah Al-Hibri, eds., *Technology and Human Affairs* (St. Louis: C. V. Mosby, 1981), and Larry A. Hickman, ed., *Philosophy, Technology, and Human Affairs* (College Station, Tex.: IBIS Press, 1985).

Hommes, Jakob. 1955. *Der technische Eros: Das Wesen der materialistischen Geschictsauffassung* [Technological eros: The essence of the materialist interpretation of history]. Freiburg: Herder. xi, 519 pp.

Hood, Webster F. 1982. "Dewey and Technology: A Phenomenological Approach." *Research in Philosophy and Technology,* 5: 189–207.

Horkheimer, Max. 1947. *Eclipse of Reason.* New York: Oxford University Press. vii, 191 pp. Unlike other Horkheimer books, this was written in English.

———. 1974. *Critique of Instrumental Reason.* Trans. Matthew J. O'Connell et al. New York: Seabury. x, 163 pp. German version, *Zur Kritik der instrumentallen Vernuft,* ed. Alfred Schmidt (Frankfurt: S. Fischer, 1967), includes both *Eclipse of Reason* (part 1) and *Critique of Instrumental Reason* (part 2).

Horkheimer, Max, and Theodor Adorno. 1947. *Dialektik der Aufklärung.* Amsterdam: Querido. 310 pp. First published as *Philosophische Fragmente* (New York: Institute of Social Research, 1944). English version, *Dialectic of Enlightenment,* trans. John Cumming (New York: Seabury, 1972). xvii, 258 pp.

Hottois, Gilbert. 1984a. *Pour une éthique dans un univers technicien* [For an ethics in the technical universe]. Brussels: Editions de l'Université de Bruxelles. 107 pp.

———. 1984b. *Le Signe et la technique: La Philosophie à l'épreuve de la technique* [Sign and technology: Philosophy tested by technology]. Paris: Aubier. 222 pp. For an English presentation of the thesis of this book, see Gilbert Hottois, "Aspects of a Philosophy of Technique," *Research in Philosophy and Technology* 9 (1989): 45–57.

———, ed. 1983. "Ethique et Technique" [Ethics and technology]. Theme issue of *Morale et Enseignement* (Annales de l'Institut de Philosophie et de Sciences Morales). 165 pp.

———, ed. 1987. "Questions sur la Technique" [Questions on technology]. Theme issue of *Revue Internationale de Philosophie* 41, no. 161: 151–323.

Huning, Alois. 1974. *Das Schaffen des Ingenieurs: Beiträge zu einer Philosophie der Technik* [The creativity of engineers: Contribution to a philosophy of technology]. Düsseldorf: VDI Verlag. viii, 203 pp.; 2d ed., 1978, viii, 226 pp.; 3d ed., 1987, viii, 207 pp.

———. 1979. "Philosophy of Technology and the Verein Deutscher Ingenieure." *Research in Philosophy and Technology* 2: 265–271.

Ihde, Don. 1979. *Technics and Praxis: A Philosophy of Technology.* Boston: D. Reidel. xxviii, 151 pp.

———. 1983. *Existential Technics*. Albany: State University of New York Press. ix, 190 pp.

———. 1986. *Consequences of Phenomenology*. Albany: State University of New York Press. xi, 210 pp.

———. 1990. *Technology and the Lifeworld: From Garden to Earth*. Bloomington: Indiana University Press. xiv, 226 pp.

———. 1991. *Instrumental Realism: The Interface between Philosophy of Science and Philosophy of Technology*. Bloomington: Indiana University Press. xiv, 159 pp.

———. 1993. *Philosophy of Technology: An Introduction*. New York: Paragon Press. xiii, 157 pp.

———. 1994. *Postphenomenology: Essays in the Postmodern Context*. Evanston, Ill.: Northwestern University Press. Part 1 focuses on postmodernity and technoculture.

Illich, Ivan. 1973. *Tools for Conviviality*. New York: Harper and Row. xiii, 133 pp.

———. 1981. *Shadow Work*. Boston: Marion Boyars. 152 pp.

———. 1993. *In the Vineyard of the Text: A Commentary to Hugh's "Didascalicon."* Chicago: University of Chicago Press. vi, 154 pp.

Illich, Ivan, and Barry Sanders. 1988. *ABC: The Alphabetization of the Popular Mind*. San Francisco: North Point Press. xi, 166 pp.

Note on Illich: Other works by Illich of special relevance: *Medical Nemesis: The Expropriation of Health* (New York: Pantheon, 1976), viii, 294 pp.; *Toward a History of Needs* (New York: Pantheon, 1978), xiii, 143 pp. which includes the important "Energy and Equity" that was also published as a separate volume; *H₂O and the Waters of Forgetfulness: Reflections on the Historicity of "Stuff"* (Berkeley, Calif.: Heyday Books, 1985), 92 pp.; and *In the Mirror of the Past: Lectures and Addresses, 1978–1990* (New York: Marion Boyars, 1992), 231 pp. The best general introduction to Illich is David Cayley, *Ivan Illich in Conversation* (Toronto: House of Anansi, 1992), xv, 299 pp.

Jarvie, I. C. 1966. "The Social Character of Technological Problems: Comments on Skolimowski's Paper." *Technology and Culture* 7, no. 3 (summer): 384–390. Reprinted in Mitcham and Mackey, eds., *Philosophy and Technology* (1972), pp. 50–53, along with the related "Technology and the Structure of Knowledge," pp. 54–61. Both papers are included in *Thinking about Society: Theory and Practice*, by I. C. Jarvie, pp. 314–320 and 302–313, respectively, along with "Is Technology Unnatural?" pp. 321–327 (Boston: D. Reidel, 1986).

Jaspers, Karl. 1931. *Die geistige Situation der Zeit*. Berlin: W. de Gruyter. 191 pp. English version, *Man in the Modern Age*, trans. Eden and Cedar Paul (Garden City, N.Y.: Doubleday, 1957), viii, 230 pp. This condensed version of Jaspers's three-volume *Philosophie* (Berlin: J. Springer, 1932) examines "the tension between technical mass-order and human life" in part 1, chaps. 1–3.

———. 1949. *Von Ursprung und Ziel der Geschichte*. Zurich: Artemis-Verlag. 268 pp. English version, *The Origin and Goal of History*, trans. M. Bullock (New Haven: Yale University Press, 1953), xvi, 294 pp. See especially part 2, chap. 1, sec. 2, "Modern Technology," pp. 100–127.

———. 1958. *Die Atombombe und die Zukunft des Menschen: Politisches Bewußtsein in unserer Zeit*. Munich: Piper. 506 pp. First published in part as a twenty-six-page pamphlet of a radio talk, 1957. English version, *The Future of Mankind*, trans. E. B. Ashton (Chicago: University of Chicago Press, 1961), ix, 342 pp. See especially chap. 12, "The Scientists and the 'New Way of Thinking,'" pp. 187–208.

Johnson, Deborah. 1985. *Computer Ethics*. Englewood Cliffs, N.J.: Prentice-Hall. xv, 110 pp. For a collection of more than thirty texts on the themes analyzed by and referred to in this monograph, see Deborah G. Johnson and John W. Snapper, eds., *Ethical Issues in the Use of Computers* (Belmont, Calif.: Wadsworth, 1985), ix, 363 pp.

Jonas, Hans, 1966. "The Practical Uses of Theory." In *The Phenomenon of Life: Toward a Philosophical Biology*, pp. 188–210. New York: Harper and Row. First printed, along with comments, in *Social Research* 26, no. 2 (1959): 151–166. Reprinted in Mitcham and Mackey, eds., *Philosophy and Technology* (1972), pp. 335–347.

———. 1974. *Philosophical Essays: From Ancient Creed to Technological Man*. Englewood Cliffs, N.J.: Prentice-Hall. xviii, 349 pp.

———. 1984. *The Imperative of Responsibility: In Search of an Ethics for the Technological Age*. Chicago: University of Chicago Press. xii, 255 pp. This English version, trans. David Herr and the author, combines *Das Prinzip Verantwortung: Versuch einer Ethik für die technologische Zivilisation* (Frankfurt: Insel, 1979) and *Macht oder Ohnmacht der Subjektivität? Das Leib-Seele-Problem im Vorfeld des Prinzips Verantwortung* (Frankfurt: Insel, 1981).

Jones, J., and D. G. Thornley, eds. 1963. *Conference on Design Methods*. New York: Pergamon Press. xiii, 222 pp. Papers from a Conference on Systematic and Intuitive Methods in Engineering, Industrial Design, Architecture and Communications, London, September 1962.

Jünger, Ernst. 1932. *Der Arbeiter*. Hamburg: Hanseatische Verlagsanstalt. 300 pp. Sections 44–57, "Technology as the Mobilization of the World through the *Gestalt* of the Worker," trans. James M. Vincent and Richard J. Rundell, in Mitcham and Mackey, eds., *Philosophy and Technology* (1972), pp. 269–289.

Kapp, Ernst. 1845. *Philosophie oder vergleichende allgemeine Erdkunde als wissenschaftliche Darstellung der Erdverhältnisse und des Menschenlebens nach ihrem inneren Zusammenhang* [Philosophy or comparative general geography as scientific presentation of environmental and human life through their inner relationship]. 2 vols. Braunschweig: Westermann. Vol. 1. x, 331 pp.; vol. 2, iv, 447 pp. There is a one-volume revision under the shortened title *Vergleichende allgemeine Erdkunde in wissenschaftlicher Darstellung* (Braunschweig: Westermann, 1868), xv, 704 pp.

————. 1877. *Grundlinien einer Philosophie der Technik: Zur Entstehungsgeschichte der Cultur aus neuen Gesichtspunkten* [Fundamentals of a philosophy of technology; The genesis of culture from a new perspective]. Braunschweig: Westermann. xvi, 360 pp. Facsimile reprint, with a new introduction by Hans-Martin Sass, Düsseldorf: Stern-Verlag, 1978.

Koen, Billy Vaughn. 1985. *Definition of the Engineering Method.* Washington, D.C.: American Society for Engineering Education, 75 pp.

Kovács, Gizella, and Siegfried Wollgast, eds. 1984. *Technikphilosophie in Vergangenheit und Gegenwart* [Philosophy of technology in past and present]. Berlin: Akademie-Verlag. 225 pp. Contents: Gerhard Banse, "Die 'Technikphilosophie' in der Sicht des dialektischen und historischen Materialismus" ["Philosophy of technology" from the perspective of dialectical and historical materialism]; Gizella Kovács, "Der Technikbergriff von Karl Marx und seine heutigen 'marxologischen' Kritiker" [Karl Marx's concept of technology and its contempoary "marxiological" critics]; Hans-Ulrich Wöhler, "Weltanschauliche Aspekte der Technikbetrachtung in der Periode des Manufakturkapitalismus" [Worldview aspects of the technical perspective in the period of capitalist manufacture]; Helga Petzoldt, "Zu einigen Problemen der philosophischen Lehre an deutschen Technischen Hochschulen im 19. Jahrhundert" [On some problems of philosophical teaching in German technical colleges in the nineteenth century]; Bernd Adelhoch, "'Technikphilosophie' in der *Zeitschrift des Vereins deutscher Ingenieure* in der Weimarer Republic" [Philosophy of technology in the *Zeitschrift des Vereins deutscher Ingenieure* during the Weimar Republic]; Siegfried Wollgast, "'Technikphilosophie' während der Herrschaft des deutschen Faschismus" ['Philosophy of technology' during the domination of German fascism]; Margit Rezsö, "Zur 'Technikphilosophie' Martin Heidegger" [Martin Heidegger's "philosophy of technology"]; Ernst Woit's "Spätbürgerliche 'Technikphilosophie' über Krieg und Frieden" [Late bourgeois 'philosophy of technology' in war and peace]; Imre Hronszky and János Rathmann, "Zur 'Technikphilosophie' in der BRD in den 70er Jahren" [On 'philosophy of technology' in West Germany in the 1970s].

Lafitte, Jacques. 1932. *Réflexions sur la science des machines* [Reflections on the science of machines]. Paris: Bloud et Gay. 162 pp. Reprint, with a new introduction by Jacques Guillerme, Paris: J. Vrin, 1972. viii, 123 pp. English version, *Reflections on the Science of Machines*, trans. by Lynda Grant, John Hart, and Jean LeMoyne (Computer Science Department, University of Western Ontario, London, Ontario, Canada: Mechanology Press, n.d.), xxvii, 108 pp. Translations of a small section from the book along with an article by Lafitte summarizing his conclusions, titled "On the Science of Machines" (1933), appear in Carl Mitcham, ed., "Analyses of Machines in the French Intellectual Tradition," in *Research in Philosophy and Technology* (1979) 2: 15–52.

Lange, Hellmuth. 1988. "Technikphilosophie." In *Enzyklopädie zur bürgerlichen Philosophie in 19. und 20. Jahrhundert,* ed. Manfred Buhr, pp. 527–561. Cologne: Pahl-Rugenstein.

Laudan, Rachel, ed. 1984. *The Nature of Technological Knowledge: Are Models of Scientific Change Relevant?* Boston: Reidel. vii, 145 pp.

Laymon, Ronald. 1985. "Idealization and the Testing of Theories by Experimentation." In *Experiment and Observation in Modern Science,* ed. Peter Achinstein and Owen Hannaway, pp. 147–173. Cambridge: MIT Press.

———. 1989. "Applying Idealized Scientific Theories to Engineering." *Synthese* 81, no. 3 (December): 353–371.

———. 1991. "Idealizations and the Reliability of Dimensional Analysis." In Durbin, ed., *Critical Perspectives on Nonacademic Science and Engineering* (1991), pp. 146–180.

———. 1992. "Idealizations, Externalities, and the Economic Analysis of Law." In Pitt and Lugo, eds., *The Technology of Discovery and the Discovery of Technology* (1992), pp. 87–101.

Layton, Edwin T., Jr. 1974. "Technology as Knowledge." *Technology and Culture* 15, no. 1 (January): 31–41.

———. 1976. "American Ideologies of Science and Engineering." *Technology and Culture* 17, no. 4 (October): 688–701.

Leiss, William. 1972. *The Domination of Nature.* New York: G. Braziller. xii, 242 pp.

———. 1976. *The Limits to Satisfaction: An Essay on the Problem of Needs and Commodities.* Toronto: University of Toronto Press. x, 159 pp.

———. 1990. *Under Technology's Thumb.* Montreal: McGill-Queen's University Press. xii, 169 pp.

Lenk, Hans. 1971. *Philosophie im technologischen Zeitalter* [Philosophy in the technological age]. Stuttgart: W. Kohlhammer. 174 pp.

———. 1982. *Zur sozialphilosophie der technik* [On the social philosophy of technology]. Frankfurt: Suhrkamp. 300 pp.

———, ed. 1973. *Technokratie als ideologie* [Technology as ideology]. Stuttgart: W. Kohlhammer. 238 pp.

Lenk, Hans, and Matthias Maring, eds. 1991. *Technikverantwortung: Güterabwägung—risikobewertung—verhaltenskodizes* [Technical responsibility: Assessing costs and benefits—calculating risks—determining codes of behavior]. New York: Campus. 353 pp.

Lenk, Hans, and Simon Moser, eds. 1973. *Techne, Technik, Technologie: Philosophische Perspektiven* [Techne, technique, technology: Philosophical perspectives]. Pullach bei Munich: Dokumentation. 247 pp. The lead contribution is Moser, "Kritik der tradionellen Technikphilosphie" (1973).

Lenk, Hans, and Günter Ropohl. 1976. *Technische Intelligenz im systemtechnologischen Zeitalter* [Technical intelligence in a systems technology age]. Düsseldorf: VDI Verlag. x, 138 pp.

———. 1979. "Toward and Interdisciplinary and Pragmatic Philosophy of Technology: Technology as a Focus for Interdisciplinary Reflection and Systems Research." Trans. Cyn Klohr and Carl Mitcham, in *Research in Philosophy and Technology* 2: 15–52. This is a revised and expanded version of

"Praxisnahe Technikphilosophie: Entwicklung und Aktualität der interdisziplinaren Technologiediskussion," in Zimmerli, ed., *Technik, oder: Wissen wir, was wir tun?* (1976), pp. 104–145.

———, eds. 1978. *Systemtheorie als Wissenschaftsprogramm* [Systems theory as a scientific program]. Königstein: Athenäum. 271 pp.

———, eds. 1987. *Technik und Ethik* [Technology and ethics]. Stuttgart: Philipp Reclam. 333 pp.

Levinson, Paul. 1988. *Mind at Large: Knowing in the Technological Age. Research in Philosophy and Technology,* suppl. 2. xviii, 271 pp.

Ley, Hermann. 1961. *Dämon Technik?* Berlin: Deutscher Verlag Wissenschaften. 428 pp.

Losonsky, Michael. 1990. "The Nature of Artifacts." *Philosophy* 65, no. 251 (January): 81–88.

McGinn, Robert E. 1978. "What Is Technology?" *Research in Philosophy and Technology* 1: 179–197. Reprinted in Hickman, ed., *Technology as a Human Affair* (1990), pp. 10–25.

———. 1991. *Science, Technology, and Society.* Englewood Cliffs, N.J.: Prentice-Hall. xvii, 302 pp.

MacKenzie, Donald, and Judy Wajcman, eds. 1985. *The Social Shaping of Technology: How the Refrigerator Got Its Hum.* Milton Keynes, England: Open University Press. viii, 327 pp.

McLean, George F., ed. 1964. *Philosophy in a Technological Culture.* Proceedings of the Workshop on Philosophy in a Technological Culture, Catholic University of America, June 1963. Washington, D.C.: Catholic University of America Press. xv, 438 pp.

McLuhan, Marshall. 1964. *Understanding Media: The Extensions of Man.* New York: McGraw-Hill. vii, 359 pp. 2d ed., 1965, xiii, 364 pp.

McLuhan, Marshall, and Eric McLuhan. 1988. *Laws of Media: The New Science.* Buffalo, N.Y.: University of Toronto Press. xi, 252 pp.

Man, Science, Technology: A Marxist Analysis of the Scientific and Technological Revolution. 1973. Compiled by three institutes: Institute of History of Natural Sciences and Technology, Institute of Philosophy (the first two belonging to the Academy of Sciences of the former USSR), and the Institute of Philosophy and Sociology (from the former Czechoslovak Academy of Sciences). Czechoslovakia: Academia (Czechoslovak Academy of Sciences Press). 387 pp.

Mander, Jerry. 1991. *In the Absence of the Sacred: The Failure of Technology and the Survival of the Indian Nations.* San Francisco: Sierra Club Books. 446 pp. See also Mander's important *Four Arguments for the Elimination of Television* (New York: William Morrow, 1978).

Marcuse, Herbert. 1964. *One-Dimensional Man: Studies in the Ideology of Advanced Industrial Society.* Boston: Beacon Press. xvii, 260. pp.

———. 1969. *An Essay on Liberation.* Boston: Beacon Press. x, 91 pp.

Marx, Karl. 1867. *Das Kapital.* Vol. 1, *Der Produktionsprocess des Kapitals.* Hamburg: Otto Meissner. 2d ed., 1873; 3d ed., ed. Frederick Engels, 1883; 4th ed., ed. Frederick Engels, 1890. In Karl Marx and Friedrich Engels, *Werke,* vol. 23 (Berlin: Dietz, 1973), 955 pp. English version, *Capital,* vol. 1, *A Critical Analysis of Capitalist Production,* trans. Samuel Moore and Edward Aveling and ed. Frederick Engels (New York: International Publishers, 1967), xii, 807 pp. (Because of changes authorized by Marx himself in the French edition published serially between 1872 and 1875, and further revisions introduced by Engels in German and the English translation, English and German chapter numbers do not correspond after chap. 3.)

Note on the text of "Das Kapital": Das Kapital consists of three volumes, the second two prepared by Friedrich Engels after Marx's death in 1883. Vol. 2, *Der Cirkulationsprocess des Kapitals* [The process of the circulation of capital], appeared in 1885; vol. 3, *Der Gesammtprocess der kapitalistischen Produktion* [The process of capitalist production as a whole] in 1894. However, according to Engels's own "Preface to the English Edition" (1886), the "first book is in a great measure a whole in itself, and [ranks] as an independent work" (trans., 1967, p. 5). In vol. 1 Marx's core technical philosophy of technology can be found in chaps. 12 and 13, "Teilung der Arbeit und Manufaktur" and "Maschinerie und große Industrie"; English version, chaps. 14 and 15, "Division of Labor and Manufacture" and "Machinery and Modern Industry."

Mayz Vallenilla, Ernesto. 1974. *Esbozo de una crítica de la razón técnica* [Outline of a critique of technical reason]. Caracas: Universidad Simón Bolívar. 249 pp. The title essay is reprinted in *Ratio Technica* (Caracas: Monte Avila, 1983), 278 pp.

———. 1990a. *Fundamentos de la meta-técnicia* [Foundations of metatechnology]. Caracas: Monte Avila. 152 pp.

———. 1990b. "Presente y futuro de la humanidad." In Mitcham and Peña Borrero, eds., *El nuevo mundo de la filosofía y la tecnología* (1990), pp. 283–291. English version, "Present and Future of Humanity," trans. by Luis Castro Leiva and Carl Mitcham, in Mitcham, ed., *Philosophy of Technology in Spanish-Speaking Countries* (1993), pp. 249–258.

Mesthene, Emmanuel. 1970. *Technological Change: Its Impact on Man and Society.* Cambridge: Harvard University Press. 127 pp.

Mitcham, Carl, ed. 1993. *Philosophy of Technology in Spanish Speaking Countries. Philosophy and Technology,* vol. 10. Boston: Kluwer. xxxvi, 320 pp.

Mitcham, Carl, and Jim Grote. 1978. "Current Bibliography in the Philosophy of Technology: 1973–1974." *Research in Philosophy and Technology* 1: 297–390.

———. 1981. *Current Bibliography in the Philosophy of Technology: 1975–1976.* Theme issue of *Research in Philosophy and Technology* 4: 1–297.

———. 1983. "Current Bibliography in the Philosophy of Technology: 1977–1978—The Primary Sources." *Research in Philosophy and Technology* 6: 231–296.

————, eds. 1984. *Theology and Technology: Essays in Christian Analysis and Exegesis*. Lanham, Md.: University Press of America. ix, 523 pp. Contents: Mitcham, "Technology as a Theological Problem in the Christian Tradition"; Mitcham and Grote, "Aspects of Christian Exegesis." Part 1, "Basic Approaches": George E. Blair, "Faith outside Technique"; Wilhelm E. Fudpucker, "Through Technological Christianity to Christian Technology"; Terry J. Tekippe, "Bernard Lonergan: A Context for Technology"; André Malet, "The Believer in the Presence of Technique"; Egbert Schuurman, "A Christian Philosophical Perspective on Technology." Part 2, "Exegeses of the Christian Traditon": Jacques Ellul, "Technique and the Opening Chapters of Genesis"; Ellul, "The Relation of Man to Creation according to the Bible"; Charles Mabee, "Biblical Hermeneutics and the Critique of Technology"; P. Hans Sun, "Notes on How to Begin to Think about Technology in a Theological Way"; Ernest Fortin, "Augustine, the Arts, and Human Progress"; Paul T. Durbin, "Thomism and Technology: Natural Law Theory and the Problems of a Technological Society"; Willis Dulap, "Two Fragments: Theological Transformation of Law, Technological Transformation of Nature"; George Grant, "Justice and Technology"; Douglas John Hall, "Toward an Indigenous Theology of the Cross"; Thomas Berry, "The New Story: Meaning and Value in the Technological World"; George W. Shield, "Process Theology and Technology"; Frederick Sontag, "Technology and Theodicy"; Albert Borgmann, "Prospects for the Theology of Technology." Part 3: "Select Bibliography of Theology and Technology."

Mitcham, Carl, and Alois Huning, eds. 1986. *Philosophy and Technology II: Information Technology and Computers in Theory and Practice*. Boston Studies in the Philosophy of Science, vol. 90. Boston: D. Reidel. xxii, 352 pp. Selected proceedings of an international conference held in New York, September 3–7, 1983.

Mitcham, Carl, and Robert Mackey. 1973. *Bibliography of the Philosophy of Technology*. Chicago: University of Chicago Press. xvii, 205 pp. First published as a special issue of *Technology and Culture* 14, no. 2, pt. 2 (April 1973): S1–S205. Paperback reprint, Ann Arbor, Mich.: Books on Demand, 1985. Author index from *Research in Philosophy and Technology* 4 (1981): 243–297, added 1986. For subsequent bibliographies in this series see Mitcham and Grote, "Current Bibliography" (1978, 1981, 1983).

————, eds. 1972. *Philosophy and Technology: Readings in the Philosophical Problems of Technology*. New York: Free Press. ix, 399 pp. Paperback reprint, with revised select bibliography, 1983. xii, 403 pp. Contents: Mitcham and Mackey, "Technology as a Philosophical Problem." Part 1, "Conceptual Issues": James K. Feibleman, "Pure Science, Applied Science, and Technology"; Henryk Skolimowski, "The Structure of Thinking in Technology"; I. C. Jarvie, "The Social Character of Technological Problems: Comments on Skolimowski's Paper"; I. C. Jarvie, "Technology and the Structure of Knowledge"; Mario Bunge, "Toward a Philosophy of Technology"; Lewis Mumford, "Technics and the Nature of Man"; Jacques Ellul, "The Technological Order." Part 2, "Ethical and Political Critiques": Emmanuel G. Mesthene, "Technology and Wisdom"; Emmanuel G. Mesthene, "How Technology Will Shape the Future"; Günther Anders, "Commandments in the Atomic Age"; Richard M. Weaver, "Humanism in an Age of Science and Technology"; C. S. Lewis, "The Abolition of Man"; Nathan Rotenstreich, "Technology and Politics"; C. B. Macpherson, "Democratic Theory: Ontology and Technology"; Yves R. Simon, "Pursuit of Happiness and Lust for Power in Technological Society"; George Grant, "Technology and Empire." Part 3, "Religious Critiques": Nicholas Berdyaev, "Man and Machine"; Eric Gill, "Christianity and the Machine Age"; R. A. Buchanan, "The Churches in a Changing World"; W. Norris Clarke, "Technology and Man: A Christian View"; Lynn White Jr., "The Historical Roots of Our Ecological Crisis." Part 4, "Two Existentialist Critiques": Ernst Jünger, "Technology as the Mobilization of the World through the *Gestalt* of the Worker"; José Ortega y Gasset, "Thoughts on Technology." Part 5, "Metaphysical Studies": Friedrich Dessauer, "Technology in Its Proper Sphere"; Hans Jonas, "The Practical Uses of Theory"; Webster F. Hood, "The Aristotelian versus the Heideggerian Approach to the Problem of Technology."

Mitcham, Carl, and Margarita M. Peña Borrero, eds., with Elena Lugo and James Ward. 1990. *El nuevo mundo de la filosofía y la tecnología* [The new world of philosophy and technology]. Proceedings of the First Interamerican Congress on Philosophy of Technology, University of Puerto Rico in Mayagüez, 5–9 October, 1988. University Park, Pa.: STS Press, Willard 133. vii, 330 pp.

Note on Mitcham: Mitcham's "Philosophy of Technology," in *A Guide to the Culture of Science, Technology, and Medicine,* ed. Paul T. Durbin, pp. 282–363 (New York: Free Press, 1980), paperback edition (1984) with bibliographic update, pp. 672–674, is an otherwise uncited state-of-the-art survey. A condensed and refocused version can be found in Mitcham's entry "Philosophy of Technology," in *Encyclopedia of Bioethics,* ed. Warren T. Reich, 3: 1638–1643 (New York: Free Press, 1978).

Moser, Simon. 1973. "Kritik der traditionellen Technikphilosophie." In Lenk and Moser, eds., *Techne, Technik, Technologie* (1973), pp. 11–81. Revised version of "Zur Metaphysik der Technik" in *Metaphysik einst und jetzt,* by Simon Moser, pp. 231–94 (Berlin: De Gruyter, 1958). Partial English version of the original, "Toward a Metaphysics of Technology," trans. William Carroll, Carl Mitcham, and Robert Mackey, *Philosophy Today* 15, no. 2 (summer 1971): 129–156. Two deleted sections are available from Mitcham on request.

Moser, Simon, and Alois Huning, eds. 1975. *Werte und Wertodnungen in Technik und Gesellschaft* [Values and the order of values in technology and society]. Düsseldorf: VDI Verlag.

———. 1976. *Wertpräferenzen in Technik und Gesellschaft* [Value preferences in technology and society]. Düsseldorf: VDI Verlag. 134 pp.

Mumford, Lewis. 1934. *Technics and Civilization.* New York: Harcourt Brace. 495 pp. Reprinted with a new, unpaginated introduction, 1963. All quotations are from the 1963 edition.

————. 1952. *Art and Technics.* New York: Columbia University Press. 162 pp. Paperback reprint, 1960.

————. 1961. *The City in History: Its Origins, Its Transformations, and Its Prospects.* New York: Harcourt, Brace and World. xi, 657 pp.

————. 1964. "Authoritarian and Democratic Technics." *Technology and Culture* 5, no. 1 (winter): 1–8.

————. 1967, 1970. *The Myth of the Machine.* Vol. 1, *Technics and Human Development.* Vol. 2, *The Pentagon of Power.* New York: Harcourt Brace Jovanovich. 342 and 496 pp.

Note on Mumford: Helpful surveys of Mumford's voluminous work are available in Elmer S. Newman, *Lewis Mumford: A Bibliography, 1914–1970* (New York: Harcourt Brace Jovanovich, 1971), and Jane Morley, *On Lewis Mumford: An Annotated Bibliography* (Philadelphia: Seminar on Technology and Culture, University of Pennsylvania, 1985). See also Donald L. Miller, *Lewis Mumford: A Life* (Pittsburgh: University of Pittsburgh Press, 1989), and Thomas P. Hughes and Agatha C. Hughes, eds., *Lewis Mumford: Public Intellectual* (New York: Oxford University Press, 1990).

Murray, Patrick. 1982. "The Frankfurt School Critique of Technology." *Research in Philosophy and Technology* 5: 223–248.

Ortega y Gasset, José. 1939. "Meditación de la técnica." In *Ensimismamiento y alteracion* (Buenos Aires: Espasa-Calpe). In *Obras completas*, 1st ed., 5: 317–375 (Madrid: Revista de Occidente, 1945–1947). This series of lectures was first translated by Helene Weyl as "Man the Technician" and included in *Toward a Philosophy of History*, pp. 87–161 (New York: W. W. Norton, 1941), a volume subsequently reprinted as *History as a System and Other Essays toward a Philosophy of History.* The translation was revised by Edwin Williams as "Thoughts on Technology" for Mitcham and Mackey, eds., *Philosophy and Technology* (1972), pp. 290–313, but even the revision is inadequate. Both translations delete substantial portions of the text, unnecessarily alter format, and contain errors in phrasing and terminology.

————. 1952. "Der Mythus des Menschen hinter der Technik" [The myth of humanity outside technology]. In *Mensch und Raum*, ed. Otto Bartning, pp. 111–117 (Darmstadt: Neue Darmstädter Verlagsanstalt). A Spanish translation (no Spanish original exists), is "El mito del hombre allende de la técnica," in *Obras completas*, 9: 617–624.

Petroski, Henry. 1985. *To Engineer Is Human: The Role of Failure in Successful Design.* New York: St. Martin's Press. xiii, 247 pp.

————. 1986. *Beyond Engineering: Essays and Other Attempts to Figure without Equations.* New York: St. Martin's Press. xii, 256 pp.

————. 1990. *The Pencil.* New York: Alfred Knopf. 448 pp.

————. 1992. *The Evolution of Useful Things.* New York: Alfred Knopf. xi, 288 pp.

Philosophy and Technology. 1983–. Vol. 1 (1983), ed. Paul T. Durbin and Friedrich Rapp; vol. 2 (1986), ed. Carl Mitcham and Alois Huning; vol. 3 (1987) and vol. 4 (1988), ed. Paul T. Durbin; vol. 5 (1989), ed. Edmund F. Bryne and Joseph C. Pitt; vol. 6 (1989) and vol. 7 (1990), ed. Paul T. Durbin; vol. 8 (1991), ed. Paul T. Durbin; vol. 9 (1992), ed. Langdon Winner; vol. 10 (1993), ed. Carl Mitcham. A series of the Society for Philosophy and Technology. Boston: D. Reidel, now Kluwer.

Pitt, Joseph C. 1987. "The Autonomy of Technology." In *Technology and Responsibility, Philosophy and Technology*, vol. 3, ed. Paul T. Durbin, pp. 99–114 (Boston: Kluwer). Reprinted in Gayle L. Ormiston, ed., *From Artifact to Habitat: Studies in the Critical Engagement of Technology* (Bethlehem, Pa.: Lehigh University Press, 1990), pp. 117–131.

———. 1988. "'Style' and Technology." *Technology in Society* 10: 447–456.

———. 1992. *Galielo, Human Knowledge, and the Book of Nature: Method Replaces Metaphysics*. Boston: Kluwer. xvi, 201 pp.

Pitt, Joseph C., and Elena Lugo, eds. 1991. *The Technology of Discovery and the Discovery of Technology*. Blacksburg: Society for Philosophy and Technology, Department of Philosophy, Virginia Polytechnic. 519 pp. Proceedings of the sixth international conference of the Society for Philosophy and Technology, University of Puerto Rico at Mayagüez, March 1991.

Polanyi, Michael. 1953. *Pure and Applied Science and Their Appropriate Forms of Organization*. Occasional Pamphlet, no. 14. Oxford: Society for Freedom in Science, December. Reprinted in *Dialectica* 10, no. 3 (1956): 231–242.

———. 1954. "Skills and Connoisseurship." In *Àtti del Congrèsso di Studi Metòdologici* pp. 381–395 (Turin: Taylor).

———. 1958. "Skills." In *Personal Knowledge: Towards a Post-critical Philosophy* pp. 49–65 (Chicago: University of Chicago Press).

Popper, Karl R. 1945. *The Open Society and Its Enemies*. Vol. 1, *The Spell of Plato*. Vol. 2, *The High Tide of Prophecy: Hegel, Marx, and the Aftermath*. London: Routledge and Kegan Paul. Vol. 1, vii, 268 pp.; vol. 2, v, 352 pp. 2d ed. (Princeton, N.J.: Princeton University Press, 1950), 2 vols. in one, xii, 732 pp. 3d rev. ed. (London: Routledge and Kegan Paul, 1957), xi, 322 pp. and v, 391 pp. 4th rev. ed., (Princeton, N.J.: Princeton University Press, 1963), xi, 351 pp. and v, 420 pp. Also available as Harper Torchbook. 5th rev. ed. (Princeton, N.J.: Princeton University Press, 1966), xi, 361 pp. and v, 420 pp.

Pylyshyn, Zenon, ed. 1970. *Perspectives on the Computer Revolution*. Englewood Cliffs, N.J.: Prentice-Hall. xx, 540 pp.

Quintanilla, Miguel Angel. 1989. *Tecnología: Un enfoque filosófico* [Technology: A philosophical perspective]. Madrid: FUNDESCO. 141 pp. For an English version of chap. 5 see "The Design and Evaluation of Technologies: Some Conceptual Issues," trans. Susan Frisbie and Belén García, in Mitcham, ed., *Philosophy of Technology in Spanish Speaking Countries* (1993), pp. 173–195.

Rapp, Friedrich. 1981. *Analytical Philosophy of Technology*. Trans. Stanley R. Carpenter and Theodor Langenbruch. Boston: D. Reidel. xiv, 199 pp. Original German, *Analytische Technikphilosophie* (Freiburg: Karl Alber, 1978), 226 pp.

———. 1982. "Philosophy of Technology." In *Contemporary Philosophy: A New Survey*, vol. 2, *Philosophy of Science*, ed. Guttorm Fløistad, pp. 361–412. The Hague: Martinus Nijhoff.

———. 1985. "Humanism and Technology: The Two Cultures Debate." *Technology in Society* 7, no. 4: 423–435.

———, ed. 1974. *Contributions to a Philosophy of Technology: Studies in the Structure of Thinking in the Technological Sciences*. Boston: D. Reidel. xiii, 228 pp. Contents: Ladislav Tondl, "On the Concepts of 'Technology' and 'Technological Sciences'"; Mario Bunge, "Technology as Applied Science"; Joseph Agassi, "The Confusion between Science and Technology in the Standard Philosophies of Science"; John O. Wisdom, "The Need for Corroboration: Comments on J. Agassi's Paper"; Joseph Agassi, "Planning for Success: A Reply to J. O. Wisdom"; John O. Wisdom, "Rules for Making Discoveries: Reply to J. Agassi"; Henryk Skolimowski, "The Structure of Thinking in Technology"; I. C. Jarvie, "The Social Character of Technological Problems: Comments on Skolimowski's Paper"; Friedrich Rapp, "Technology and Natural Science—a Methodological Investigation"; M. I. Mantell's "Scientific Method—a Triad"; Eberhard Jobst, "Specific Features of Technology in Its Interrelation with Natural Science"; Dieter Teichmann, "On the Classification of the Technological Sciences"; Tadeusz Kotarbinski, "Instrumentalization of Actions"; Morris Asimov, "A Philosophy of Engineering Design"; R. J. McCrory, "The Design Method—a Scientific Approach to Valid Design"; A. D. Hall, "Three-Dimensional Morphology of Systems Engineering"; "The Role of Experiments in Applied Science—Letters to the Editor," by A. J. S. Pippard, W. A. Tuplin, E. McEwen, and Your Reviewer; F. V. Lazarev and M. K. Trifonova, "The Role of Apparatus in Cognition and Its Classification."

———, ed. 1990. *Technik und Philosophie* [Technology and philosophy]. *Technik und Kultur*, vol. 1. Düsseldorf: VDI Verlag. xviii, 338 pp. Contents: For review, see Carl Mitcham, "German Philosophy of Technology," in *Research in Philosophy and Technology*, vol. 13 (forthcoming 1993). Part 1, "Entwicklung der Technikphilosophie" [The Development of the Philosophy of Technology], includes: Alois Huning, "Der Technikbegriff" [The concept of technology]; Alois Huning, "Die philosophische Tradition" [The philosophical tradition]; Alois Huning, "Deutungen vom 19. Jahrhundert bis zur Gegenwart" [Interpretations from the nineteenth century to today]; Friedrich Rapp, "Geistesgeschichtliche Voraussetzungen der modernen Technik" [Historicophilosophical presuppositions of modern technology]. Part 2, Günter Ropohl, "Technisches Problemlösen und soziales Umfeld" [Technical problem solving and social environment]. Part 3, "Technik und Verantwortung" [Technology and responsibility]: Friedrich Rapp, "Die zwei Kulturen: technische und humanistische Rationalität" [The two cultures: Technical and humanistic rationality]; Friedrich Rapp, "Sachzwänge und Wertentscheidungen" [Material constraints and value decisions]; Ernst Oldemeyer, "Geschichtlicher

Wertwandel" [Historical value change]; Hans Lenk, "Verantwortungsdifferenzierung und Systemkomplexität" [Responsibility differentiation and systems complexity]; Friedrich Rapp, "Möglichkeiten und Grenzen der Technikbewertung" [Possibilities and limits of technology assessment]; Walther Ch. Zimmerli, "Spezifische Problembereiche" [Specific problem areas]. Part 4, "Die Ambivalenz der Technik" [The ambivalence of technology]: Friedrich Rapp, "Utopien und Antiutopien" [Utopias and antiutopias]; Friedrich Rapp, "Die technische Weltzivilisation" [The technological world civilization]; Friedrich Rapp, "Die Leistungen der Technik und ihr Preis" [The achievements of technology and their cost].

Research in Philosophy and Technology. 1978–. Vol. 1 (1978) to vol. 8 (1985), vol. 9 (1988) to present. Greenwich, Conn.: JAI Press. From 1978 to 1985, a series of the Society for Philosophy and Technology, ed. Paul T. Durbin; independent afterward, ed. Frederick Ferré, 1985–1994.

Reuleaux, Franz. 1875. *Theoretische Kinematik: Grundzüge einer Theorie des Maschinenwesens.* Braunschweig: F. Vieweg. English version, *The Kinematics of Machinery: Outlines of a Theory of Machines,* trans. and ed. Alex B. W. Kennedy (London: Macmillan, 1876), 662 pp. Reprinted, with new introduction by Eugene S. Ferguson (New York: Dover, 1963).

Richta, Radovan, ed. 1967. *Civilizace na rozcesti.* Prague: Svoboda. 3d expanded ed., 1969. 412 pp. English version, *Civilization at the Crossroads,* trans. Marian Šlingová (White Plains, N.Y.: International Arts and Sciences Press, 1969), 371 pp.

Riessen, Hendrik van. 1949. *Filosofie en techniek.* Kampen: J. H. Kok. 715 pp.

Rogers, G. F. C. 1983. *The Nature of Engineering: A Philosophy of Technology.* London: Macmillan. 105 pp.

Ropohl, Günter. 1975. *Systemtechnik: Grundlagen und Anwendung* [Systems technology: Foundations and applications]. Munich: Hanser. xvi, 356 pp.

———. 1985. *Die unvollkommene Technik* [Imperfect technology]. Frankfurt: Suhrkamp. 277 pp.

———. 1991. *Technologische Aufklärung: Beiträge zur Technikphilosophie* [Technological enlightenment: Contribution to the philosophy of technology]. Frankfurt: Suhrkamp. 264 pp.

Note on Ropohl: In addition to his own monographs, Ropohl is coauthor and coeditor with Hans Lenk of a number of key volumes listed above under Lenk.

Rothenberg, David. 1993. *Hand's End: Technology and the Limits of Nature.* Berkeley: University of California Press. xix, 256 pp.

Sachsse, Hans. 1987. *Anthropologie der Technik* [Anthropology of technology]. Braunschweig: Vieweg. vi, 291 pp.

———, ed. 1974–1976. *Technik und Gesellschaft* [Technology and society]. Vol. 1, *Literaturführer* [Leading literature] (1974), 309 pp. Vol. 2, *Texte: Technik in der Literatur* [Texts: technology in literature] (1976), 260 pp. Vol. 3,

Selbstzeugnisse der Techniker: Philosophie der Technik [Personal testimonies of technologists: Philosophy of technology]. (1976), 260 pp. Munich: Dokumentation.

Sanmartín, José. 1987. *Los nuevos redentores: Reflexiones sobre la ingeniería genética, la sociobiología y el mundo feliz que nos prometen* [The new saviors: Reflections on genetic engineering, sociobiology, and the happy world they promise us]. Barcelona: Anthropos. 207 pp.

——. 1990. *Tecnología y futuro humano* [Technology and human future]. Barcelona: Anthropos. 158 pp.

Note on Sanmartín: For representative English articles by Sanmartín see "Alternatives for Evaluating the Effects of Genetic Engineering on Human Development," in *Broad and Narrow Interpretations of Philosophy of Technology, Philosophy and Technology,* vol. 7, ed. Paul T. Durbin, pp. 153–166 (Boston: Kluwer, 1990); and "From World3 to the Social Assessment of Technology: Remarks on Science, Technology, and Society," and "Genethics: The Social Assessment of the Risks and Impacts of Genetic Engineering," both in *Philosophy of Technology in Spanish Speaking Countries, Philosophy and Technology,* vol. 10 ed. Carl Mitcham, pp. 197–209, 211–225 (Boston: Kluwer, 1993).

Sass, Hans-Martin. 1980. "Man and His Environment: Ernst Kapp's Pioneering Experience and His Philosophy of Technology and Environment." In *German Culture in Texas: A Free Earth,* ed. Glen E. Lich and Dona B. Reeves, pp. 82–101. Boston: Twayne.

Schaub, James H., and Karl Pavlovic, eds. 1983. *Engineering Professionalism and Ethics.* New York: John Wiley. xv, 559 pp.

Schirmacher, Wolfgang. 1983. *Technik und Gelassenheit: Zeitkritik nach Heidegger* [Technology and detachment: Contemporary criticism after Heidegger]. Freiburg: Karl Alber. 274 pp.

——. 1990. *Ereignis Technik* [Event technology]. Vienna: Passagen-Verlag. 245 pp.

Schön, Donald A. 1983. *The Reflective Practitioner: How Professionals Think in Action.* New York: Basic Books. x, 374 pp.

Schuurman, Egbert. 1972. *Techniek en toekomst.* Assen: Van Gorcum. English version, *Technology and the Future: A Philosophical Challenge,* trans. Herbert Donald Morton (Toronto: Wedge, 1980), xxii, 4343 pp.

Shrader-Frechette, Kristin S. 1980. *Nuclear Power and Public Policy: The Social and Ethical Problems of Fission Technology.* Boston: D. Reidel: rev. ed., 1983. xx, 178 pp.

Note on Shrader-Frechette: Further works on science policy from Shrader-Frechette: *Science Policy, Ethics, and Economic Methodology* (Boston: D. Reidel, 1985), xv, 321 pp.; *Risk Analysis and Scientific Method* (Boston: D. Reidel, 1985), x, 232 pp.

Simon, Herbert A. 1981. *The Sciences of the Artificial*. 2d ed., rev. Cambridge: MIT Press. xiii, 247 pp. The second edition adds three new chapters to the original (1969) four.

————. 1983. *Reason in Human Affairs*. Stanford: Stanford University Press. viii, 115 pp.

Note on Simon: Among this economics Nobel Prize (1978) winner's extensive writings relevant to philosophy of technology are *The Shape of Automation for Men and Management* (New York: Harper and Row, 1965), xv, 111 pp.; *Administrative Behavior: A Study of Decision-Making Processes in Administrative Organizations*, 3d ed. (New York: Free Press, 1976), 364 pp.; *Models of Discovery* (Boston: D. Reidel, 1977), esp. secs. 3, "The Logic of Imperatives," and 4, "Complexity," pp. 135–261; and *Models of Bounded Rationality*, 2 vols. (Cambridge, Mass.: MIT Press, 1982). The coauthored (with Allen Newell) *Human Problem Solving* (Englewood Cliffs, N.J.: Prentice-Hall, 1972) is also important. Simon's autobiography, *Models of My Life* (New York: Basic Books, 1991), is a good introduction.

Simondon, Gilbert. 1958. *Du Mode d'existence des objets techniques* [On the mode of the existence of technical objects]. Paris: Aubier-Montaigne. Reprinted with plates, 1969 and 1989. 265 pp.

Skolimowski, Henryk. 1966. "The Structure of Thinking in Technology." *Technology and Culture* 7, no. 3 (summer): 371–383. Reprinted in Mitcham and Mackey, eds., *Philosophy and Technology* (1972), pp. 42–49.

————. 1968. "On the Concept of Truth in Science and in Technology." In *Proceedings of the XIV International Congress of Philosophy* (Vienna, September 2–9, 1968), 2: 553–559. Vienna: Herder.

————. 1970–1971. "Problems of Truth in Technology." *Ingenor* (College of Engineering, University of Michigan) 8 (winter): 5–7, 41–46.

Note on Skolimowski: More books by Skolimowski: *Eco-Philosophy: Designing New Tactics for Living* (Salem, N.H.: Marion Boyars, 1981), 117 pp.; *Technology and Human Destiny* (Madras, India: University of Madras, 1983), xii, 139 pp.; *Living Philosophy: Eco-Philosophy as a Tree of Life* (London: Arkana, 1992), 254 pp.

Spengler, Oswald. 1931. *Der Mensch und die Technik: Beitrag zu einer Philosophie des Lebens*. Munich: C. H. Beck. 61 pp. English version, *Man and Technics: A Contribution to a Philosophy of Life*, trans. Charles F. Atkinson (New York: Alfred A. Knopf, 1932), 104 pp. See also Oswald Spengler, *The Decline of the West*, vol. 2, *Perspectives of World History*, trans. Charles Francis Atkinson (London: Allen and Unwin, 1928), chap. 14, "The Form-world of Economic Life (B), The Machine."

Staudenmaier, John M., S.J. 1985. *Technology's Storytellers: Reweaving the Human Fabric*. Cambridge: MIT Press. xxiii, 282 pp.

Stork, Heinrich. 1977. *Einführung in die Philosophie der Technik* [Introducing the philosophy of technology]. Darmstadt: Wissenschaftliche Buchgesellschaft. 189 pp.

Stover, Carl F., ed. 1963. *The Technological Order*. Detroit: Wayne State University Press. xii, 280 pp. First published as a theme issue of *Technology and Culture* 3, no. 4 (fall 1962).

Tierney, Thomas F. 1993. *The Value of Convenience: A Genealogy of Technical Culture*. SUNY Series in Science, Technology, and Society. Albany: State University of New York Press. x, 281 pp.

"Toward a Philosophy of Technology." 1966. *Technology and Culture* 7, no. 3 (summer): 301–390. A symposium that includes Lewis Mumford, "Technics and the Nature of Man"; James K. Feibleman, "Technology as Skills"; Mario Bunge, "Technology as Applied Science"; Joseph Agassi, "The Confusion between Science and Technology in the Standard Philosophies of Science"; John O. Wisdom, "The Need for Corroboration: Comments on Agassi's Paper"; Henryk Skolimowski, "The Structure of Thinking in Technology"; I. C. Jarvie, "The Social Character of Technological Problems: Comments on Skolimowski's Paper." Four of these papers are included in Mitcham and Mackey, eds., *Philosophy and Technology* (1972); four (two of them different) can also be found in Rapp, ed., *Contributions to a Philosophy of Technology* (1974).

"Toward a Philosophy of Technology." 1971. *Philosophy Today* 15, no. 2 (summer): 75–156. Contents: Hans Jonas, "The Scientific and Technological Revolutions"; Carl Mitcham and Robert Mackey, "Jacques Ellul and the Technological Society"; Donald Brinkmann, "Technology as Philosophic Problem"; and Simon Moser, "Toward a Metaphysics of Technology." The last two essays are translations.

Tuchel, Klaus, ed. 1967. *Herausforderung der Technik: Gesellschaftliche Voraussetzunger und Wirkungen der technischen Entwicklung* [Challenge of technology]. Bremen: Carl Schünemann. 317 pp.

Ullrich, Otto. 1977. *Technik und Herrschaft: Vom Hand-Werk zur verdinglichten Blockstruktur industrieller Produktion* [Technology and domination: From handicraft to reified modular industrial production]. Frankfurt: Suhrkamp. 484 pp. For a brief version of the argument in English, see "Technology," in *The Development Dictionary: A Guide to Power as Knowledge*, ed. Wolfgang Sachs, pp. 275–287 (London: Zed Books, 1992).

Ure, Andrew. 1835. *The Philosophy of Manufactures, or An Exposition of the Scientific, Moral, and Commercial Economy of the Factory System of Great Britain*. London: Charles Knights. 480 pp. Reprint, New York: Augustus Kelley, 1967. Partial reprint in *The Philosophy of Manufactures: Early Debates over Industrialization in the United States*, ed. Michael Brewster Folsom and Steven D. Lubar, pp. 365–388 (Cambridge: MIT Press, 1982).

Van Melsen, Andrew G. 1961. *Science and Technology*. Pittsburgh: Duquesne University Press. ix, 373 pp.

Van Poolen, Lambert J. 1989. "A Philosophical Perspective on Technological

Design." *International Journal of Applied Engineering Education* (U.K.) 5, no. 3: 319–329.

Van Riessen, Hendrik. 1949. *Filosofie en techniek* [Philosophy and technology]. Kampen: J. H. Kok. 715 pp. For a bibliography of Van Riessen's work and translations of two of his essays—"Technology and Culture" (1951) and "The Structure of Technology" (1961)—see Donald Morton, ed., "Symposium: Hendrik Van Riessen and Dutch Neocalvinist Philosophy of Technology," in *Research in Philosophy and Technology*, (1979) 2: 293–340.

Vincenti, Walter. 1990. *What Engineers Know and How They Know It*. Baltimore: Johns Hopkins University Press. viii, 326 pp.

Vries, M. J. de, N. Cross, and D. P. Grant, eds. 1993. *Design Methodology and Relationships with Science*. Series D, Behavioural and Social Sciences, vol. 71. Boston: Kluwer. vii, 327 pp.

Walker, Timothy, 1831. "In Defense of Mechanical Philosophy." *North American Review* 31 (July): 122–136. Reprinted in *Readings in Technology and American Life*, ed. Carroll W. Pursell Jr., pp. 67–77 (New York: Oxford University Press, 1969), and in *The Philosophy of Manufactures: Early Debates over Industrialization in the United States*, ed. Michael Brewster Folsom and Steven D. Lubar, pp. 295–304 (Cambridge: MIT Press, 1982).

Wartofsky, Marx. 1979. "Philosophy of Technology." In *Current Research in Philosophy of Science*, ed. Peter D. Asquith and Henry E. Kyburg Jr., pp. 171–184 (East Lansing, Mich.: Philosophy of Science Association).

Weizenbaum, Joseph. 1976. *Computer Power and Human Reason: From Judgment to Calculation*. San Francisco: W. H. Freeman. xii, 300 pp.

Weston, Anthony. 1989. "Ivan Illich and the Radical Critique of Tools." *Research in Philosophy and Technology* 9: 171–182.

Wiener, Norbert. 1947. "A Scientist Rebels." *Atlantic Monthly* 179, no. 1 (January): 46. Also printed in *Bulletin of the Atomic Scientists* 3, no. 1 (January 1947): 31. Wiener reiterates his position in "A Rebellious Scientist after Two Years," *Bulletin of the Atomic Scientists* 4, no. 11 (November 1948): 339–340.

———. 1948. *Cybernetics, or Control and Communication in the Animal and the Machine*. Cambridge: MIT Press. 194 pp. 2d ed., 1961. xvi, 212 pp.

———. 1950. *The Human Use of Human Beings: Cybernetics and Society*. New York: Houghton Mifflin. 241 pp. 2d ed., Garden City, N.Y.: Doubleday, 1954. 199 pp. See also "Some Moral and Technical Consequences of Automation," *Science* 131, no. 3410 (May 6, 1960): 1355–1358.

———. 1964. *God and Golem, Inc.: A Comment on Certain Points Where Cybernetics Impinges on Religion*. Cambridge: MIT Press. ix, 99 pp.

Note on Wiener: Two other related books: Wiener's novel, *The Tempter* (New York: Random House, 1959), 242 pp.; *Invention: The Care and Feeding of Ideas* (Cambridge: MIT Press, 1993), xxiv, 159 pp. For biography and analysis, see Wiener's two volumes of autobiography—*Ex-Prodigy: My Childhood and Youth* (New York: Simon and Schuster, 1953) and *I Am a Mathematician: The*

Later Life of a Prodigy (Garden City, N.Y.: Doubleday, 1956)—and Steve J. Heims, *John von Neumann and Norbert Wiener: From Mathematics to the Technologies of Life and Death* (Cambridge: MIT Press, 1980).

Winner, Langdon. 1977. *Autonomous Technology: Technics-out-of-Control as a Theme in Political Thought.* Cambridge: MIT Press. x, 386 pp.

————. 1986. *The Whale and the Reactor: A Search for Limits in an Age of High Technology.* Chicago: University of Chicago Press. xiv, 220 pp.

————. 1991. "Upon Opening the Black Box and Finding It Empty: Social Constructivism and the Philosophy of Technology." In Pitt and Lugo, eds., *The Technology of Discovery and the Discovery of Technology* (1991), pp. 503–519. Abridged reprint, *Science, Technology, and Human Values* 18, no. 3 (summer 1993): 362–378.

Zimmerli, Walther Christopher. 1986. "Who Is to Blame for Data Pollution? On Individual Moral Responsibility with Information Technology." In Mitcham and Huning, eds., *Philosophy and Technology II* (1986), pp. 291–305.

————, ed. 1976. *Technik, oder Wissen wir, was wir tun?* [Technology, or Do we know what we are doing?]. Basel: Schwabe. 210 pp.

————, ed. 1989. *Herausforderung der Gesellschaft durch den technischen Wandel: Informationstechnologie und Sprache, Biotechnologie, Technikdiscussion im Systemvergleich* [The challenge to society of technical change: Information technology and language, biotechnology, technical discussion in systems comparison]. Düsseldorf: VDI-Verlag. 277 pp.

Zimmerman, Michael E. 1990. *Heidegger's Confrontation with Modernity: Technology, Politics, Art.* Bloomington: Indiana University Press. xxvii, 306 pp.

Zschimmer, Eberhard. 1914. *Philosophie der Technik: Vom Sinn der Technik und Kritik des Unsinns über die Technik* [Philosophy of technology: Concerning the meaning of technology and criticisms of the meaninglessness of technology]. Jena: E. Diederichs. 184 pp. 2d ed., 1919, 166 pp. 3d ed., completely revised, with a new subtitle, *Einfuhrung in die technische Ideenwelt* [Introduction to the world of technical ideas] (Stuttgart: F. Enke, 1933), viii, 76 pp.

Note on Zschimmer: Further works by Zschimmer are *Philosophie der Technik* [Philosophy of technology] (Berlin: E. J. Mittler, 1917), 22 pp., a pamphlet publishing a lecture; *Technik und Idealismus* [Technology and idealism] (Jena: Jenaer Volksbuchhandlung, 1920), 31 pp., another pamphlet from a lecture; *Deutsche Philosophen der Technik* [German philosophers of technology] (Stuttgart: F. Enke, 1937), 115 pp., essays on Kapp, Eyth, Eduard von Mayer, Ulrich Wendt, DuBois-Reymond, and Viktor Engelhardt (Nazi influence is indicated by the exclusion of Dessauer); "Vom Wessen des technischen Schaffens" [On the essence of technological creation], *Zeitschrift für Deutsche Philosophie* (Stuttgart) 6 (1940): 231–238.

INDEX

• • • •

This index is to the main body of the text and its notes, excluding preface and references. It is exhaustive with respect to authors and editors, some of whose names are more complete here than in their textual occurrence. Other names, subjects, and technical (especially foreign-language) terms are selectively indexed for the main body of the text but not the notes.